Sustainable Development and Resilience of Tourism

Deepak Chhabra • Navin Atal
Alka Maheshwari
Editors

Sustainable Development and Resilience of Tourism

Wellbeing and Quality of Life Perspectives

Springer

Editors
Deepak Chhabra
School of Community Resources
and Development, Watts College of Public
Service and Community Solutions
Arizona State University
Phoenix, AZ, USA

Navin Atal
Department of Internal Medicine
Max Super Speciality Hospital
Delhi, Delhi, India

Alka Maheshwari
Amity Institute of Travel and Tourism
Amity University
Noida, Uttar Pradesh, India

ISBN 978-3-031-63144-3 ISBN 978-3-031-63145-0 (eBook)
https://doi.org/10.1007/978-3-031-63145-0

This Springer imprint is published by the registered company Springer Nature Switzerland AG
The registered company address is: Gewerbestrasse 11, 6330 Cham, Switzerland

Preface

Crises in the form of a pandemic, wars, or in some other form will continue to test communities and institutions across the globe. Such disruptions can generate adverse outcomes by dissolving or diluting initiatives aimed at restoring some kind of new normalcy within the tourism systems. Ongoing core need is for innovative and resilient strategies to plan products and services as tourism systems are constantly in a state of flux (Lew, 2020). Resilience refers to an acquired capacity that holds potential to resist disruptions. It is the ability to recover and thrive from traumatic situations (Harms et al., 2018). Liu et al., (2020) describe resilience as a personal attribute that can buffer people from negative situations while developing the skills to manage them. According to McCartney et al. (2021), resilience-based research needs to look beyond reactive strategies to overcome a crisis and promote capacity-building efforts to minimize future vulnerabilities.

Floating fixity is a defining feature of the contemporary environment. In other words, the systems might accomplish equilibrium for a brief period but will move again to accommodate/resolve health and socio-economic issues as and when they gravitate. Adaptiveness and resilience (of the self and that of others and the extended environment) are key to help stay anchored/grounded as our human systems adapt to unpredictable sway of uncontrollable factors such as the pandemic and the war in Ukraine. From the demand side, significance of sustainable and wellness consumption is becoming increasingly important (Saarinen, 2021; Um & Yoon, 2021) to buffer shocks of magnitude. This implies that it has become crucial to design and develop tourism strategically and sustainably by seeking trade-offs between environmental, economic, social, and health/wellness goals of different societies across the globe (Streimikiene et al., 2020). In fact, quality of life (QOL) of host communities has become more pronounced due to the pandemic, as local communities display more keenness to become involved in sustainable tourism planning and development decisions associated with the shaping of tourism in a manner that their overall wellness and QOL prioritizations are not compromised. The concept of the life quality is largely regarded as satisfying life, happiness and prosperity, wellbeing, and comfort, and reflects the level of the realization of human needs (Więczkowska et al., 2011: 1).

Concisely, the present and future success of sustainable tourism calls for a deep-rooted paradigm shift by embracing four pillars of sustainability. With so many environmental, political, economic, and social changes observed during the contemporary times, relevancy and integration of the health sector as one core dimension of human sustainability has become paramount. In fact, "the inclusion of health as the fourth pillar for sustainability has direct implications not only for the national health systems but also at the level of governance and formulation of public policies" (Marques et al., 2020, p. 1). Structuring of healthy societies need to be guided by innovative, evolving, and inspirational knowledge. Growing emphasis can also be evidenced in the 2030 agenda, of United Nations, for sustainable development. It aims at promoting peace, emphasize on health and wellbeing. Furthermore, the UN-Tourism's (formerly called UN-WTO) slogan for "tourism raises its voice for peace" is a call for employing tourism as a pillar of peace to support the path of dialogue and non-violence. Health has clearly become an increasing cause of concern in war zones. Wellness of the humanity at large can make an impactful contribution toward promoting a resilient planet. According to Marques et al., "the popular expression of 'I'm in good health, the rest is secondary', holds zero meaning today" (2020, p. 2).

The focus of this book proposal is timely. It examines progress in sustainable and resilient development of, particularly, niche forms of tourism from quality of life, health, and wellbeing perspectives. It also aims to offer deep and innovative insights on the supporting role of non-pharmaceutical interventions (such as herbs) to enhance resilience and wellbeing of visited and visiting communities. A comparative and critical discourse of the role of allopathic medicine is also vital in modern medicine, from both a core and supplementary perspective, particularly, from eudaimonic and physiologically restorative standpoints (Haywantee, 2021). Allopathic system of medicine focuses on the source of disease and treatment of that cause to cure the body and mind. The treatment is based on evidence-based scientific research. Allopathy uses surgery, radiation, clinical trials, and approved medications to treat patients. Additionally, allopathic treatment provides preventive care through antibiotics, vaccinations, and screenings. These medicines are governed and restricted under the drug laws of countries. The herbal system is in comparison based on ancient wisdom (Atal, 1979). Herbs and herbal medicines have been the mainstay of treatment since ancient times. Even with the modernized technology today, the time-tested medicinal plants have not lost their importance. In fact, their popularity has surged in recent times. The perception of a safe and affordable herbal treatment can be attributed to its popularity in many countries across the world (Atal & Khurana, 2016). The faith and belief in herbs are strengthened by the application of modern science and technology that have tested the safety and efficacy of many herbal drugs and proven their robusticity (Atal et al., 1981). Now in the modern era, herbal medicines have become subjects of research and pathways to groundbreaking discoveries. Practitioners today accentuate the need for hybrid treatments.

The potential of conscientious and healthful consumerism, in enhancing the quality of life and holistic wellness of visiting and visited communities, is an emerging field of study (Chhabra, 2021, 2022; Dwyer, 2022; Seo et al., 2021; Uysal &

Sirgy, 2023). This book makes an important contribution toward this line of inquiry. It will aid future researchers to nurture the present repository of knowledge and offers valuable insights to academicians, practitioners, and policymakers. The focus is not only timed well, but it also holds tremendous potential to become an important point of reference for future studies on sustainable development of tourism and transformative quality of live frameworks.

Harmonious collaboration between numerous stakeholders has become crucial to plan pathways for integrated development. In fact, the pandemic has resulted in generating the need to add one more key set of stakeholders to the sustainable development equation: health and wellness experts and related hospitality services. There is a need for a critical discourse to examine the usefulness of integrating health as an essential pillar of sustainable development in tourism. The second editor of the book is a well-known physician from India. In his chapter, co-authored with late Dr. Atal, he offers useful insights on history and healing potential of herbs. The second author, late Dr. Atal, was recognized by the United Nations for the discovery of several medicinal remedies from herbal plants including the groundbreaking "Rizorine" that offers an herbal remedy for tuberculosis.

Insights on the healing potential of Ayurveda is discussed by Atal and Atal (2024) in this book. Ayurveda, with its emphasis on holistic wellbeing, offers alternative treatments that aim to enhance adaptiveness and resilience (Deshmukh, 2023). Through personalized approaches based on an individual's unique constitution or doshas (fundamental regulatory principles of the body, mind, and behavior), Ayurvedic practices seek to strengthen the body's natural ability to adapt to stressors and challenges. Herbal formulations, dietary recommendations, and mindfulness techniques are often employed to balance the self, promoting physical and mental resilience. By addressing the root causes of imbalance, ayurvedic treatments strive to build a foundation for sustained adaptability and resilience in the face of life's uncertainties (Chauhan et al., 2015; Frawley, 2000; Kumar et al., 2017).

Structure of the Book

The first chapter introduces the notions of sustainable tourism, resilience, and quality of life. The United Nations' sustainable tourism agenda, in the context of its 17 sustainable development goals (SDGs), and their alignment with tourism are traversed. The chapter discusses traversing points between SDGs, resilience, and quality of life paradigms. Chapter 2 focuses on the notion of health, wellness, and quality of life from the standpoint of tourists and local residents using the overarching umbrella of sustainability. The surging popularity of wellness tourism is critically examined, particularly in the context of its contribution to the quality of life and wellbeing of guests and host communities.

Chapter 3 by Atal and Atal offers insights on the significance of various types of herbal treatments, with a special focus on herbs found in India. The potential of herbal programs and spaces to promote herbal tourism is discussed using a variety of examples. A synopsis is offered regarding the history and progress in the development of herbal products to promote herbal tourism. This is followed by an overview of various frameworks and models employed in some countries to plan and develop a sustainable form of herbal tourism. India and its progress in herbal tourism is specifically discussed followed by suggestions for a forward-looking inclusive path to develop and promote sustainable herbal tourism.

Chapter 4 by Joson, Pyke, Delacy, and Davidson offers insights on how wellness as a tourism segment has evolved as a field of academic and practical interest. This trend is a response to the surging need for maintenance of overall wellbeing, broadly understood as a state of being or feeling that holds potential to improve life, both physically and emotionally. This can include happiness, financial stability, health, and spirituality. The authors illustrate that spas or wellness hotels, offering water-based treatments, are an important component of the wellness tourism industry. They focus on hot springs popularly patronized due to their therapeutic benefits, explore Australian tourism and wellness tourism, in context of hot springs tourism industry, and propose a set of indicators as a tool to measure the impact of hot springs in Australia. Ayazlar and Ayazlar in Chap. 5 stress on the need for enterprises to promote employee mindfulness. The authors point out that general wellbeing is important for all individuals and these include employees of the tourism industry. Optimal employee performance is possible by improving employee motivation and wellbeing. It cannot be denied that uncertainty is unavoidable and chaotic situations can happen from time to time. Efforts to provide effective solutions, to customer concerns and needs in a short time, can impose a mental and physical strain on human resources. Ayazlar and Ayazlar suggest that mindful training and attitude can counter stress-related situations and enhance the overall wellbeing of employees. The authors scrutinize the concept of mindfulness and its importance for tourism employees and offer insights on how to reduce negative emotions and increase positive emotions in tourism employees.

Chapter 6 by Gogoi and Maheshwari focuses on a unique subset of adventure tourism (motorcycle tourism). In fact, motorcycle tourism has emerged as a popular form of tourism in India. The authors share that this unique form of tourism requires professional expertise in addition to technical skilled guidance and specialized equipment. It is particularly popular in rural and remote regions of the Himalayas. However, sustainable development and promotion of this unique form of tourism is crucial and very few studies, to date, have examined motorcycle tourism from this perspective. To fill this gap, the authors offer insights on how it can meaningfully contribute to the economy of rural communities located in the remote parts of Ladakh (in the Jammu & Kashmir State, India) and discuss the most favorable technique to develop innovative initiatives that are centered on inclusivity and pro-poor guidelines. A conceptual sustainable livelihood model is suggested. Cohen, in Chap. 7, scrutinizes the sustainability of marine diving tourism. He discusses the extent to which it holds potential to generate transformative experiences that can prove to be beneficial for the diver tourists and the host communities. The author uses three pillars of sustainability (environmental, economic, and social) in the context of dive tourism. He also offers a discourse on the viability of sustainable development goals

(SDGs) and the effectiveness of management goals based on a transformative paradigm. A case study (the Hawaiian Island of Maui, which suffered an immense setback due to the devastating wildfire in August 2023) is employed to discuss the transformative potential of the sustainable marine tourism industry on host communities.

Chapter 8 by Kim focuses on off-highway vehicle (OHV) recreation from a marketing perspective. It examines whether key versions of word of mouth (WOM) significantly shape consumer behavior in terms of loyalty and spending and whether they hold potential to facilitate responsible behavior. By exploring the relationship between consumer behavior and traditional WOM and eWOM, Kim stresses on the significance of integrating traditional WOM with social networking platforms. By offering comparative/integrated insights on the impact of WOM on local spending behavior of OHV visitors, the author shares important implications from a sustainable marketing perspective.

Chapters 9 and 10 are penned by Pham, Vogt, and Andereck and Pham, Andereck, and Vogt respectively. The spotlight is on the challenges of overtourism for popular destinations. The authors specifically focus on two key concepts: increase in visitations throughout the year and annoyance of local residents to tourism development as it overwhelms the social and environmental thresholds of a community. In Chap. 9, a review of the overtourism notion is offered. A toolkit of indicators is presented to measure the "size of tourism economy, employment, tourism visitation, development levels, media reach, and stakeholder's perspectives." Chapter 10 delves deeper into the causes of overtourism and its solutions. The authors share a broad spectrum of solutions such as tools and management/development techniques, from academic and practitioner standpoints, to mitigate crowding and other adverse impacts from overtourism. Many of these solutions are grounded in sustainability and resilience principles and also related to destination capacity management. An integrated overtourism ecosystem model is suggested that is specifically centered on stakeholders. The authors call for developing a tourism plan as a roadmap for communities to identify, manage, and mitigate externalities associated with overtourism.

Chapter 11 discusses the manner in which the work of all contributors is aligned with a broad spectrum of SDGs (sustainable development goals) proposed by the United Nations. It examines the notions of transformation, transformative experiences, and transformative learning and contextualizes them from a wellness/wellbeing standpoint. Deliberations are offered on the association between transformative tourism and wellbeing/quality of life. Chapter 12 presents a wellbeing-centered sustainable tourism roadmap going forward. Contemporary trends, their challenges, and opportunities are presented. Key highlights from the contributors are presented. An effort is made to integrate the wealth of information from the book and contemporary trends. Moving forward, a sustained transformative quality of life paradigm is proposed.

In summary, as evidenced in the preceding paragraphs, this book offers a broad range of topics, challenges, and solutions associated with resilience, transformation, and sustainable development of tourism to improve and sustain quality of life of visited and visiting communities. Its enriching content will aid future researchers to nurture the present repository of knowledge and will offer valuable insights to academician, practitioners, and policymakers. It is not only timed well, it also holds tremendous potential to become an important point of reference for future studies on sustainable development of tourism, particularly from a health/wellbeing and transformative quality of live perspectives.

Phoenix, AZ, USA Deepak Chhabra
Delhi, India Navin Atal
Noida, UP, India Alka Maheshwari

References

Atal, C. K. (1979). A breakthrough in drug bioavailability—A clue from age-old wisdom of Ayurveda (pp. 483–484). *IDMA Bulletin*

Atal, C. K., Zutshi, U., & Rao, P. (1981). Scientific evidence on the role of Ayurvedic herbals on bioavailability of drugs. Journal of ethnopharmacology 4(2), 229–232.

Atal, N., & Khurana, P. (2016). *Guest lecture at VAMSCON, Vardhman Mahavir Medical College, Safdarjung hospital*. https://youtu.be/jncrDaMChIk?si=2yida8ItfYl3NYTt

Chauhan, A., Semwal, D. K., Mishra, S. P., & Semwal, R. B. (2015). Ayurvedic research and methodology: Present status and future strategies. *Ayurveda 36*(4), 364.

Chhabra, D. (2021). Transformative perspectives of tourism: Dialogical perceptiveness. *Journal of Travel & Tourism Marketing*, *38*(8), 759–768.

Chhabra, D. (2022). Transformational model of well-being for serious travellers. *International Journal of Spa and Wellness*, *5*(2), 129–146.

Deshmukh, A. (2023). *How Ayurveda is the new medical tourism attraction*, Updated Feb, 2023. https://travel.economictimes.indiatimes.com/news/tourism/wellness/how-ayurveda-is-the-new-medical-tourism-attraction/98043429

Dillette, A. K., Douglas, A. C., & Andrzejewski, C. (2019). Yoga tourism–A catalyst for transformation? *Annals of Leisure Research*, *22*(1), 22–41.

Dwyer, L. (2022). Destination competitiveness and resident well-being. *Tourism Management Perspectives*, *43*, 100996.

Frawley, D. (2000). *Ayurvedic healing: a comprehensive guide*. Lotus Press.

Harms, P. D., Brady, L., Wood, D., and Silard, A. (2018). Resilience and wellbeing. In E. Diener, S. Oishi, L. Tay. (Eds.), *The handbook of wellbeing*. DEF Publishers.

Haywantee, R. (2021). Body-mind medicine interventions in COVID-19 place confinement for mental, physical and spiritual wellbeing. *OBM Integrative and Complementary Medicine*, *6*(2), 1–1.

Kumar, S., Dobos, G. J., & Rampp, T. (2017). The significance of ayurvedic medicinal plants. *Journal of Evidence-Based Complementary & Alternative Medicine*, *22*(3), 494–501.

Lehto, X. Y., Brown, S., Chen, Y., & Morrison, A. M. (2006). Yoga tourism as a niche within the wellness tourism market. *Tourism Recreation Research*, *31*(1), 25–35.

Lew, A. (2020). The global consciousness path to sustainable tourism: a perspective paper. *Tourism Review.*

Liu, C. H., Zhang, E., Wong, G. T. F., Hyun, S., & Hahm, H. (2020). Factors associated with depression, anxiety, and PTSD symptomatology during the COVID-19 pandemic: Clinical implications for U.S. young adult mental health. *Psychiatry Research,* 290, 113172. https://doi.org/10.1016/j.psychres.2020.113172

Marques, I., Serrasqueiro, Z., & Nogueira, F. (2021). Managers' competences in private hospitals for investment decisions during the COVID-19 Pandemic. *Sustainability*, *13*(4), 1757.

McCartney, G., Pinto, J., & Liu, M. (2021). City resilience and recovery from COVID-19: The case of Macao. *Cities*, 112, 103130. https://doi.org/10.1016/j.cities.2021.103130

Saarinen, J. (2021). Is being responsible sustainable in tourism? Connections and critical differences. *Sustainability*, *13*(12), 6599.

Streimikiene, D., Svagzdiene, B., Jasinskas, E., & Simanavicius, A. (2021). Sustainable tourism development and competitiveness: The systematic literature review. *Sustainable Development*, *29*(1), 259–271.

Seo, K., Jordan, E., Woosnam, K. M., Lee, C. K., & Lee, E. J. (2021). Effects of emotional solidarity and tourism-related stress on residents' quality of life. *Tourism Management Perspectives*, *40*, 100874.

Um, J., & Yoon, S. (2021). Evaluating the relationship between perceived value regarding tourism gentrification experience, attitude, and responsible tourism intention. *Journal of Tourism and Cultural Change*, *19*(3), 345–361.

Zielińska-Więczkowska, H., Kędziora-Kornatowska, K., & Ciemnoczołowski, W. (2011). Evaluation of quality of life (QoL) of students of the University of Third Age (U3A) on the basis of socio-demographic factors and health status. *Archives of gerontology and geriatrics*, *53*(2), e198–e202.

Acknowledgments

Deepak Chhabra:

Recent times call for a shift in the manner quality of life and sustainability are scrutinized. Health and wellness have become a crucial part of the quality of life, sustainable tourism, and resilience discourses. This book will add to this new wave of discourses and other emerging concepts in the ocean of knowledge. I am enormously grateful to numerous colleagues, scholars, and wellness institutions for their unwavering assistance and support. I am also indebted to my dogs Ebony and Sania for patiently watching over me during extended hours of work. Finally, I wish to dedicate this book to my father, Dr. G.S. Chhabra, and my mother, Sharanjit Walia, who are no more with me but they continue to inspire me through the fond memories I will always hold of them.

Navin Atal:

I am happy to be involved in this important book, which provides wisdom, knowledge, and information in wellness tourism. With the support of my co-editors, it became easier to complete the write up in the desired shape as we see it.

I am thankful to all our scientists and physicians from ancient times till today who have brought a smile to the faces of countless diseased and suffering people. It is indeed remarkable to see the ancient wisdom of herbal health practitioners, who have bestowed upon mankind a priceless gift of the herbal medicinal system. Even today in modern times, herbal medicines are used, by people across the globe, for healing purposes.

I am grateful to my father, late Dr Chand Kumar Atal, a towering figure in his field, for being a role model and guiding light to me throughout his lifetime. He has continued to inspire a generation of scientists in the field of medicinal plants. I therefore dedicate my chapter to my father for the knowledge he imparted to me since childhood in the field of medicinal plants.

I am grateful to my mother, Mrs Saroj Atal, my spouse Mrs Ruchita Atal, and my children Aman and Amrit, for their constant support in my work. Lastly, I also thankfully acknowledge all my patients who have always trusted me in my treatment and supported me in my profession.

Alka Maheshwari:

I extend my deepest gratitude to my incredible support system, starting with my husband, whose unwavering encouragement and understanding have been my pillars throughout this academic journey. His patience and belief in my pursuits have been a constant source of inspiration.

To my parents, whose sacrifices and encouragement laid the foundation for my academic pursuits, I am profoundly thankful. Their enduring support and belief in my capabilities have been a driving force behind my accomplishments.

My heartfelt appreciation goes to my son, whose boundless love and understanding have brought joy and perspective to my academic endeavors. His presence has been a motivating force, reminding me of the importance of balance and family amidst the rigors of academia.

I also wish to acknowledge the vast academic experience that has shaped my journey. The mentors, colleagues, and students I have encountered have enriched my understanding and passion for my field. Each academic milestone has been a collaborative effort, and I am grateful for the collective wisdom and camaraderie that have defined my professional trajectory.

Contents

About the Editors

Deepak Chhabra, corresponding editor, teaches in the School of Community Resources and Development and holds the position of Senior Global Futures Scientist in the Julie Ann Wrigley Global Futures Laboratory at Arizona State University, USA. Her research interests include authenticity and authentication of heritage; social and economic viability of different forms of tourism; and wellness and wellbeing through alternate healing/preventive therapeutic settings and programs. She has authored several books and published more than 60 articles in leading journals in the travel and tourism field. She serves on several advisory boards and is also currently serving as the "Section (Cultural Heritage and Authenticity in Tourism) Editor" for the *Frontiers in Sustainable Tourism* journal.

Navin Atal is currently working as a senior consultant physician at Max Super Speciality Hospital in Delhi (India) and specializes in Internal Medicine. He has a special interest in diabetes mellitus and treats all types of diabetic patients and diabetic complications. During and throughout the difficult first Covid wave, he was responsible for managing Covid-19-infected diabetic patients admitted to the Pulmonology Department at Max Super Speciality Hospital. His areas of interest also include medicinal plants and bioenhancers derived from medicinal plants. He has treated numerous tuberculosis patients with bioenhanced antitubercular formulation since 2009, the year in which the formulation was released in the market. He has delivered a talk as a guest speaker on bioenhancers at Vamscon 2016 (Vardhman Mahavir medical college) in Delhi, India. He has also published in EDPA (East Delhi Physicians Association) medical journal for 4 years, and many articles for the EDPA medical journal. He also holds the position of Managing Director of the Chand Kumar Atal Foundation and Atal Museum in Gurugram (NCR region of Delhi, India).

Alka Maheshwari is working as an Associate Professor in the Amity Institute of Travel and Tourism at Amity University (Noida, India). She holds a PhD in community-based tourism and her research areas of interest include community-based tourism, tourism marketing, sustainable tourism, tourism policy and planning, and tourism products and services. She has published articles in numerous journals at the national and international level. She boasts over three decades of enriching experience in academia, including a distinguished 17-year tenure at South Delhi Polytechnic for Women, where she ascended from lecturer to the esteemed position of Head of the Department. During her remarkable journey, she also had the privilege of representing her institute at Lancashire University, Blackburn, as a team member for UKIERI.

Chapter 1
Insights on Sustainable Tourism, Resilience, and Quality of Life Notions

Deepak Chhabra

Abstract This chapter first scrutinizes the dynamic notion of sustainable tourism. This is followed by insights on the significance of the United Nations Sustainable Development Goals for the tourism system. SDG roadmap by the G20 Tourism Working Group is featured with insights on sustainable pathways adopted by several green tourism case studies. Furthermore, attention is drawn to the significance of the resilience concept. The chapter closes with a discussion of the critical association between the notions of resilience, SDGs, and well-being.

Keywords Sustainable tourism · Sustainable Development Goals · Resilience · Well-being

Defining Sustainable Tourism

There is a conceptual dilemma with regard to the definition of sustainable tourism as on one hand it focuses on developing and promoting tourism and, on the other hand, it requires viable measures and minimization of environmental and social externalities (Bramwell, 2015; Liburd et al., 2022; Ruhanen et al., 2015; Sharpley, 2020; Weaver, 2009). As several scholars postulate, an overarching goal of sustainable tourism is to maintain a natural balance as a way of gauging the best intervention mechanisms to achieve economic viability and equilibrium between utilizing resources for today while ensuring their preservation for future generations (Bricker et al., 2013; Fennell, 2018; Fennell & Cooper, 2020; Holden, 2005; Mihalic, 2016). According to Liburd et al., sustainable tourism development should prioritize scrutiny and management of complex and "dynamic interrelationships, collaboration, and stewardship and engagement with civic society" (2022, p. 2302).

D. Chhabra (✉)
School of Community Resources and Development, Watts College of Public Service and Community Solutions, Arizona State University, Tempe, AZ, USA
e-mail: Deepak.chhabra@asu.edu

D. Chhabra et al. (eds.), *Sustainable Development and Resilience of Tourism*,
https://doi.org/10.1007/978-3-031-63145-0_1

The aforementioned complexities can be evidenced in the manner UN Tourism (UNWTO, 2024) has defined sustainable tourism. According to the UN Tourism (formerly known as UNWTO), sustainable tourism aims to consider current and future socio-economic and environmental impacts of tourism in addition to meeting the needs of visitors, the tourism industry, local communities, and the environment. Sustainable tourism should strive to:

1. Make optimal use of environmental resources that constitute a key element in tourism development, maintaining essential ecological processes and helping to conserve natural heritage and biodiversity.
2. Respect the sociocultural authenticity of host communities, conserve their built and living cultural heritage and traditional values, and contribute to inter-cultural understanding and tolerance.
3. Ensure viable, long-term economic operations, providing socio-economic benefits to all stakeholders that are fairly distributed, including stable employment and income-earning opportunities and social services to host communities, and contributing to poverty alleviation (UNWTO, 2023a).

Several scholars propose that the development of sustainable tourism should involve all key stakeholders guided/administered by effective leadership to ensure equitable participation and coordination (Butler, 1998; Fennell & Cooper, 2020; Mihalic, 2016; Ruhanen et al., 2015). The sustainable tourism path evolves and needs continuous monitoring and intervention to ensure appropriate measures are in place at the appropriate time so that interruptions can be avoided or minimized at the right time. It should strive towards generating high levels of tourist satisfaction while promoting mindful tourism practices and overall well-being of the guests and the hosts.

Almost two decades ago, Doppell (2003) suggested a "Wheel of Change Toward Sustainability" paradigm by identifying common flaws or blunders that are likely to impede an organization's progress towards sustainable development. The paradigm highlighted five issues: patriarchal thinking that can trigger a false sense of security, ambiguity towards cause and effect dynamics; poor information, failure to take a systematic/institutional view of sustainability, and lack of clear vision of sustainability. Doppell (2003) questioned the relevance of top-down approach towards governance and management of tourism. Today, global catastrophes such as the pandemic and devastating effects of climate change call for co-creative, inclusive, and more localized integration of different stakeholders of tourism. Transparency and accountability are critical. A myopic approach can restrict measures towards long-term viability. In other words, a systems approach is critical to manage sustainable planning and development of tourism. Aligned with Doppell's work, several recent scholars have called for a systems approach that take interdependencies and feedback loops into consideration (Fennell & Cooper, 2020; Hall, 2019; Liburd et al., 2022; Ruhanen et al., 2015; Uysal & Sirgy, 2023). Proactive approach by staying informed of both positive and negative impacts on the host community, being educated on sustainable practices, and an institutional approach guided by a clear vision/mission statement are critical elements. Proactive strategies are required so that important agents, of change, can coordinate in a systematic manner to achieve sustainable goals.

In the post-pandemic times, several studies report that economic sustainability continues to dominate the agenda of local residents and other important stakeholders of tourism (Dwyer, 2022; G€ossling et al., 2020; Kaygalak-Celebi et al., 2022; Movono & Hughes, 2022; Scheyvens & Cheer, 2022; Stone & Nyaupane, 2018). Acknowledging the economic potential of tourism, Uysal and Sirgy note that "governments through-out the world are investing heavily in the development, promotion, and operation of tourism destinations. Both public policy and government expenditures seem to be increasingly focused on establishing a tourism economy" while, at same time, striving to enhance the QOL (quality of life) of residents of popular destinations (2023, p. 3). The United Nations Sustainable Development Goals (SDGs) are regarded as the most viable path to guide sustainable development of tourism and ensure enhancement of the QOL of, particularly, the host community and the tourists.

Approximately 159 countries endorse the SDGs through their memberships. More specifically, through the SDGs, the United Nations is playing a key role in "advancing knowledge and tourism policies worldwide, advocating for responsible tourism and promoting tourism as a driving force towards economic growth, inclusive development and environmental sustainability" (UNWTO, 2023b). The organization facilitates enactment of "global code of ethics" in the tourism industry so that socio-economic benefits of tourism can be optimized and possible externalities minimized. The next section delves in to the SDGs in a detailed manner.

United Nations SDGs

Boluk and Rasoolimanesh (2022) write that the SDGs were particularly developed to address the most urgent obstacles associated with reducing poverty, safeguarding the planet, and ensuring equitable growth and success by year 2030 (UN, 2016). At the 2019 SDG summit, the world leaders acknowledged the critical need to speed up tasks associated with sustainability to address contemporary global issues of magnitude. The outcome was a "Decade of Action," plan that stresses on the need to fast-track and marshal sustainability solutions at the global, local, and individual levels (UN, 2020). In the year 2016, the United Nations SDGs' plan launched 17 goals based on 169 targets and an initial list of 230 markers. It is a comprehensive framework designed to seek solutions to economic, social, and environmental issues so that an inclusive and sustainable future can be planned for the future generations. Several inherent contradictions between the different goals have been reported by some scholars, particularly in regard to socio-economic and environmental sustainability indicators (Dwyer, 2022; Voola et al., 2022). Having said that, on a positive note, these SDGS use universal language and aim at strengthening collaborations between a broad spectrum of stakeholders to collectively address pressing challenges impeding the sustainable initiatives. The SDG agenda has been welcomed by tourism stakeholders across the globe (Laimer, 2017; Dwyer, 2022; G€ossling & Hall, 2019). The United Nations World Tourism Organization (now called United Nations Tourism) recognizes the tremendous potential of the tourism industry to contribute to the success of the SDGs and has taken numerous initiatives to strengthen partnerships (Dwyer, 2022) between a broad spectrum of stakeholders such as "other UN agencies, national governments, development banks, private sector actors" (UNWTO, 2018).

The UN Tourism (formerly known as UNWTO) recognizes that critical role of the tourism industry in embracing the SDGS and achieving desirable outcomes (UNWTO, 2023b). Tan et al. (2023) offer an overview of how the tourism industry can align itself with the SDGs and shape a sustainable path going forward. For the most part, these SDGs are not treated separately but as a set of interconnecting goals, supporting the key pillars of sustainability, integrated in an umbrella framework (Nunkoo et al., 2023). Tan et al. (2023) offer an SDG agenda for the tourism industry. It is argued that from a social standpoint, the tourism industry can promote an inclusive environment and advocate global solutions to resolve challenges associated with societies across the world such as inequality and marginalization. More conducive tourism and alternative forms of tourism can be advocated such as pro-poor tourism, social tourism, solidarity tourism, volunteer tourism, and emotional peace tourism. Economic optimization strategies expand opportunities to expand jobs, generate more taxes, and promote health and wellness-related initiatives. Infrastructure improvements in destinations and renovation projects hold potential to enhance the local quality of life. Backward and forward local supply chain linkages can be strengthened (Hall, 2019; Movono & Hughes, 2022). Environmental measures can safeguard against damage to the wildlife and the fragile natural resources. Furthermore, Tan et al. (2023) opine that collaborative consumption is a key feature of the sharing economy and it is flourishing today. Evidently the tourism industry offers a promising platform to bring several SDGs to fruition. Like any other business, success requires carefully crafted scientifically supported strategies associated with "strategic and operational management, supported by professional evidence-based tourism planning and development and empowered by smart tourism tools" (Tan et al., 2023, p. 306). It is important to avoid externalities associated with over-indulgence and overuse such as in overtourism scenarios (Butler & Dodds, 2022; Walmsley et al., 2022; Wall, 2020).

The year 2017 was celebrated as the "International Year of Sustainable Tourism for Development," and it was recognized that the tourism holds tremendous potential and can play a critical role in supporting the SDGs such as SDG 12 (Responsible Consumption and Production), SDG 8 (Decent Work and Economic Growth), and SDG 14 (Life Below Water) (UN, 2020; UNWTO, 2023b). Extant literature recognizes the challenges associated with mobilizing these goals and other relevant SDGs as initiatives were and are being planned to utilize the code of ethics to guide sustainability (e.g., Boluk et al., 2019; Bramwell et al., 2017; Hall, 2019). However, the capability of the tourism and hospitality industry to expedite sustainable development and aid in the accomplishment of the SDGs has been questioned by some scholars and practitioners (Dwyer, 2022; Scheyvens, 2018). Part of the issue can be attributed to the lack of unanimity with regard to the connotations of sustainable tourism and a broad range of sustainable indicators which has led to unorganized and inconstant management of tourism impacts (Butler, 1998).

The catastrophic impact of COVID-19 on the tourism industry has generated a critical need to strategically build resilient communities and businesses, and SDGs can perform an inspiring role in this regard. As Boluk writes "the global framework acts as a reminder that the status quo must continue to be challenged. Specifically, the growth trajectory supported by capitalism is explicitly recognized and supported in SDG 12 Responsible Consumption and Production. Clearly, it is important to recognize that the business-as-usual neoliberal ideology will not help us in this moment, as we think through overcoming COVID-19 and its resulted impacts" (2022, p. 3). Several scholars recognize that the pandemic has altered the outlook and performance of key stakeholders of tourism such as residents, tourists, and businesses (Adie et al., 2022; Liburd et al., 2022; Scheyvens & Cheer, 2022; Singh & Aggarwal, 2013; Sharpely, 2020). Therefore, "understanding these changes to establish a more sustainable future for tourism and positive contribution to the SDGs is paramount and should be considered in the process of tourism recovery" (Boluk & Rasoolimanesh, 2022, p. 4). In the current times, it has also been noted that some countries are simultaneously dealing with dual issues of poverty and climate change (Tan et al., 2023; Walker et al., 2019). Furthermore, as the application of technology accelerated during the pandemic, the significant role of technology in tourism in accomplishing the SDGs is acknowledged by numerous studies (e.g., Ali et al., 2020; Azmat et al., 2023; Glyptou, 2022; Uysal & Sirgy, 2023). Therefore, it is important to devise ways in which technology can be utilized to integrate SDGS in tourism. As a significant step in this direction, the next section shares the initiatives of the G20 Tourism Working Group at the 2023 G20 summit, in India. The Group of Twenty (G20) is an assemblage of 20 of the globe's leading economies. It was formed in 1999 and was envisaged as a group that would join the most important developed and developing economies to deliberate on economic and commercial stability across the world (McBride et al., 2023).

G20 Tourism Working Group

G20 is an intergovernmental forum comprising of a group of 19 autonomous countries, the European Union, and the African Union. Its key goals are to work towards resolution of issues associated with the world economy such as global economic stability, abating climate change and sustainable growth. The Ministry of Tourism in India organized the fourth G20 Working Tourism group forum in Goa (a state on the southwestern coast of India) to develop a pathway for more sustainable, resilient, and inclusive tourism. Five priority areas were identified: green tourism, digitalization, skills, tourism micro- and small MSMEs (micro small and medium enterprises), and destination management. One outcome of the summit is the Goa Roadmap for tourism. UNWTO, as the knowledge partner of India's G20 Presidency, has developed the "Goa Roadmap (named after the place where the summit took place) for Tourism as a Vehicle for Achieving the Sustainable Development Goals." The Roadmap is built around the six priority areas identified by the Tourism Working Group under the leadership of India's G20 Presidency:

- Green development, climate action, and LIFE (lifestyle for environment):
- Accelerated inclusive and resilient growth
- Accelerating progress on the SDGs
- Technological transformation and digital public infrastructure
- Multilateral institutions for the twenty-first century and three Fs—food, fuel, and fertilizers, and women-led development (G20 2023, p. 10)

Priority areas under green tourism include sustainable infrastructure and management of resources, lending support to land/people and local inhabitants, developing sustainable and inclusive economies guided by credible data and innovation, and supporting actions steered by visitors to promote responsible consumption. The next priority area is digitalization, and the focus is to extend digital products/services during the entire journey of the visitor, formulating policies and guidelines for digitalization of tourism, building control panels to oversee activities under travel for "LIFE." The third priority area is focused on "SKILL" where the emphasis is on building competencies for reorienting strategies for resilient tourism to ensure different needs of the tourism industry are met in addition to developing future-oriented training prospects. "MSMEs" is the fourth priority area which focuses on offering resources to traditional "MSMEs" to enable them to stay relevant and offering a conducive environment to help them grow and be successful and encouraging initiatives managed to support the "MSMEs." The fifth priority area is "DESTINATION MANAGEMENT," and the aim of this area is to position the DMOs (destination marketing organizations) as system organizers to create innovative business models (G20 2023).

The summit conducted a survey, of G20 members, to identify ways in which tourism can contribute to top Sustainable Development Goals. The key SDGs identified for consideration for the suggested roadmap were (G20 2023, p. 11):

- SDG 8: Decent Work and Economic Growth
- SDG 9: Industry, Innovation, and Infrastructure
- SDG 11: Sustainable Cities and Communities
- SDG 12: Responsible Consumption and Production
- SDG 17: Partnership for the Goals

Other related SDGS mentioned by the G20 countries are SDG 1 (No Poverty), SDG 2 (Zero Hunger), SDG 3 (Good Health and Well-being), SDG 4 (Quality Education), SDG 5 (Gender Equality), SDG 6 (Clean Water and Sanitation), SDG 7 (Affordable and Clean Energy), SDG 8 (Decent Work and Economic Growth), SDG 9 (Industry, Innovation, and Infrastructure), SDG 10 (Reduced Inequality), SDG 11 (Sustainable Cities and Communities), SDG 12 (Response Consumption and Production), SDG 13 (Climate Action), SDG 14 (Life Below Water), SDG 15 (Life on Land), SDG 16 (Peace, Justice, and Strong Institutions), and SDG 17 (Partnerships).

The G20 Roadmap report includes a set of objectives and recommended actions for each of the identified areas. It highlights the interconnection between different priorities and their distinct contributions to the SDGs. It also features a collection of case studies that illustrate how the aforementioned priorities are being advanced by the G20 countries. As an instance, Table 1.1 illustrates roadmap for Tadoba-Andhari Tiger Reserve in India. Tadoba-Andhari Tiger Reserve, also known as the "Tadoba National Park" is the oldest and largest park in the state of Maharashtra in India. The word "Tadoba" is taken from "Tadoba" or "Taru" God who is worshipped by the local tribe and "Andhari" is obtained from the river "Andheri" which flows through the park. Table 1.2 offers a brief overview of roadmaps pertaining to other studies across the globe. A list of relevant SDGS is included to help formulate a forward path of action. For instance, the ADAPTUR Project is an ecosystem-based adaptation to climate change in the tourism sector, and its key components are economic risk analysis, communication strategy, strengthening of public and private sector cooperation, capacity building, planning and implementation of EbA (ecosystem-based adaptation) solutions with the involvement of all key stakeholders, gathering financial resources from the private and public sector, and rallying for development of policy on the regional and national scale (Valera, 2021). This project is funded by the German Ministry of Environment and launched in three popular tourism destinations in Mexico.

The case study in Germany focuses on offering assistance to the tourism industry to build, execute, and record effective decline in greenhouse gas (GHG) which is aligned with the targets set up by the national climate protection agenda. The goal

Table 1.1 Tadoba-Andhari Tiger Reserve, India

Challenge: India aimed to involve the local community in the tourism value chain along with conservation efforts for biodiversity at the forest service.
Initiative: The main features of the program related to community involvement include:
Development of homestays in nearby villages
Training of local villagers as tour guides, in storytelling, identification of animal calls, and sensitization towards tourists
Readiness assessment programs
Enabling sense of ownership amongst locals to ensure conservation
Creating alternate activities and experiences
Enhancing income of locals
Impact: Today, the Tadoba-Andhari Tiger Reserve is regarded as one of the world's most preferred tiger reserve destinations, with over 80 tigers in the reserve and 200 in the larger landscape. Thanks to the functional connectivity of the reserve with other protected areas in the central Indian tiger landscape, tigers in Tadoba are now found restocking the gene pools of protected areas such as Navegaon Nagzira, Umred Karhandla, and Tipeshwar and forests as far afield as Kawal, Nagarjun Sagar, and Indravati.
Relevant SDGs: 1,8,10,11,14,15, and 17

Source: G20 (2023, p. 76)

Table 1.2 Roadmap for green tourism case studies across the globe

Country	Case study	Relevant SDGs
Germany	Conception and piloting of the German Fund for Tourism	9 and 13
Italy	Sustainable Tourism Fund	8, 11, and13
Spain	Transitioning to Circular Economy	8, 9, and 11
Mauritius	Sustainable Island Mauritius	3, 6, 7, 8, 9, 12, 13, 14, 15, and 17
India	Responsible Tourism Mission	
Mexico	Adaptur Project	1, 2, 7, 8, 9, 10, 11, 13, and 17
Brazil	Waterfront Project	8, 12, 13, and 14

Source: G20 (2023, pp. 76–91)

is to increase execution of activities geared towards climate protection. The Sustainable Tourism Fund initiative in Italy aims to broaden infrastructure upgrades in a manner that minimizes environmental impacts of tourism such as reducing overcrowding, supporting eco-sustainable tourism that offers tools to support sustainable services and accommodations. The Smart Tourism Destination initiative in Spain is focused on charting a well-rounded vision for integrating technology in tourism to promote digital transformation in tourism destinations. The challenge undertaken by the country was to position itself as a digital-friendly country to enhance tourism experiences and services with the help of integrating technology by launching a network of smart tourism destinations. The Sustainable Island Mauritius case study is focused on reducing tourism-related externalities on the environment and facilitating use of sustainable products and services. This initiative was launched by the Ministry of Tourism to fostering sustainable tourism by illustrating and advancing self-supporting tools to ensure sustainable protocol was followed throughout the tourism supply chain.

The goal of the Adaptur Project in Mexico is to build useful and well-organized measures for ecosystem acclimatization metrics in collaboration with the public and private sectors. This initiative aims to unite the private and public sectors to offer tools to the tourism enterprises in their initiative to safeguard ecosystem services and natural resources. The Waterfront Project in Brazil is geared towards an organized localization of regulations to distribute power with local towns and cities and enable local communities to act on decisions related to their lands by empowering them.

Building resilience has become a key goal in the post-COVID times. Extant literature confirms that the sustainability of a tourism destination is dependent on its resilient capacity. In other words, it is a core factor in improving sustainability initiatives of different stakeholders of tourism. Resilience and sustainability are related closely albeit their attributes make them distinct from each other. While sustainability is commonly described in the context of a trade-off of economic, sociocultural, and environmental factors, and environment resilience "is the intervening circle that absorbs external stresses and supports sustainability" (Bertella, 2022, p. 146). Sustainable tourism is supported by the triple bottom-line pillars (Mihalic, 2016), and scholars recognize that these three dimensions can often conflict with each other. For instance, natural tourism relies on the natural environment but can generate externalities for the same (Matthew & Nimmi 2021; Tahiri et al., 2022). This implies that resilience is about seeking a change, while sustainability strategies are about pursuing a state of stability. Nevertheless, resilience continues to be a crucial tenet of sustainability as noted by scholars focusing on risk management and tourism sustainability. According to Espiner et al. (2017), resilience acts a mediating factor or as a facilitating cushion for sustainability. In the next section, the notion of resilience is presented, and its key models are examined in the context of their alignment with the SDGs, from a tourism standpoint.

Resilience

Resilience can be described as capability of a feature to return to its original shape after being subject to a disturbance or intervention (Amore et al., 2018; Bertella, 2022; Espinar et al., 2017; Holling, 2001; Prayag, 2023; Westman, 1978). This feature can pertain to a specific environment, community, landscape, or an ecosystem. Adaptive or socio-ecological systems mindset has been frequently used to position the notion of resilience in the context of tourism (Bertella, 2022). Prayag (2023) writes that resilience in the new normal post-COVID times is depicted as a life-saving mechanism for enterprises. The ongoing interventions of unpredictable and sudden catastrophic events (such as calamities, wars, COVID-19, climate change, and political/social turmoil) across the globe are fueling a complicated, susceptible, and volatile business landscape that can lead to high stress levels and apprehension at multiple (individual/organizational and community) levels. As a resolution mechanism, "resilience has become a powerful discourse for multiple stakeholders to reassure their constituents that the present and the future is under control and predictable" (Prayag, 2023, p. 513).

Prayag (2023) delineates resilience into three categories: engineering and ecological resilience and socio-ecological system resilience. Bouncing back has been related to the notion of engineering resilience. In other words, a system can return to its original state of equilibrium after a disturbance, and it is about understanding the extent to which it can be shaped to reach its original state (Amore et al., 2018; Espinar et al., 2017). In contrast, ecological resilience is about understanding the magnitude of disruption that a system can tolerate before it spins to a new state of equilibrium. In the context of tourism, the industry leans more towards achieving a state of ecological resilience (Holland et al., 2022; Holladay & Powell, 2013; Westman, 1978). In other words, resilient tourism systems do not strive to return to the original state but strive to strengthen adaptive capacity to absorb change and embrace a new symmetrical state. It is about interactions and adaptations to change. The socio-ecological system resilience standpoint is associated with the extent to which a system has the capacity to self-organize itself, attain new knowledge, and embrace innovation (Folke, 2006). In other words, as explained by Holling and

Gunderson (2002), from this perspective, this type of resilience is about dealing "with change while continuing to renew and develop" (Prayag, 2023, p. 514). As reiterated by Sterk et al. (2017), this standpoint highlights the significance of "environment actions from the grass-roots level," and a collaborative approach is required to scrutinize and organize chaos and disturbances in a system, particularly, due to unexpected catastrophes. Multifaceted and dynamic resilience models are crucial to aid in developing proactive and reactive strategies to deal with catastrophic interventions. The next section offers an overview of some notable resilience models.

Resilience Models

Holland et al. (2022) are of the view that an integrated enterprise-driven conservation development (ICD) model holds potential to enhance resilience of local communities. It supports strong linkages between source of livelihood and conservation of biodiversity (Stone & Nyaupane, 2018). The authors tested the robusticity of this model at the Maasai Mara National Reserve in Kenya, which is heavily reliant on wildlife tourism for livelihood. It is noted that overtourism can compromise conservation initiatives and resilience of host communities because of non-inclusive governance structures and inequitable benefits at the Reserve. Resilience of the community serves as a progressive bridge between tourism and conservation and disproportionate accumulation of income compromised social capital of the Reserve community. The authors illustrated that the ICD model can ensure synergistic benefits for the people and the community as a whole and offer a blueprint to build a strong resilient community.

Notably, most resilience frameworks are multidimensional and are shaped by socio-ecological factors. As an instance, Amore et al. (2018) use a multiple-level paradigm to illustrate the roles and exchanges between various stakeholders at different points in developing resilience of destinations. Hall (2018) ascertains communications between multiple types of resilience operating at different stages while being guided by the "panarchy model of adaptive cycle." Panarchy approach implies that adaptive systems are organized in layered gradings across space and time, and each stage can be in a different stage on the hierarchical ladder (Hall, 2018). Pragyag (2023) adds to these models by detecting different attributes of resilience and inducing agents that hold potential to build and shape resilience at different levels of a tourism system. Some common points include robustness (to ensure quick restoration through efficient processes and procedures), vulnerability (ability to identify sensitive areas which can be severely impacted by stress and disruption), risk perceptions and management (factors that can offer deep insights into the relationship between resilience and vulnerability and best practices to navigate through adverse situations), adaptive capacity (such as through psychological resilience, flexibility, and ability to adapt in chaotic times), and absorption to change (using a holistic approach and taking elements and sub-elements of a system into consideration) (Prayag, 2023).

Prayag further writes that "tourism systems already have characteristics that can limit or assist in resilience building. Thus, resilience and tourism system characteristics can interact in complex and unintended ways, requiring clarity in the intended goals for resilience building at all levels of the system and leadership in ensuing change management processes" (2023, p. 515). Key shapers and features at each level include:

- Thorough grasp of scale, intensity, and duration of the adversity
- Awareness that equilibrium in a system at the holistic and sub-level is transitory which implies that the resilient building journey is as crucial as its consequences and it is important to understand the evolutionary nature of systems
- Planning and preparedness helps to adopt a proactive approach, with contingency plans, towards long-term survival
- Path dependencies associated with economic prosperity and overtourism are easy pitfalls; governance structures are the center point of a system

By highlighting these factors and influencers, Prayag (2023) points out that it is the interaction and integration between the tourism system and its resilient features that determine or help with its optimal stability. Prayag's model illustrates both top-down and bottom-up impacts of different categories and stages of resilience, for instance, at the individual and organizational level. Individual level is more shaped by psychological resistance, employee resilience, entrepreneur resilience, tourist resilience, and resident resilience.

Bertella (2022) suggests a resilience model based on elasticity, hysteresis, and malleability and argues that these are the core characteristics of resilience and help to explain the manner in which a community might respond to a disturbance and the learning path it will embrace to strengthen itself. Elasticity refers to the quick way a system responds to a disturbance and restores itself to a state of equilibrium. It assumes that the system has the ability to bounce back to its original state. Single-loop learning approach enables a system to resolve errors and perform better my modifying behavior. The other two features, hysteresis and malleability are reliant on sustainable learning journeys. According to Rios et al. (2017), hysteresis implies that a system undergoes change after being subject to a shock, and it never returns to its original state although it tries to return as close to its original equilibrium position as possible. Malleability, on the other hand, refers to the effortlessness with which a system embraces a permanently altered equilibrium state (Westman, 1978). In summary, the core message is that the shock opens learning pathways as a destination's system attempts to reinvent itself (Amore et al., 2018; Holladay & Powell, 2013; Prayag, 2023; Westman, 1978).

Bertella's (2023) model also features on learning pathway towards resilient sustainable futures and illustrates the linkage between resilience and sustainability. It includes pre-crisis situation, shock reactions to the crisis, and weak and strong alternative future scenarios, and reactions and plans shape these paths. The first path "is characterized by elasticity and includes behavioral episodic changes that respond to the crisis with initiatives for community safety and recovery. The second path is a single-loop learning path. It is characterized by hysteretic reactions that reconsider assumptions, such as the desirability of mass tourism and the possibility of devel oping target-specific market segments through innovation. Multiple loop learning characterizes the third path. Here, the reconsidered assumptions are epistemic and involve worldviews that can be related to strong sustainability" (2023, p. 155). In a nutshell, a community's response to a crisis can suggest the manner in which a destination will embrace sustainability in the future. Bertalla write that "discussions after covid-19 outbreak showed some new themes and signs of different types of resilience and learning. Theoretically, …… the oft-cited journeys towards sustainability can be better conceptualized as a set of journeys. Focusing on the community level and viewing such journeys as learning paths has the advantage of placing the focus in the learning processes necessary for understanding and framing issues relevant to the future of destinations" (2023, p. 156). Seeing sustainability as an array of journeys towards different futures can prove to be beneficial by prompting deliberations on societal responsibility and inspiring innovative people to act.

Resilience, SDGs, and Well-Being

As pointed out earlier, human well-being is increasingly being recognized as a critical element of sustainable development plans (Costanza et al., 2020; Dwyer, 2020, 2022; Kanbur Patel & Stiglitz, 2018; Roy, 2021; Stone & Nyaupane, 2018). In fact, the notion of a well-being economy has taken center stage thereby stressing on the significance of managing economic strategies in a manner that human well-being is augmented, by placing people and their well-being at the core of governance policies and decision-making processes (Roy, 2021). Several authors have postulated that the end goal behind developing destinations for tourism purpose should be to boost the well-being of residents, the definition of which should encompass the entire population instead of the stakeholders of tourism (Crouch & Ritchie, 2012; Berbekova et al., 2021).

Dwyer (2020) aptly points out that "the thrust of argument is to support the use of well-being metrics alongside economic, social and environmental indicators to capture changes in resident 'quality of life' associated with industry development, and to better assess and design policy to support sustainable development" (Adler & Seligman, 2016). Here the 'resident' notion extends beyond tourism stakeholders to include the entire population of a region (Dwyer, 2020). As more and more scholars are highlighting the significance of resolving issues related to well-being (Smith & Diekmann, 2017; Hartwell et al., 2018), more attention needs to shift towards scrutinizing and strategizing SDG targets to ensure favorable outcomes can be accomplished (Dwyer, 2022).

From an economic standpoint, recent trends promote the concept of "well-being economy" placing the wellness and overall health on the priority agenda in policy formulation and policy-making processes. There is a consensus that the key goal of economics should be to facilitate human well-being (Dwyer, 2022; Kanbur Patel & Stiglitz, 2018; Roy, 2021). However, in reality, tourism growth is shadowed by substantial economic, social, and environmental externalities (Dwyer, 2018), and therefore, assessment of tourism success in the attainment of SDG is worsened by the fact that success in one SDG might be at the cost of another SDG (Laimer, 2017). For instance, economic growth that brings revenue and reduces poverty is likely to deteriorate the natural landscape through overcrowding and pollution. Another potential conflict can be between the SDG aimed at protecting ecosystem and the ones addressing hunger-related issues and improving health.

In view of the aforementioned conflicts, Dwyer (2022) writes that some types of tourism development are more likely to obstruct the progress of SDGs and the key issue to resolve is arrive at a trade-off point. The SDG agenda is silent on that front and "it has been argued an aggregate indicator of overall progress within the SDG 2030 agenda is required to determine the relative contribution of each of the SDGs and the effects of their interactions with each other (Qasim, 2017)" (Dwyer, 2022, p. 5). The resolution of this dilemma can lie in developing an umbrella goal that focuses on "sustainable well-being" which refers to prosperity, inclusivity, and sustained initiatives that can promote enhanced quality of life (Costanza et al., 2016). It is important to remember that sustainable well-being goal is a complicated target, and trade-off between various values is likely to remain a contentious issue. In resolution of this dilemma, Dwyer suggests that "a useful option is to determine the well-being outcomes of the various strategies used to achieve the SDGs. Tourism strategies can achieve well-being outcomes for destination residents directly through policy initiatives" (2022, p. 5).

Next chapter offers insights into the notions of health, well-being, and wellness. It stresses on the need to modify the conventional triple bottom-line pillars of sustainability by adding a fourth (health) pillar.

References

Adie, B. A., Amore, A., & Hall, C. M. (2022). Just because it seems impossible, doesn't mean we shouldn't at least try: The need for longitudinal perspectives on tourism partnerships and the SDGs. *Journal of Sustainable Tourism, 30*(10), 2282–2297.

Adler, A., & Seligman, M. (2016). Using wellbeing for public policy: Theory, measurement, and recommendations. *International Journal of Wellbeing, 6*(1), 1–35.

Ali, A., Rasoolimanesh, S. M., & Cobanoglu, C. (2020). Editorial. *Journal of Hospitality and Tourism Technology, 11*(2), 177–181.

Amore, A., Prayag, G., & Hall, C. M. (2018). Conceptualizing destination resilience from a multilevel perspective. *Tourism Review International, 22*(3–4), 235–250.

Azmat, F., Lim, W. M., Moyeen, A., Voola, R., & Gupta, G. (2023). Convergence of business, innovation, and sustainability at the tipping point of the sustainable development goals. *Journal of Business Research, 167*, 114170.

Berbekova, A., Uysal, M., & Assaf, A. (2021). Toward an assessment of quality of life indicators as measures of destination performance. *Journal of Travel Research, 61*(6), 1424–1436.

Bertella, G. (2022). Discussing tourism during a crisis: Resilient reactions and learning paths towards sustainable futures. *Scandinavian Journal of Hospitality and Tourism, 22*(2), 144–160.
Bertella, G. (2023). Care-full academic activism for sustainable transformations in tourism. *Current Issues in Tourism, 26*(2), 212–223.
Boluk, K. A., & Rasoolimanesh, S. M. (2022). Introduction to the special issue on "Deepening our understandings of the roles and responsibilities of the tourism industry towards the United Nations Sustainable Development Goals (SDGs)". *Tourism Management Perspectives, 41*, 100944.
Boluk, K., Cavaliere, C., & Higgins-despoilees, F. (2019). A crtical framework for interrogating the United Nations sustainable development goals 2030 agenda in tourism. *Journal of Sustainable Tourism, 27*(7), 847–864.
Bramwell, B. (2015). Theoretical activity in sustainable tourism research. *Annals of Tourism Research, 54*, 204–218.
Bramwell, B., Higham, J., Lane, B., & Miller, G. (2017). Twenty-five years of sustainable tourism and the Journal of Sustainable Tourism: Looking back and moving forward. *Journal of Sustainable Tourism, 25*(1), 1–9.
Bricker, K., Black, R., & Cottrell, S. (2013). *Sustainable tourism & the millennium development goals: Effecting positive change*. Jones and Bartlett.
Butler, R. W. (1998). Sustainable tourism: Looking backwards in order to progress? In C. M. Hall & A. Lew (Eds.), *Sustainable tourism: A geographical perspective* (pp. 25–34). Addison Wesley Longman.
Butler, R. W., & Dodds, R. (2022). Overcoming overtourism: a review of failure. *Tourism Review, 77*(1), 35–53.
Costanza, R., Fioramonti, L., & Kubiszewski, I. (2016). The UN sustainable development goals and the dynamics of well-being. *Frontiers in Ecology and the Environment, 14*(2), 20–22.
Costanza, R., Erickson, J. D., Farley, J., & Kubiszewski, I. (Eds.). (2020). *Sustainable wellbeing futures: A research and action agenda for ecological economics*. Edward Elgar Publishing.
Crouch, I. G., & Ritchie, B. J. R. (2012). Destination competitiveness and its implications for host-community QOL. In M. Uysal, R. Perdue, & M. J. Sirgy (Eds.), *Handbook of tourism and quality-of-life research: Enhancing the lives of tourists and residents of host communities* (pp. 491–514). Springer.
Doppell, B. (2003). Overcoming the seven sustainability blunders. *The Systems Thinker, 14*(5), 2–7.
Dwyer, L. (2018). Saluting while the ship sinks: The necessity for tourism paradigm change. *Journal of Sustainable Tourism, 26*(1), 29–48.
Dwyer, L. (2020). Tourism development and sustainable well-being: A beyond GDP perspective. *Journal of Sustainable Tourism, 28*, 1–18.
Dwyer, L. (2022). Tourism contribution to the SDGs: Applying a well-being lens. *European Journal of Tourism Research, 32*, 3212–3212.
Espiner, S., Orchiston, C., & Higham, J. (2017). Resilience and sustainability: A complementary relationship? Towards a practical conceptual model for the sustainability–resilience nexus in tourism. *Journal of Sustainable Tourism, 25*(10), 1385–1400.
Fennell, D. A. (2018). *Tourism ethics*. Channel View.
Fennell, D. A., & Cooper, C. (2020). *Sustainable tourism: Principles, context and practices*. Channel View.
Folke, C. (2006). Resilience: The emergence of a perspective for social–ecological systems analyses. *Global Environmental Change, 16*(3), 253–267.
G20 (2023). Goa roadmap for tourism as a vehicle for achieving the sustainable development goals: https://www.ilo.org/sites/default/files/wcmsp5/groups/public/@ed_dialogue/@sector/documents/genericdocument/wcms_887681.pdf. Accessed January 2024.
G€ossling, S., & Hall, C. M. (2019). Sharing versus collaborative economy: How to align ICT developments and the SDGs in tourism? *Journal of Sustainable Tourism, 27*(1), 74–96.
G€ossling, S., Scott, D., & Hall, C. M. (2020). Pandemics, tourism and global change: A rapid assessment of COVID-19. *Journal of Sustainable Tourism, 29*(1), 1–20.

Glyptou, K. (2022). Operationalising tourism sustainability at the destination level: A systems thinking approach along the SDGs. *Tourism Planning & Development, 21*, 1–27.

Hall, C. M. (2018). Resilience theory and tourism. In *Resilient destinations and tourism* (pp. 34–47). Routledge.

Hall, C. M. (2019). Constructing STD: The 2030 Agenda and the managerial ecology of sustainable tourism. *Journal of Sustainable Tourism, 27*(7), 1044–1060.

Hartwell, H., Fyall, A., Willis, C., Page, S., Ladkin, A., & Hemingway, A. (2018). Progress in tourism and destination wellbeing research. *Current Issues in Tourism, 21*(16), 1830–1892.

Holden, A. (2005). Achieving a sustainable relationship between common pool resources and tourism: The role of environmental ethics. *Journal of Sustainable Tourism, 13*(4), 339–352.

Holladay, P. J., & Powell, R. B. (2013). Resident perceptions of social–ecological resilience and the sustainability of community-based tourism development in the Commonwealth of Dominica. *Journal of Sustainable Tourism, 21*(8), 1188–1211.

Holland, K. K., Larson, L. R., Powell, R. B., Holland, W. H., Allen, L., Nabaala, M., et al. (2022). Impacts of tourism on support for conservation, local livelihoods, and community resilience around Maasai Mara National Reserve, Kenya. *Journal of Sustainable Tourism, 30*(11), 2526–2548.

Holling, C. S. (2001). Understanding the complexity of economic, ecological, and social systems. *Ecosystems, 4*, 390–405.

Holling, C. S., & Gunderson, L. H. (2002). Resilience and adaptive cycles. In *Panarchy: Understanding transformations in human and natural systems* (pp. 25–62). Island Press.

Kanbur, R., Patel, E., & Stiglitz, J. E. (2018). Sustainable development goals and the measurement of economic and social progress. In J. E. Stiglitz, J.-P. Fitoussi, & M. Durand (Eds.), *For good measure: Advancing research on well-being metrics beyond GDP* (pp. 33–48). OECD Publishing.

Kaygalak-Celebi, S., Ozeren, E., & Aydin, E. (2022). The missing link of the Sustainable Development Goals (SDGs) in tourism: A qualitative research on Amsterdam Pride. *Tourism Management Perspectives, 41*, 100937.

Laimer, P. (2017). Tourism indicators for monitoring the SDGs. In *Sixth UNWTO international conference on tourism statistics, measuring sustainable tourism* (pp. 1–33).

Liburd, J., Duedahl, E., & Heape, C. (2022). Co-designing tourism for sustainable development. *Journal of Sustainable Tourism, 30*(10), 2298–2317.

Mathew, P., & Nimmi, P. (2021). Sustainable tourism development: Discerning the impact of responsible tourism on community well-being. *Journal of Hospitality and Tourism Insights, 5*(5), 987–1001.

McBride, J., Siripurapu, A., & Berman, N. (2023). What does the G20 do? https://www.cfr.org/backgrounder/what-does-g20-do. Accessed in December 2023.

Mihalic, T. (2016). Sustainable-responsible tourism discourse—Towards 'responsustable' tourism. *Journal of Cleaner Production, 111*, 461–470.

Movono, A., & Hughes, E. (2022). Tourism partnerships: Localizing the SDG agenda in Fiji. *Journal of Sustainable Tourism, 30*(10), 2318–2332.

Nunkoo, R., Sharma, A., Rana, N. P., Dwivedi, Y. K., & Sunnassee, V. A. (2023). Advancing sustainable development goals through interdisciplinarity in sustainable tourism research. *Journal of Sustainable Tourism, 31*(3), 735–759.

Prayag, G. (2023). Tourism resilience in the 'new normal': Beyond jingle and jangle fallacies? *Journal of Hospitality and Tourism Management, 54*, 513–520.

Qasim, M. (2017). Sustainability and wellbeing: a scientometric and bibliometric review of the literature. *Journal of Economic Surveys, 31*(4), 1035–1061.

Rios, L. A., Rachinskii, D., & Cross, R. (2017). A model of hysteresis arising from social interaction within a firm. *Journal of Physics: IOP: Conference Series, 811*(1), 012011.

Roy, M. J. (2021). Towards a 'Wellbeing Economy': What can we learn from social enterprise? In *The new social and impact economy: An international perspective* (p. 269). Springer.

Ruhanen, L., Weiler, B., Moyle, B. D., & McLennan, C. L. J. (2015). Trends and patterns in sustainable tourism research: A 25-year bibliometric analysis. *Journal of Sustainable Tourism, 23*(4), 517–535.

Scheyvens, R. (2018). Linking tourism to the sustainable development goals: A geographical perspective. *Tourism Geographies, 20*(2), 341–342.

Scheyvens, R., & Cheer, J. M. (2022). Tourism, the SDGs and partnerships. *Journal of Sustainable Tourism, 30*(10), 2271–2281.

Sharpley, R. (2020). Tourism, sustainable development and the theoretical divide: 20 years on. *Journal of Sustainable Tourism, 28*(11), 1932–1946.

Singh, H., & Aggarwal, N. (2013). Achieving sustainable development goals through elevating socio-economic status. *Competitiveness Review: An International Business Journal, 23*(4/5), 398–407.

Smith, M. K., & Diekmann, A. (2017). Tourism and wellbeing. *Annals of Tourism Research, 66*, 1–13.

Sterk, M., van de Leemput, I. A., & Peeters, E. T. (2017). How to conceptualize and operationalize resilience in socio-ecological systems? *Current Opinion in Environmental Sustainability, 28*, 108–113.

Stone, M. T., & Nyaupane, G. P. (2018). Protected areas, wildlife-based community tourism and community livelihoods dynamics: Spiraling up and down of community capitals. *Journal of Sustainable Tourism, 26*(2), 307–324.

Tahiri, A., Kovaçi, I., & Trajkovska Petkoska, A. (2022). Sustainable tourism as a potential for promotion of regional heritage, local food, traditions, and diversity—case of Kosovo. *Sustainability, 14*(19), 12326.

Tan, G., Nunkoo, R., & Farmaki, A. (2023). Tourism 2030 and the contribution to the sustainable development goals: The tourism review viewpoint. *Tourism Review, 78*(2), 293–231.

United Nations (UN). (2016). Transforming our world: The 2030 agenda for sustainable development. Retrieved August 24, 2020, from https://sustainabledevelopment.un.org/content/documents/21252030%20Agenda%20for%20Sustainable%20Development%20web.pdf

United Nations (2020). Global indicator framework for the Sustainable Development Goals and targets of the 2030 Agenda for Sustainable Development Retrieved from: https://unstats.un.org/sdgs/indicators/indicators-list/

UNWTO (2018). Tourism and the sustainable development goals: journey to 2030. Retrieved from: https://www.eunwto.org/doi/pdf/10.18111/9789284419401

UNWTO. (2023a). *Sustainable development*. Retrieved on December 2023: https://www.unwto.org/sustainable-development

UNWTO. (2023b). *G20 tourism and SDGs dashboard*. Retrieved December 2023: https://tourism4sdgs.org/g20india/

UNWTO. (2024). *UNWTO becomes "UN Tourism" to Mark A New Era for global sector*. Retrieved January 24 from: https://www.unwto.org/news/unwto-becomes-un-tourism-to-mark-a-new-era-for-global-sector?utm_source=news&utm_medium=crm

Uysal, M., & Sirgy, M. J. (2023). Prologue: Tourism and quality of life (QoL) research II. In *Handbook of tourism and quality-of-life research II: Enhancing the lives of tourists, residents of host communities and service providers* (pp. 1–8). Springer International Publishing.

Valera, D. (2021). *ADAPTUR: Mainstreaming Ecosystem-based Adaptation (EbA) in the tourism sector as a strategy for NDC implementation in Mexico*. Retrieved in November 2023 from: https://panorama.solutions/en/solution/adaptur-mainstreaming-ecosystem-based-adaptation-eba-tourism-sector-strategy-ndc

Voola, R., Carlson, J., Azmat, F., Viet Ngo, L., Porter, K., & Sinha, A. (2022). Re-imagining marketing scholarship in the era of the UN sustainable development goals. *Australasian Marketing Journal, 30*(2), 97–106.

Walker, J., Pekmezovic, A., & Walker, G. (2019). *Sustainable development goals: Harnessing business to achieve the SDGs through finance, technology and law reform*. Wiley.

Wall, G. (2020). From carrying capacity to overtourism: A perspective article. *Tourism Review, 75*(1), 212–215.
Walmsley, A., Koens, K., & Milano, C. (2022). Overtourism and employment outcomes for the tourism worker: Impacts to labour markets. *Tourism Review, 77*(1), 1–15.
Weaver, D. (2009). Reflections on sustainable tourism and paradigm change. In S. G€ossling, M. Hall, & D. Weaver (Eds.), *Sustainable tourism futures: Perspectives on systems, restructuring and innovations* (pp. 33–40). Routledge.
Westman, W. E. (1978). Measuring the inertia and resilience of ecosystems. *Bioscience, 28*(11), 705–710.

Chapter 2
Intersection Between Health, Wellness, and Quality of Life Notions

Deepak Chhabra

Abstract This chapter focuses on the notion of health, wellness, and quality of life from the standpoint of tourists and local residents. Sustainability is the over-arching umbrella under which these terms and the quality of life phenomena are scrutinized. First, the concepts of health and wellness/well-being are described. This is followed by a critical discourse on the surging popularity of wellness tourism. Last, deliberations are offered regarding the association between quality of life and wellness.

Keywords Health · Wellness · Quality of life · Wellness tourism

Health

Health can be defined as "a state of complete physical, mental and social well-being and not merely the absence of disease or infirmity" (WHO, n.d.). In other words, health is a goal and wellness is a path to accomplish it (Meera & Vinodan, 2019). The WHO (World Health Organization) promotes high quality of health based on principles such as holistic well-being, equitable access to health resources, collaboration between people and the government, predominant focus on children's health, accurate dissemination of knowledge and ongoing feedback from the people, and active role of the government in ensuring accurate/meaningful health and social protocol is in place. Table 2.1 presents a detailed breakdown of these principles.

Until recent times, health and well-being were primarily featured in the social dimension of sustainability. However, in the post-COVID times, several scholars have recommended extending the sustainability paradigm to include human health as the fourth pillar (Marques et al., 2022). This viewpoint can be attributed to the growing public health problems. Recent studies argue that most of public health

D. Chhabra (✉)
School of Community Resources and Development, Watts College of Public Service and Community Solutions, Arizona State University, Phoenix, AZ, USA
e-mail: Deepak.chhabra@asu.edu

D. Chhabra et al. (eds.), *Sustainable Development and Resilience of Tourism*,
https://doi.org/10.1007/978-3-031-63145-0_2

Table 2.1 Principles of Health

Health is a state of complete physical, mental, and social well-being and not merely the absence of disease or infirmity.
The enjoyment of the highest attainable standard of health is one of the fundamental rights of every human being without distinction of race, religion, political belief, economic, or social condition.
The health of all peoples is fundamental to the attainment of peace and security and is dependent on the fullest cooperation of individuals and states.
The achievement of any state in the promotion and protection of health is of value to all.
Unequal development in different countries in the promotion of health and control of diseases, especially communicable disease, is a common danger.
Healthy development of the child is of basic importance.
The ability to live harmoniously in a changing total environment is essential to such development.
The extension to all peoples of the benefits of medical, psychological, and related knowledge is essential to the fullest attainment of health.
Informed opinion and active cooperation on the part of the public are of the utmost importance in the improvement of the health of the people.
Governments have a responsibility for the health of their peoples which can be fulfilled only by the provision of adequate health and social measures.

Source: WHO (2006)

issues can be reduced or resolved with the assistance of preventive measures and a proactive wellness agenda (Atal & Atal, 2024). As the demand for health tourism soars, it is important to ensure that it is promoted and experienced in a sustainable manner. Sustainable health tourism can also be referred as tourism that aims at physical and mental health through medical and wellness activities which hold potential to enhance both physical and mental health (Meera & Vinodan, 2019).

Health received scant attention in early tourism studies. Until the 1980s, medical services and facilities did not focus on tourists per se. However, in recent times, healthcare has become big business with travelers willing to travel far to other countries for health-related services and medical treatment and for the purpose of staying and enhancing health (Goodrich & Goodrich, 1987; Mueller & Kaufmann, 2001). The parameters of health tourism have cognitively expanded, and people are keen to travel to improve or maintain health, and this has steered toward "a commercial phenomenon" geared toward "improving health tourism services" (Hall, 2011). Examples of health-related travel include travel by tourists for physical examination associated with disease prevention or cosmetic surgery. Faroldi, Fabi, Vettori, Gola et al. write that "the recent evolution of European health policies and tourism market trends highlights the increasing attention to health promotion strategies along with healthcare tourism innovation" (2019, p. 405).

Health tourism can be considered a canopy consisting of medical tourism and wellness tourism. The former refers to medical treatment as the main travel motivation, while the latter is about travel to prevent health-related issues (Gulyas & Molnar, 2023; Mueller & Kaufman, 2001). Mental health crisis has been noted by several studies, particularly during the pandemic and in the post-COVID times.

Some scholars have argued that existing medical treatments often distance the patient from outdoor activities which can, in fact, help in curing by facilitating connection with nature (Walker et al., 2017). For instance, Walker et al. (2017) postulate that walking and outdoor adventure activities and engaging in art should be considered important health and well-being resources (Walker et al., 2017). Outdoor adventure activities instill deep connection with nature and inspire appreciation which can facilitate pro-environmental behavior and promote sustainable societies (Brymer et al., 2009). In fact, a significant correlation between engaging with the environment and sustainable behavior has been reported by several studies (Hanna et al., 2015; Klaniecki et al., 2018; Rosa et al., 2018).

Well-Being and Wellness Notions

While wellness and well-being notions have been used interchangeably, they have different connotations. Table 2.2 presents an overview of similarities and dissimilarities between the wellness and well-being terms, based on the Global Wellness Institute. Well-being is also often considered a proxy for "happiness" (e.g., McCabe & Johnson, 2013) and that which reduces stress. The Global Wellness Institute (GWI 2024) is a nonprofit organization with an overarching aim to empower wellness across the globe by "educating public and private sectors about preventive health and wellness." It is pointed out that wellness is about indulging in activities, decision-making and developing a lifestyle that promotes a complete state of health.

Besides the description offered by the GWI, several scholars have made an effort to conceptualize the notion of human well-being (HWB). For instance, Summers et al. (2012) delineate the subjective and objective domains of well-being. While the former relates to benefits such as health, place identity, and inspiration, the latter comprises of tangible and social traits offering fundamental human needs in addition to supporting economic and environmental requirements that enhance the overall welfare of the society (Bryce et al., 2016; Nowak-Olejnik et al., 2022). Interaction with nature has been specifically noted to enhance psychological and physical

Table 2.2 Similarities and dissimilarities between wellness and well-being

Wellness	Well-being and happiness
Relates to intention, action, activities	Perceptions of a state of being
Has a prominent physical dimension	Has a prominent mental/emotional dimension
Consumers associate it with healthy lifestyles, choices, and market offerings	Citizens associate it with feelings of satisfaction and sense of fulfillment
Common usage in business/private industry context (e.g., wellness sector)	Common usage in policy/government context (e.g., well-being budgeting)
Useful concept for measuring industry, size and discussing business opportunities (e.g., wellness economy)	Useful for measuring individual/citizen welfare (e.g., global happiness index) and for organizing policies and interventions

Source: GWI (2020, 2024)

health by reducing stress, depression, anxiety, blood pressure, obesity, and diabetes, while improving immune function, happiness, cognitive development, and social behavior (McDougall et al., 2020; Willis, 2015). Well-being is commonly associated with human flourishing and refers to "feeling or being well" and "being away" to participate in activities that enhance lifestyle (Gatrell, 2013). The HWB (human well-being) notion postulates that well-being leads to good health and vice versa to reach a state of holistic well-being from psychological, physiological, and social perspectives (Park et al., 2011; Riva & Curtis, 2016). It is a static state and can fluctuate based on communication and bonds with other people or with inanimate objects (such as material space, the natural environment, other nonhuman species, etc.) (Walker et al., 2017).

Wellness

In tracing the history of wellness, the pioneer work of Dunn, Ardell, and Hettler is worthy of attention. Dunn (1957) described wellness as a holistic term encompassing physical, mental, cultural, spiritual, and social domains. The historical contribution of Hettler's (1976) six-dimensional model also stands capturing social, physical, spiritual, occupational, intellectual, and emotional dimensions. The social dimension refers to amicable relationship with the environment, overall well-being of the community, and minimization of conflicts. The physical dimension emphasizes on the prevention of harmful practices and advocacy of activities that boost health, such as routine physiological activity and a healthful regime. The spiritual domain emphasizes on building a value system and forbearance toward others whereas "the occupational dimension encompasses the profession, a career that suits the individual's personality and is in harmony with the personal values and interests," so that it "becomes a source of happiness." The identification of intellectual problems and solving them in the most effective way significantly decreases stress. The emotional dimension draws attention to the importance of an optimistic perspective (Gulyas & Molnar, 2023, p. 173).

As indicated earlier, the Global Wellness Institute defines wellness as "the active pursuit of activities, choices and lifestyles that lead to a state of holistic health" GWI (2020). See Table 2.3 for a more detailed description of wellness.

GWI also delineates wellness into multiple dimensions such as the following:

- Physical (nourishing a healthy body through exercise nutrition, sleep, etc.)
- Mental (engaging the world through learning, problem-solving, creativity, etc.)
- Emotional (being aware of, accepting and expressing our feelings, and understanding the feelings of others)
- Spiritual (searching for meaning and higher purpose in human existence)
- Social (connecting and engaging with others and our communities in meaningful ways)
- Environmental (fostering positive inter-relationships between planetary health and human actions, choices, and well-being) (GWI, 2020, n.d.a, b)

Table 2.3 The meaning of wellness

Wellness is a modern word with ancient roots. The key tenets of wellness as both preventive and holistic can be traced back to ancient civilizations from the East (India, China) to the West (Greece, Rome). In nineteenth-century Europe and the United States, a variety of intellectual, religious, and medical movements were developed in parallel with conventional medicine. With their focus on holistic and natural approaches, self-healing and preventive care, these movements have provided a firm foundation for wellness today. Wellness-focused and holistic modalities have gained more visibility since the 1960s/1970s under the writings and thought leadership of an informal network of US physicians and thinkers (such as Halbert Dunn, Jack Travis, Don Ardell, Bill Hettler, and others). As these have evolved, proliferated, and gone mainstream, they have informed the healthy-living, self-help, self-care, fitness, nutrition, diet, and spiritual practices that have become a flourishing wellness movement in the twenty-first century.
There are two important aspects to this definition. First, wellness is not a passive or static state but rather an "active pursuit" that is associated with intentions, choices, and actions as we work toward an optimal state of health and well-being. Second, wellness is linked to holistic health—that is, it extends beyond physical health and incorporates many different dimensions that should work in harmony.
Wellness is an individual pursuit—we have self-responsibility for our own choices, behaviors, and lifestyles—but it is also significantly influenced by the physical, social, and cultural environments in which we live.
Wellness is often confused with terms such as health, well-being, and happiness. While there are common elements among them, wellness is distinguished by not referring to a static state of being (i.e., being happy, in good health, or a state of well-being). Rather, wellness is associated with an active process of being aware and making choices that lead toward an outcome of optimal holistic health and well-being.

Source: GWI (2020)

GWI shares a wellness scale that stretches from sickness to optimal well-being. At one end of the spectrum are sick patients who engage with medicine to treat their illness, and on the other end, there are people who are proactively engaged in self-responsible and preventive behavior. To optimize their vitality, they embrace attitudes and lifestyles that protect them from disease and enhance their health, quality of life, and well-being. It is important to note that wellness is not the same as healthcare. As shared by GWI, the "healthcare systems use a *pathogenic* and reactive approach, focused on causes, consequences, diagnosis and treatment of diseases and injuries. In contrast, wellness is a *salutogenic* and proactive approach, focused on prevention, healthy lifestyles and the pursuit of optimal wellbeing" (GWI, 2020, n.d.a, b).

Wellness Tourism

From a tourism standpoint, wellness is about targeting a robust state of health and offering a path to accomplish goals which strive toward holistic health (Dillette et al., 2021). Majeed and Kim (2023) refer to wellness tourism as voyages that enhance health and well-being (HWB) of people and shape the anatomy of the body, mind, and soul. Mueller and Kaufman define wellness as "a state of health featuring the harmony of body, mind and spirit, with self-responsibility, physical fitness/beauty care, nutrition/healthy diet, relaxation (de-stressing)/meditation, mental activity/education and environmental sensitivity/social contacts, as fundamental elements" (2001, p. 6). Examples of wellness-related travel include visiting a retreat that offers healing services, spa, or resort for holiday trip (DeMicco, 2017; Kotur, 2022; Thal & Hudson, 2019). Initially, wellness tourism was considered a proxy for spa tourism (Laing & Weiler, 2007; Smith et al., 2009). Recent studies argue that wellness tourism is a broad term and encompasses more than spa-related facilities and services.

Wellness tourism has played a significant role in the accelerated growth of the wellness industry (Voigt et al., 2011). It was estimated to generate $919 billion in 2022 (GWI, 2020). According to Gulyas and Molnar, "increasing stress, a fast-paced lifestyle, changing the traditional community structures, attention to health preservation and disease prevention, and, in many cases, skepticism toward Western medicine" have led to the surge in demand for wellness services (2023, p. 172). As mentioned earlier, wellness tourism is often described as a subset of health tourism (Lee & Kim, 2015). According to He et al. (2023), wellness has health-related connotations and refers to complete harmony between the body, mind, and spirit. UNWTO's report titled "Exploring Health Tourism," in 2018 recognized the significance of wellness and medical tourism. Acknowledging the unprecedented growth of this market, it points out that the most important market segment driving this trend is the baby boomer generation (born between 1946 and 1964). Several studies have proven that wellness contributes to better quality of life. People are striving for "holistic self-care at all stages of their lives" (Patterson & Balderas-Cejudo, 2022, p. 3).

Although most studies describe wellness as a multidimensional construct, they differ in the manner they describe the wellness attributes. For instance, Adams (2003) refers to four fundamental tenets of wellness: it is multidimensional, it focuses on identifying roots of wellness instead of reasons of illness, it also emphasizes on harmony, and it has relative and perceptual underpinnings. Liao et al. (2023) stress on the multidimensional nature of personal health and write that it is a blend of physiological and psychological health and social and environment factors. Koncul defines it as a state where it is possible to experience overall "harmony between mental, physical, spiritual or biological health" (2012, p. 530).

Different motivations for wellness travel have been identified. Tourists seek wellness travel to maintain good health and wellness and some travel to recover from stress/fatigue or a sickness. Wellness tourists desire to fulfill numerous aspects of well-being such as through immersing in educational, esthetic, and escapist type of experiences (Luo et al., 2018). According to He et al. "educational experience refers to customers' intellectual growth and horizons broadening through active engagement in a wellness destination. Esthetic experience refers to tourists' enjoyment induced by an appreciation of art or beauty of local landscapes, which may consequentially enhance their spiritual life. Escapist experience is typically obtained by breaking away from routine life and discovering novelty" (2023, p. 1118). Voigt et al.'s (2011) typology categorizes wellness tourists into three key groups: day visitors to a spa, lifestyle visitors to resorts, and visitors patronizing spiritual retreats. Extant literature notes that wellness tourists patronize traditional health remedies and alternative health treatments, such as indigenous and complementary medicine to treat or avoid disease (Global Wellness Institute, 2020). Gulyas and Molnar (2023) give credit to the Wellness Report by Ardell and Ardell for raising public awareness toward wellness. The report presents a five-dimensional wellness model consisting of nutritional consciousness, anxiety management, ecological mindfulness, physical strength/robustness, and the core feature, responsibility for the self (Ardell, 1977). For the most part, wellness tourists are seeking physiological and/or psychological health through restoration.

Wellness tourism is reliant on the service industry that offers supporting activities and services to support healthy lifestyles in daily life. Wellness destinations depart from the mainstream destinations, and their appeal is, for the most part, based on therapeutic and/or spiritual landscapes (He et al., 2023). Several studies have recognized that tourism offers an opportunity to step out of the daily life and the work environment and explore/experience novel things. Leisure activities improve psychological well-being (such as altruist feelings of volunteer tourists and intellectual satisfaction by learning something novel); some scholars have employed the bottom-up tourism benefit spillover theory to argue that the various aspects of life shape life satisfaction and these domains include work, leisure, travel, personal health, social life, and family (Peters, 2000).

The goal of wellness tourism is predominantly "to link individual health with social and personal dimensions" and taking personal lifestyles of people into consideration (Liao et al., 2023, p. 2). Scholars from a variety of disciplines (for instance, psychology, geography, tourism, occupational health), studying conventional health treatments (such as allopathic medical treatment) and alternative health treatments (such as traditional and complementary medicine) (Atal & Atal, 2024), recognize the potential of wellness tourism to promote HWB in individuals (Majeed et al., 2019; Smith & Diekmann, 2017). Therapeutic landscapes such as parks, riverside areas, beach, forests, hill stations, and mountain retreats are sought by tourists to enhance their HWB. Tourists try to distance themselves away from the stresses of work and mundaneness of everyday life and seek happiness to optimize HWB (Chen et al., 2014).

Majeed and Kim point out that, "due to the risk of chronic disease …, the emergence of therapeutic centers, natural and built therapeutic landscapes and the pressing need for safe and reliable health treatments across the globe, people have started to transform their mindset toward using wellness treatments with a blend of different health treatments and tourist attractions" (2023, p. 864). The authors examine connections between conventional medical treatments, alternative health treatments, and tourist attractions and determine the manner in which they complement each other to satisfy HWB and wellness expectations of tourists. Another point the authors make is that although "alternative health treatments overshadow the functioning of wellness tourism in the recent context, the integral contribution of conventional medical treatments to maintain individual HWB cannot be set aside…….

Conventional medical treatments and alternative health treatments overlap, determining tourist HWB where tourism plays a synergistic role" (2023, p. 864).

Wellness tourists, evidently, expect a mix of health treatments, to optimize HWB, in the contemporary era. Wellness tourism, therefore, refers to a mix of conventional medical treatments, alternative health treatments (such as herbal treatments), and tourist attractions (Atal & Atal, 2024). Recent trends illustrate commonalities between wellness tourism and medical tourism that offer access, to tourists, to a broad spectrum of HWB treatments. Unique attributes of wellness tourism from tourists' standpoint can be used to craft effective marketing strategies to promote holistic wellness tourism. Healthcare hospitality holds potential to play a key role in promoting holistic wellness tourism. Undeniably, the hospitality industry has emerged as a key stakeholder of wellness tourism (Cormany, 2017; Evans et al., 2009; Park et al., 2022). For HWB optimization, wellness tourism matrix can serve as a roadmap to effectively comprehend health conditions and needs of a visitor to ensure access to appropriate treatments and alternative health remedies along with access to therapeutic natural spaces. Majeed and Kim (2023) present four quadrants based on two dimensions (type of treatment desired and health condition): wellness treatment vs seeking health restoration by visiting tourist attractions and critical health conditions versus noncritical health conditions.

Figure 2.1 presents a modified version of this matrix by delimiting tourist attractions to therapeutic natural and spiritual landscapes that hold potential to serve as spaces for relaxation and inner transformation. It is argued that tourist attractions are a broad term and do not clearly specify places and settings that have healing

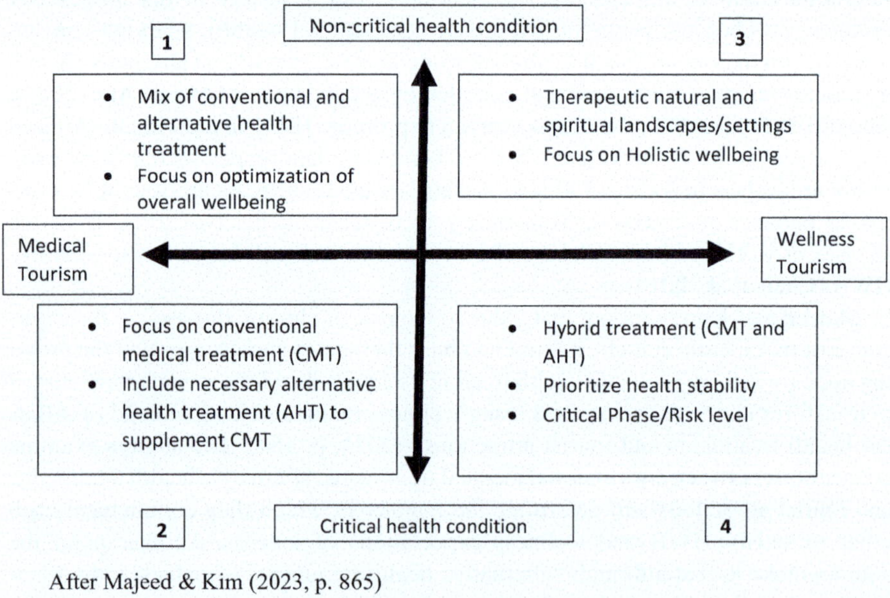

Fig. 2.1 Wellness Tourism Matrix (After Majeed and Kim (2023, p. 865))

properties. The matrix delineates health tourism into medical tourism and wellness tourism and presents four scenarios of required treatment and remedies based on the condition. Wellness and medical tourism destinations can offer a hybrid package to visitors who prefer conventional treatment and are open to receiving supplementary with alternative health treatments. Visitors in critical health condition can be categorized as medical visitors. The third quadrant refers to conventional medical facilities for treatment but with ability to offer/arrange for alternative medical treatment such as physical therapies, herbal healing, and engagement with the natural surroundings. Medical visitors in this category will prioritize a destination that specifically offers high-quality conventional medical treatment. Quadrant 4 also requires hybrid treatment where the medical visitors specifically seek an alternative health treatment and the choice of location is based on availability of both types of treatments. Quadrants 1 and 3 relate to wellness tourists who might have an ailing but nonemergency health condition (in the case of quadrant 1) or seek a destination for preventive reasons supplemented with a desire to seek optimal flow (B-cognition prompted by disoriented or out of the ordinary experiences and/or positive emotions) and transformative experiences.

Tourists visiting for treatment of critical disease conditions can be offered optimal opportunities for HWB by coordinating with doctors and health facilities of the visited destination (Atal & Atal, 2024). Alternative treatments such as exercises, herbal remedies, and activities that offer immersive experiences in the natural therapeutic landscapes can also be offered (Atal & Atal, 2024). Examples of natural landscapes, which offer therapeutic benefits include beaches, mountains, forests, and gardens include walking and gazing. Tourists who do not need critical medical treatment can be offered health services such as normal health checkups, yoga, massage, etc. while offering visits to natural spaces with therapeutic and restorative features. This strategy can enhance the overall visiting experience.

A mix of conventional medical treatments, alternative health treatments, and therapeutic natural landscapes (such as herbal gardens/arboretums) can "ensure tourists' health stability" (Atal & Khurana, 2016). This approach may be adopted when the chance of being cured is limited but the health condition still allows tourists to travel or walk, such as during the treatment of end-stage cancer. Wellness tourism service providers may position their resources in the four quadrants and offer tailored wellness/medical packages based on the health and condition of the visitor. This type of roadmap can "ensure the management of optimal tourist HWB in the diversified realms of wellness tourism" (Majeed & Kim, 2023, p. 865). Destinations need appropriate infrastructure and natural and human resources to offer and manage a combination of different treatments in addition to the existence of natural and therapeutic spaces that hold potential to relax the wellness tourists. Furthermore, "to efficiently satisfy the unprecedented HWB needs of wellness tourists, wellness and medical tourism destinations need to prepare themselves for the advanced and interconnected healthcare infrastructure that spans conventional healthcare systems, alternative healthcare systems and hospitality and tourism. An effective marketing strategy may reflect tourist ideas and preferences for healthfulness to confirm their health beliefs" (Majeed & Kim, 2023, p. 866). Such a marketing strategy can decrease cognitive dissonance and increase perceptions of well-being as well as stimulate behavioral intention (Atal, N. Personal Communication, October 2023).

Clearly, a strategic coordination between local wellness tourism stakeholders is crucial. For instance, involvement is required from retreat centers, hotels and resorts, hospitals, alternative health treatment centers, local attractions, the local government, and local residents who are not involved in offering the medical or alternative treatment services (Atal, N. Personal Communication, December 2023). As Majeed and Kim note, "there is much to do to streamline the linkages among conventional health practices, alternative health practices and tourist attractions to deliver promising HWB experiences to tourists according to tourist health conditions" (2023, p. 866). The healing potential of serene and natural therapeutic landscapes situated near a body of water (such as a river or ocean) or mountains should not be underestimated.

As pointed out earlier, tourism focusing on nature and outdoors often stimulates bonding with nature and can enhance well-being while facilitating more environment-friendly attitude and behavior. For instance, Hanna et al. (2019) argue that outdoor adventure tourism is a sustainable form of tourism that holds potential to inform and advance engagement with and interest in sustainability, mental health, and well-being. Wellness resources help to buffer individuals from stresses of the contemporary society and the fast pace of lifestyle much of which is shaped by digitalization (Feng et al., 2021). A multidimensional wellness tourism destination model is shared by Phuthong et al. (2022) to identify factors that make a destination competitive from a wellness standpoint. These include the environment of the destination, its infrastructure and capacity, its resources, potential to offer innovative products and services, visitor-friendly policies, and wellness-friendly strategies and arrangements. Polat and Koseoglu (2023) scrutinize intellectual path of wellness tourism based on review of documented literature. Research review by the authors identified several clusters of research (please see Table 2.4). It can be noted that most research focuses on wellness services and tourist motivations and satisfaction. The impact of wellness tourism on the health and well-being of local residents, however, has received scant attention.

Polat and Koseoglu (2023) note that wellness tourists of post-pandemic times seek good sleep and mindfulness besides health- and fitness-related products and services, nutrition, and beauty. As noted by several studies, wellness tourism was considered synonymous with spa tourism in early times, and review of research in health tourism shows natural resources as one of the recurring keywords. Other aspects of research focus is on psychological benefits such as stress relief, relaxation, and/or spiritual benefits.

Majeed and Kim (2023) note that visitor demand for healthcare and tourist attractions are influenced by variety of health treatments available which include "traditional and alternative medical therapies, and attractions, to enhance health and well-being" (2023, p. 863). Most studies have focused on travel motivations of wellness tourists (cluster 6 in Table 2.4). For instance, Mueller and Kauffmann (2001) share that the key motive of wellness tourists is to promote health and prevent

Table 2.4 Topics associated with wellness tourism research

Clusters	Well tourism
Cluster 1	Competitiveness of wellness tourism (green): most articles in this cluster focus on research associated with ecological resources, cultural components, distinct service packages, authenticity, innovation, scrutiny of strategic decisions, infrastructure, specialized staff, their retention, and quality of services. Other topics such as culinary or gastronomy services and their linkages with wellness can feature in this cluster
Cluster 2	Wellness services and wellness philosophy (red): a similarity index can be generated between wellness dimensions and wellness services. As an instance, the spiritual aspect and physiological aspects of yoga, meditation, Ayurveda, mindfulness, and religion-related pilgrimages are all exemplified by a deep connection with the inner self of a person. The social and the physical (be it active, e.g., fitness, or passive, e.g., spa services) dimensions
Cluster 3	Quality (gold): research focusing on the quality appraisal on the demand side. Topics within this cluster include medical wellness, spas, and service establishments adjacent to hot springs or thermal water springs
Cluster 4	Sustainability and digitalization (light blue) include evaluation of natural resources such as misuse of hot springs, carbon dioxide emissions, outlook toward ecotourism and eco-consciousness, environmental conservation, and preservation of biodiversity. Research on smart and virtual tourism and artificial intelligence is also included
Cluster 5	Analysis of COVID-19 effects (purple) also focuses on designing possible solutions. The adverse effects, the present tenseness, speedy responsiveness, and strategic tactics collectively play a role in mitigating the negative effects and promoting safety
Cluster 6	Examining motivation and marketing activity (dark blue): different cohorts based on age and other demographic characteristics can have different motivations and preferences. This calls for tailored marketing and communication strategies. A broad wide range of topics and investigation techniques are noted. Several researches occupied the examination of marketing activities effects; some of them highlight the role of: positive messages, active presence on social media platforms, and overcoming language barriers. These are only a few examples of the activities that influence the efficiency of marketing activity. In the case of research methodology used, there is a wide range of methods used, for example eye-tracking, servqual, netnography, etc. Connectedness between wellness tourism research and marketing-related perspectives are noted in terms of consumer behavior, and quality assessment of services
Cluster 7	Loyalty and measurement of customer satisfaction (orange): this cluster includes satisfaction assessment, prior experience, favorable emotions, and experiences, impact of satisfaction on behavioral loyalty such as repeat visit

Source: Polat and Koseoglu (2023)

health-related issues by purchasing services that enhance physiological health, diet, mediation, education, and care about how one looks. Based on their study on wellness tourism (WT) in Switzerland, Mueller and Kaufman are of the view that wellness services should be organized based on the target markets. The authors identify five clusters based on knowledge related to WT: yoga and spiritual part of WT and motivations, conceptual framework of WT, market segmentation of WT, spa part of WT, and the contribution of vacations on the sense of well-being (2023, p. 1867). Worthy of mention is the "Wheel of Wellness and Prevention" by Sweeney and Witmer (1991) which illustrates that wellness is the subject of interaction between six tasks of life (self-direction, work, leisure, spirituality, love, and friendship) and seven life forces (family, education, religion, community, government, media, business/industry) with spirituality epicenter (1991, p. 1870).

One popular form of wellness tourism is yoga tourism (cluster 2). From a spiritual tourism standpoint/potential, yoga tourism stimulates introspection and harmony between mind and body which leads to social transformation by enhancing psychological, spiritual, and emotional capital (Ponder & Holladay, 2013). People are motivated by the desire to seek self-authenticity and enhanced physiological and mental health (Chhabra, 2022). Career progress, search for spirituality, and regulation of positive emotions are noted to be other motivational factors (Bowers & Cheer, 2017). Quest for spirituality is driven by motivational factors such as "transcendence, indulgence, physical health and appearance" (Voigt et al., 2011, p. 27). In the case of yoga tourism and spiritual tourism, satisfaction of motivations augments destination loyalty and increases participation rate (Kim et al., 2017). In fact, many studies contend that yoga tourism hold potential to be a transformative force for individuals and communities. Ponder and Holladay (2013) used the human development theory referred to as "spiral Dynamics" while explaining the transformative power of yoga tourism. The theory is based on the assumption "that individuals and communities advance to higher levels by gaining knowledge and experience while solving life's complexities and becoming more egalitarian, libertarian, conscious. In this context, yoga tourism helps tourists rise to the top of the spiral by increasing their sense of social justice and consciousness" (2013, p. 1870). Next section offers a discourse on wellness as an important feature of sustainability.

Wellness and Sustainability

Sustainable promotion and use of wellness resources by a destination is equally important to ensure that the local residents benefit from this type of tourism from the perspective of overall quality of life and well-being. Tutberidze writes that "wellness tourism is a global phenomenon; its implementation is predominantly localized in the resorts with the appropriate natural conditions and infrastructure" (2021, p. 1). Undoubtedly, the natural environment offers resources that wellness tourism is reliant on such as therapeutic landscapes, natural beauty, and herbs (Hojcska & Szabo, 2021). These resources are not infinite and often become vulnerable toward exploitation (Printz-Markó & Molnar, 2020). Wendri et al. (2021) write that the success of wellness tourism is reliant on health elements, public services and overall infrastructure, service offerings, and quality subject to human resource efficiencies.

Significance of pursuing a sustainable approach has been recognized due to economic potential of wellness tourism. The tipping point of concern has been commodification and overuse of natural and cultural resources to promote wellness tourism (Sgroi, 2020). As pointed out by Rosato et al. (2021), the crucial presence of anthropogenic factors for wellness tourism can impact the life of local residents, natural resources, and the ecosystem in general. Wellness tourism, if strategically developed and promoted, holds potential to boost health and well-being of not just the tourists but also the local residents thereby elevating their quality of life (QOL). As pointed out by Gulyas and Molinar, the role of management is crucial, especially in the context of destination management, supply chain management followed by "knowledge management and information management, relationship management, brand management and strategic planning, modern management tools and sustainable tourism management" (2023, p. 186).

Research on quality of life (QOL) and well-being in tourism and hospitality has received unprecedented attention in the past few decades. Recent research inquiries have specifically focused on the impact of tourism on the QOL of life of residents, tourists, and employees (Uysal & Sirgy, 2023; Wang et al., 2021, 2022). Extant literature notes that health and well-being are significant determinants of QOL. In fact, some scholars refer to well-being as a proxy for QOL. The next section offers key highlights from recent studies on QOL particularly from a wellness and well-being standpoint.

Quality of Life

QOL today has become a major area of research (Uysal & Sirgy, 2023; Uysal et al., 2016; Vogt et al., 2020; Woo et al., 2015). In destinations across the globe, local residents are expressing concerns and keenness to become involved in decision-making process to address the negative impact of tourism on the QOL. Organizational justice theory postulates that satisfaction levels increase if residents perceive that they have been treated fairly (Lovelock et al., 1998). Likewise, employees behave in a positive manner if they feel they have received equitable treatment (Blodgett et al., 1993). This view will be covered in a detailed manner by Ayazlar and Ayazlar (2024).

According to Wang et al. (2022), antecedents of QOL of residents include economic status, social relationships and perceptions of tourism impacts (e.g., Chi et al., 2017), emotions (favorable as well non-favorable) toward those hosting Airbnbs (Suess et al., 2021), educational benefits, perceived employment rate (e.g., Roehl, 1999), financial performance, and community satisfaction levels (Suess et al., 2018). Other emerging perspectives are focused on equity issues (i.e., residents' perception of equitable distribution of tourism benefits can generate a higher level of support for tourism) and health. The QOL indicators, unique to a specific place, should reflect both subjective and objective QOL assessments (e.g., see Table 2.4). The subjective indicators relate to perceived state of happiness of residents and approval of some QOL dimensions, associated with germane impacts of tourism and general satisfaction with life. As an instance, measures of satisfaction metrics of residents based on intangible, societal, wellness, safety, and emotional QOL are influenced by the perceived impacts of tourism such as economic, socio/cultural, and environmental (Chi et al., 2017; Kim et al., 2013; Uysal & Sirgy,

2023). The tangible QOL dimensions can encompass "economic, educational, health, environmental, safety, and security conditions, leisure and recreational opportunities in the destination, etc. A variety of proxies can be used to measure each of the abovementioned domains" (Berbekova et al., 2024, p. 37).

Recent scholars have attempted to identify predictors of well-being of residents, tourists, and employees (Baker & Kim, 2020; Wang et al., 2022). Wang et al. (2022) note that a plethora of studies have employed two approaches to gauge QOL: residents' subjective assessment (judged from a cognitive and emotional perspective) of their well-being and use of objective QOL dimensions such as income, unemployment rate, state of health and overall education level, and crime incidences (Meng et al., 2010). Earlier studies noted that most variables measuring residents' sense of well-being were focused on economic dimensions, perceived positive/negative impacts of tourism, and sociodemographic attributes of residents. Recent studies assess value of tourism beyond economic gains and extend to intangible benefits.

Vogt et al. encapsulates that QOL is determined by a multitude of factors such as "business attraction and retention, social services, safety, utilities, transportation planning, building permitting, recreation and parks, and tourism management and marketing" (2020, p. 1). The authors opine that the best approach to counter negative impact is to integrate sustainability into the "select or comprehensive community plans, be socially just, conserve resources, invest in new technologies, and be fiscally responsible" (2020, p. 1) not only from residents' standpoint but also in tourists' QOL.

Humans across the globe share a common desire to be healthy through the life span. As stated by Vogt et al. "health conditions matter in the places we live, work, play, and vacation. The design of places for residency, employment, and leisure require health considerations of safety, clean air and water, and greenery. Governments oversee and regulate both built and unbuilt contexts to ensure public health and wellbeing" (2020, p. 3). Parks are patronized by residents and tourists for mental, physical, and social benefits. Several studies have illustrated that engagement with parks and other nature-based landscapes improve physical and psychological health and support social capital. Undeniably, "land cover and density of development play a role in tourism through the attractiveness of a place and how built and natural environments together create a unique setting. Tourism development often occurs in fragile and rare ecosystems such as coastal areas, forests, mountains, and deserts as developers aim to create a designed environment, one different than home" (Vogt et al., 2020, p. 6).

Although it is believed that the planet Earth is resilient, scientists warn that there is a resiliency threshold. The Earth needs to heal and restore itself to continue to support needs of human beings (Palmer et al., 2005). Recent studies call for the key stakeholders of tourism to conduct research and develop intelligent decision-making skills to ensure costs do not surpass gains and residents' needs are safeguarded and not compromised while satisfying the desires of tourists (Vogt et al., 2020). Destinations committed to sustainable planning are more prone to developing an equitable and resilient natural ecosystem and society. Planning, management, and marketing of tourism should be rooted in sustainability principles and equitable representations of all key stakeholders because these are more likely to offer solutions at the local, regional, and global levels (Bricker, 2018; Sharpely, 2009; Vogt et al., 2020).

It is important to offer priority to local residents when strategies are being designed to enhance QOL of a visited community. They should not feel crowded out due to tourists, nor should they have to compete for local services. For instance, they also enjoy patronizing local restaurants, souvenir shops, festivals, parks, and other historical and natural attractions (Andereck & Nyaupane, 2011). Long-term viability and sustainability of tourism relies on local resident support. Extant literature acknowledges that the key goal of tourism development should be to improve QOL and well-being of host communities (Berbekova et al., 2024). As mentioned in the previous chapter, sustainable tourism agenda in the post-pandemic times need to integrate a well-being component, and one initiative can be to apply the capitals approach to promote revised public policies that give precedence to the well-being of local residents (Dwyer, 2021).

Berbekova et al. (2024) suggest that the development and implementation of QOL indicators can guide tourism policies aimed at improving the well-being and QOL of not just stakeholders but also the appeal of the place to visit and live. As pointed out by Berbekova et al., "there is always a benefit to achieving a more accurate public policy formulation when policymakers fully consider the importance and consequences of economic and QOL elements. Having informed public policy formulation should also help sustaining local support and ensuring effective and sustainable use of resources" (2024, p. 34). QOL of life needs to be integrated in tourism performance and sustainability metrics. The QOL tools can be used to shape public policy and tourism development plans. Along similar lines, Uysal et al. (2020) argue that dogmas with regard to access and sustainability need to guide destinations to plan tourism experiences of high-quality and design strategies to ensure well-being of all tourism stakeholders. The authors postulate that such an approach will not only aid in enhancing well-being of tourists and residents but will also make destinations more inhabitable.

Conclusion

This chapter describes the notion of health, wellness, and quality of life from the standpoint of tourists and the local residents. Insights are offered on the surging popularity of wellness tourism, and a modified wellness tourism matrix is proposed. This is followed by an overview of the quality of life phenomenon. A summary of various dimensions of QOL is shared. Last, this chapter emphasizes on the important role of sustaining QOL to promote holistic wellness and sustainable tourism. Next chapter, by Atal and Atal, delves deeper into alternative health treatments by focusing on herbal healing and herbal tourism.

References

Adams, T. (2003). *The power of perceptions: Measuring wellness in a globally acceptable, philosophically consistent way*. Wellness Management.

Ardell, D. B. (1977). High level wellness strategies. *Health Education, 8*(4), 2–2.

Andereck, K. L., & Nyaupane, G. P. (2011). Exploring the nature of tourism and quality of life perceptions among residents. *Journal of Travel Research, 50*(3), 248–260.

Atal, N. & Atal, C. (2024). Healing potential of herbs and herbal tourism. In sustainable development and resilience of tourism: wellbeing and quality of life perspectives. D. Chhabra, N. Atal & A. Maheshwari (Eds.) Springer: New York. In Press.

Atal, N., & Khurana, P. (2016). *Guest lecture at VAMSCON*. Vardhman Mahavir Medical College, Safdarjung Hospital. https://youtu.be/jncrDaMChIk?si=2yida8ItfYl3NYTt

Ayazlar, R. & Ayazlar, G. (2024). The Importance of Mindfulness for Tourism Employees. In Sustainable development and resilience of tourism: wellbeing and quality of life perspectives, D. Chhabra, N. Atal & A. Maheshwari (Eds.) Springer: New York. In Press.

Baker, M. A., & Kim, K. (2020). Dealing with customer incivility: The effects of managerial support on employee psychological well-being and quality-of-life. *International Journal of Hospitality Management, 87*, 102503.

Berbekova, A., Assaf, A. G., & Uysal, M. (2024). Interdisciplinary approach to tourism demand modeling: Quality of life indicators. *Journal of Hospitality & Tourism Research*. https://doi.org/10.1177/10963480241229238

Blodgett, J. G., Granbois, D. H., & Walters, R. G. (1993). The effects of perceived justice on complainants' negative word-of-mouth behavior and repatronage intentions. *Journal of Retailing, 69*(4), 399–428.

Bowers, H., & Cheer, J. M. (2017). Yoga tourism: Commodification and western embracement of eastern spiritual practice. *Tourism Management Perspectives, 24*, 208–216.

Bricker, K. (2018). Positioning sustainable tourism: Humble placement of a complex enterprise. *Journal of Park and Recreation Administration, 36*(1).

Bryce, R., Irvine, K. N., Church, A., Fish, R., Ranger, S., & Kenter, J. O. (2016). Subjective well-being indicators for large-scale assessment of cultural ecosystem services. *Ecosystem Services, 21*, 258–269. https://doi.org/10.1016/j.ecoser.2016.07.015

Brymer, E., Downey, G., & Gray, T. (2009). Extreme sports as a precursor to environmental sustainability. *Journal of Sport & Tourism, 14*(2–3), 193–204.

Chen, Y., Hicks, A., & While, A. E. (2014). Quality of life and related factors: a questionnaire survey of older people living alone in Mainland China. *Quality of Life Research, 23*, 1593–1602.

Chi, C. G. Q., Cai, R., & Li, Y. (2017). Factors influencing residents' subjective well-being at World Heritage Sites. *Tourism Management, 63*, 209–222.

Chhabra, D. (2022). Transformational model of well-being for serious travellers. *International Journal of Spa and Wellness, 5*(2), 129–146.

DeMicco, F. J. (Ed.). (2017). *Medical tourism and wellness: Hospitality bridging healthcare (H2H)*. CRC Press.

Dillette, A. K., Douglas, A. C., & Andrzejewski, C. (2021). Dimensions of holistic wellness as a result of international wellness tourism experiences. *Current Issues in Tourism, 24*(6), 794–810.

Dunn, H. L. (1957). Points of attack for raising the levels of wellness. *Journal of the National Medical Association, 49*(4), 225.

Dwyer, L. (2021). Resident well-being and sustainable tourism development: The 'capitals approach. *Journal of Sustainable Tourism*. https://doi.org/10.1080/09669582.2021.1990304

Evans, J. D., Crooks, V. A., & Kingsbury, P. T. (2009). Theoretical injections: On the therapeutic aesthetics of medical spaces. *Social Science & Medicine, 69*(5), 716–721.

Faroldi, E., Fabi, V., Vettori, M. P., Gola, M., Brambilla, A., & Capolongo, S. (2019). Health tourism and thermal heritage: Assessing Italian spas with innovative multidisciplinary tools. *Tourism Analysis, 24*(3), 405–419.

Feng, C., Li, H., Feng, L., Yu, M., Zhang, H., Ma, Y., et al. (2021). Study on current status, problems and countermeasures of countryside health and wellness tourism in China. *Open Journal of Social Sciences, 9*(04), 212. https://doi.org/10.4236/jss.2021.94016

Gatrell, A. C. (2013). Therapeutic mobilities: Walking and 'steps' to wellbeing and health. *Health & Place, 22*, 98–106.

Goodrich, J., & Goodrich, G. (1987). Health-care tourism—an exploratory study. *Tourism Management, 8*(3), 217–222.

Gulyas, G., & Molnar, E. I. (2023). Wellness tourism management research A bibliometric analysis. *Management & Marketing, 18*(2), 172–191.

GWI. (2020). *What wellness is, wellness definitions*. Global Wellness Institute. https://globalwellnessinstitute.org/what-is-wellness/wellness-economy-definitions/

GWI (2024). Vision. Wellness Defined. Link: https://globalwellnessinstitute.org/ accessed January 2024

GWI (Global Wellness Institute). (n.d.-a). *History of wellness*. Retrieved from: https://globalwellnessinstitute.org/industry-research/history-of-wellness/

GWI (Global Wellness Institute). (n.d.-b). *What is the wellness economy?* https://globalwellnessinstitute.org/what-is-wellness/what-is-the-wellness-economy/

GWI (Global Wellness Summit). (2020). *Global wellness trends report. The future of wellness*. Available at: https://www.globalwellnesssummit.com

Hall, M. (2011). Health and medical tourism: A kill or cure for global public health? *Tourism Review, 66*(1/2), 4–15.

Hanna, P., Johnson, K., Stenner, P., & Adams, M. (2015). Foucault, sustainable tourism, and relationships with the environment (human and nonhuman). *GeoJournal, 80*(2), 301–314.

Hanna, P., Wijesinghe, S., Paliatsos, I., Walker, C., Adams, M., & Kimbu, A. (2019). Active engagement with nature: Outdoor adventure tourism, sustainability and wellbeing. *Journal of Sustainable Tourism*. https://doi.org/10.1080/09669582.2019.1621883

Hettler, B. (1976). *The six dimensions of wellness model* (pp. 1–2). National Wellness Institute.

Hojcska, Á. E., & Szabó, Z. (2021). Investigating natural treatment factors and inequalities of medicinal water institutions in the aspect of tourism in Hungary. *Geo Journal of Tourism and Geosites, 36*, 555–562. https://doi.org/10.30892/gtg.362spl01-683

Kim, E., Chiang, L., & Tang, L. (2017). Investigating wellness tourists' motivation, engagement, and loyalty: In search of the missing link. *Journal of Travel & Tourism Marketing, 34*(7), 867–879.

Kim, K., Uysal, M., & Sirgy, M. J. (2013). How does tourism in a community impact the quality of life of community residents?. *Tourism Management, 36*, 527–540.

Klaniecki, K., Leventon, J., & Abson, D. J. (2018). Human–nature connectedness as a 'treatment' for pro-environmental behavior: Making the case for spatial considerations. *Sustainability Science, 13*, 1375–1388.

Koncul, N. (2012). Wellness: a new mode of tourism. *Economic Research-Ekonomska Istraživanja, 25*, 525–534. https://doi.org/10.1080/1331677X.2012.11517521

Kotur, A. S. (2022). Exploring the wellness dimensions of wine tourism experiences: A netnographic approach. *International Journal of Wine Business Research, 34*, 608–626. https://doi.org/10.1108/ijwbr-07-2021-0040

Laing, J., & Weiler, B. (2007). Mind, body and spirit: Health and wellness tourism in Asia. In *Asian tourism: Growth and change* (pp. 379–389). Routledge.

Lee, J., & Kim, H. (2015). Success factors of health tourism: Cases on Asian tourism cities. *International Journal of Tourism Cities, 1*(3), 216–233.

Liao, C., Zuo, Y., Xu, S., Law, R., & Zhang, M. (2023). Dimensions of the health benefits of wellness tourism: A review. *Frontiers in Psychology, 13*, 1071578.

Lovelock, C. H., Patterson, P. G., & Walker, R. H. (1998). *Services marketing: Australia-New Zealand*. University Of Tasmania.

Luo, Y., Lanlung, C., Kim, E., Tang, L. R., & Song, S. M. (2018). Towards quality of life: the effects of the wellness tourism experience. *Journal of Travel & Tourism Marketing, 35*, 410–424. https://doi.org/10.1080/10548408.2017.1358236

Majeed, S., & Gon Kim, W. (2023). Emerging trends in wellness tourism: A scoping review. *Journal of Hospitality and Tourism Insights, 6*(2), 853–873.

Majeed, S., Majeed, M., & Ajike, M. A. (2019). Dry cupping therapy and the wellness management of health travelers. *Traditional Medicine Research, 4*(1), 12–24.

Marques, I. C. P., Serrasqueiro, Z., & Nogueira, F. (2022). Covid-19 and organisational development: Important signs of a new pillar for sustainability. *Social Responsibility Journal, 18*(8), 1520–1546.

McCabe, S., & Johnson, S. (2013). The happiness factor in tourism: Subjective well-being and social tourism. *Annals of Tourism Research, 41*, 42–65.

McDougall, C. W., Quilliam, R. S., Hanley, N., & Oliver, D. M. (2020). Freshwater blue space and population health: An emerging research agenda. *Science of the Total Environment, 737*, 140196. https://doi.org/10.1016/j.scitotenv.2020.140196

Meera, S., & Vinodan, A. (2019). Attitude towards alternative medicinal practices in wellness tourism market. *Journal of Hospitality and Tourism Insights, 2*(3), 278–295.

Meng, F., Li, X., & Uysal, M. (2010). Tourism development and regional quality of life: The case of China. *Journal of China Tourism Research, 6*(2), 164–182.

Mueller, H., & Kaufmann, E. L. (2001). Wellness tourism: Market analysis of a special health tourism segment and implications for the hotel industry. *Journal of Vacation Marketing, 7*(1), 5–17. https://doi.org/10.1177/135676670100700101

Nowak-Olejnik, A., Schirpke, U., & Tappeiner, U. (2022). A systematic review on subjective well-being benefits associated with cultural ecosystem services. *Ecosystem Services, 57*, 101467.

Palmer, M. A., Bernhardt, E. S., Chornesky, E. A., Collins, S. L., Dobson, A. P., Duke, C. S., et al. (2005). Ecological science and sustainability for the 21st century. *Frontiers in Ecology and the Environment, 3*(1), 4–11.

Park, B. J., Furuya, K., Kasetani, T., Takayama, N., Kagawa, T., & Miyazaki, Y. (2011). Relationship between psychological responses and physical environments in forest settings. Landsc. *Urban Planning, 102*, 24–32. https://doi.org/10.1016/j.landurbplan.2011.03.005

Park, E., Kim, W. H., & Kim, S. B. (2022). How does COVID-19 differ from previous crises? A comparative study of health-related crisis research in the tourism and hospitality context. *International Journal of Hospitality Management, 103*, 103199.

Patterson, I., & Balderas-Cejudo, A. (2022). Baby boomers and their growing interest in spa and wellness tourism. *International Journal of Spa and Wellness, 5*(3), 237–249.

Peters, M. (2000). Sociology beyond societies: Mobilities for the twenty-first century. *Sociol. Rev. 48*, 480–482. https://doi.org/10.2307/1318589

Phuthong, T., Anuntavoranich, P., Chandrachai, A., & Piromsopa, K. (2022). Developing and validating an assessment scale to measure the competitiveness of wellness destinations. *Sustainability, 14*(7), 4152.

Polat, E., & Koseoglu, M. A. (2023). Wellness tourism scholarship: A research agenda. *Journal of Hospitality and Tourism Insights, 6*(5), 1860–1889.

Ponder, L. M., & Holladay, P. J. (2013). The transformative power of yoga tourism. In *Transformational tourism: Tourist perspectives* (pp. 98–107). CABI.

Printz-Markó, E., & Molnar, E. I. (2020). Health tourism—A new system-model based on the four basic elements. In *Challenges and opportunities to develop organizations through creativity, technology and ethics: The 2019 Griffiths School of Management Annual Conference on Business, Entrepreneurship and Ethics (GSMAC)* (pp. 253–267). Springer International Publishing. https://doi.org/10.1007/978-3-030-43449-6_14

Riva, M., & Curtis, S. (2016). The significance of material and social contexts for health and wellbeing in rural England. In *Wellbeing and place* (pp. 53–72). Routledge.

Roehl, W. (1999). Quality of life issues in a casino destination. *Journal of Business Research, 44*(3), 223–229.

Rosa, C. D., Profice, C. C., & Collado, S. (2018). Nature experiences and adults' self-reported pro-environmental behaviors: The role of connectedness to nature and childhood nature experiences. *Frontiers in Psychology, 9*, 1055.

Rosato, P. F., Caputo, A., Valente, D., & Pizzi, S. (2021). 2030 agenda and sustainable business models in tourism: A bibliometric analysis. *Ecological Indicators, 121*, 106978. https://doi.org/10.1016/j.ecolind.2020.106978

Sgroi, F. (2020). Forest resources and sustainable tourism, a combination for the resilience of the landscape and development of mountain areas. *Science of the Total Environment, 736*, 139539. https://doi.org/10.1016/j.scitotenv.2020.139539

Sharpley, R. (2009). *Tourism development and the environment: Beyond sustainability?* Earthscan.

Smith, M., Puczkó, L. Á. S. Z. L. Ó., & Rátz, T. (2009). Twenty-three districts in search of a city: Budapest-the capitaless capital? In *City tourism: National capital perspectives* (pp. 201–213). CABI.

Smith, M. K., & Diekmann, A. (2017). Tourism and wellbeing. *Annals of tourism research, 66*, 1–13.

Suess, C., Baloglu, S., & Busser, J. A. (2018). Perceived impacts of medical tourism development on community wellbeing. *Tourism Management, 69*, 232–245.

Suess, C., Woosnam, K., Mody, M., Dogru, T., & Sirakaya Turk, E. (2021). Understanding how residents' emotional solidarity with Airbnb visitors influences perceptions of their impact on a community: The moderating role of prior experience staying at an Airbnb. *Journal of Travel Research, 60*(5), 1039–1060.

Summers, J. K., Smith, L. M., Case, J. L., & Linthurst, R. A. (2012). A review of the elements of human well-being with an emphasis on the contribution of ecosystem services. *Ambio, 41*, 327–340. https://doi.org/10.1007/s13280-012-0256-7

Sweeney, T. J., & Witmer, J. M. (1991). Beyond social interest: Striving toward optimum health and wellness. *Individual Psychology: Journal of Adlerian Theory, Research & Practice, 47*.

Thal, K. I., & Hudson, S. (2019). A conceptual model of wellness destination characteristics that contribute to psychological well-being. *Journal of Hospitality & Tourism Research, 43*(1), 41–57.

Tutberidze, M. (2021). The georgian resorts as a basis for wellness tourism development. *Georgian Geographical Journal, 1*(1). https://doi.org/10.52340/ggj.2021.261

Uysal, M., Sirgy, M. J., Woo, E., & Kim, H. L. (2016). Quality of life (QOL) and well-being research in tourism. *Tourism management, 53*, 244–261.

Uysal, M., Berbekova, A., & Kim, H. (2020). Designing for quality of life. *Annals of Tourism Research, 83*, 102944.

Uysal, M., & Sirgy, M. J. (2023). Prologue: tourism and quality of life (QoL) research II. In handbook of tourism and quality-of-life research II: enhancing the lives of tourists, residents of host communities and service providers (pp. 1–8). Cham: Springer International Publishing.

Vogt, C. A., Andereck, K. L., & Pham, K. (2020). Designing for quality of life and sustainability. *Annals of Tourism Research, 83*, 102963.

Voigt, C., Brown, G., & Howat, G. (2011). Wellness tourists: In search of transformation. *Tourism Review, 66*(1/2), 16–30. https://doi.org/10.1108/16605371111127206

Walker, C., Hart, A., & Hanna, P. (2017). *Building a new community psychology of mental health*. Palgrave Macmillan UK.

Wang, S., Berbekova, A., & Uysal, M. (2021). Is this about feeling? The interplay of emotional well-being, solidarity, and residents' attitude. *Journal of Travel Research, 60*(6), 1180–1195.

Wang, S., Berbekova, A., & Uysal, M. (2022). Pursuing justice and quality of life: Supporting tourism. *Tourism Management, 89*, 104446.

Wendri, I. G. M., Alistyawati, I. A., & Sudarmini, N. M. (2021, April). Human resource development model for wellness products to support sustainable tourism in Sangkan Gunung Village, Karangasem Regency, Bali Province. In *International conference on Applied Science and Technology on Social Science (ICAST-SS 2020)* (pp. 368–375). Atlantis Press. https://doi.org/10.2991/assehr.k.210424.071

Willis, C. (2015). The contribution of cultural ecosystem services to understanding the tourism-nature-wellbeing nexus. *Journal of Outdoor Recreation and Tourism, 10*, 38–43. https://doi.org/10.1016/j.jort.2015.06.002

Woo, E., Kim, H., & Uysal, M. (2015). Life satisfaction and support for tourism development. *Annals of Tourism Research, 50*, 84–97.

World Health Organization (n.d.). Constitution of the World Health Organization. Retrieved August 2023 from https://www.who.int/about/governance/constitution

Chapter 3
Healing Potential of Herbs and Herbal Tourism

Navin Atal and Chand Kumar Atal

Abstract This chapter offers insights on the significance of various types of herbal treatments, with a special focus on the herbs found in India. The potential of herbal programs and spaces to promote herbal tourism is discussed using a variety of examples. A synopsis is offered regarding the history and progress in the development of herbal products to promote herbal tourism. This is followed by an overview of various frameworks and models employed in some countries to plan and develop a sustainable form of herbal tourism. India and its progress in herbal tourism are specifically discussed followed by suggestions for a forward-looking inclusive path to develop and promote sustainable herbal tourism.

Keywords Health · Herbs · Herbal tourism · Healing systems · Ayurveda · Well-being

Introduction

Health and wellness tourism, in the context of consumption and use of traditional medicine, has witnessed exponential growth worldwide. Several scholars and practitioners have pointed out that the monotony and stress of the modern lifestyle have made people yearn for peace and tranquility at a place away from the place of residence (Coghlan & Gooch, 2011; Sheldon, 2020). Sometimes the reason for traveling can be to heal a sick body or take relief from the stresses of daily life. The

N. Atal (✉)
Max Superspeciality Hospital, Delhi, India

C. K. Atal (Deceased)
Biosciences, Omaha, Nebraska, USA

CSIR, IIIM, Jammu, J&K, India

Pharmacognosy Department in University institute of Pharmaceutical Sciences
Punjab University, Chandigarh, India

D. Chhabra et al. (eds.), *Sustainable Development and Resilience of Tourism*,
https://doi.org/10.1007/978-3-031-63145-0_3

sick are ready to travel to any corner of the world for treatment if the desired treatment is unavailable in their native place (Deka et al., 2015; Nipo et al., 2021). For instance, if Chinese treatment involves herbal medicine, then this travel can be referred as herbal tourism which is contrary to medical tourism focused on allopathy (modern medicines).

Herbal tourism, as a subset of health and wellness tourism, is becoming popular as a niche form of tourism. Extant literature (Priadarshni 2023) is recognizing it as a new tourism trend with the potential to enhance economic benefits and well-being, particularly in the post-pandemic times (Juneidi et al., 2021; Nipo et al., 2021; Widarini & Wijaya, 2022). Even before the pandemic, local governments in rural towns had begun initiating efforts as the economic, cultural, and social advantages of herbal tourism were becoming more apparent (Farasani et al., 2016; Deka et al., 2015). There are abundant untapped entrepreneurial opportunities for promoting the use of medicinal and aromatic plants in nature-based, indigenous, or wellness tourism (Kala, 2015) which are rooted in local and ancestral wisdom (Atal, 1979; Atal & Schwarting 1961; Bhutani & Gohil, 2010; Kala, 2015; Firenzuoli & Gori, 2007; Pandaya, 2007). In recent times, community herbal gardens and community herbal learning centers are increasingly promoted as key tourism attractions (Butthongdee, 2015). Countries which are spearheading this niche tourism trend are noted to be Indonesia (Juneidi et al., 2021; Utama et al., 2021; Wibawa et al., 2016), India (Deka et al., 2015), Thailand (National Thailand, 2024), and Iran (Farasani et al., 2016). In fact, Indonesia has pioneered the concept of herbal villages and progressed toward building community-empowered models by strategic planning, developing, and promoting herbal tourism using a bottom-up stakeholder collaboration approach (Utama et al., 2021).

Several herbal garden projects and herbal attractions have also mushroomed in India. The traditional use of herbal health remedies in India offers important nourishing and health-related benefits while generating income and environmental sustainability for rural communities who are willing to embrace tourism (Deka et al., 2015). This chapter predominantly focuses on the herbs in India and their potential to attract tourists seeking medical or preventive treatment or in quest to enhance knowledge of herbs. Acquiring knowledge can facilitate feelings of self-actualization and improve psychological well-being (Sheldon, 2020).

It is important to note that herbal remedies are popular worldwide and people are willing to travel far to seek herbal treatment. As an instance, people patronize Kerala from all over the world to enjoy the ayurvedic herbal oil massages (Kala, 2015; Laub, 2019). The practitioners and scholars postulate that travel for herbal treatment (i.e., herbal tourism) is triggered by a variety of reasons such as for cure, good health, supplementing nutrition, improving immunity, using herbal culinary or healthy foods, herbal cosmetics, herbal rehabilitation, herbal detoxification, and herbal weight loss (Deka et al., 2015). Moreover, botanists and traders import foreign herbs for local cultivation and utilization, besides delivering of ready-to-use herbal products for public use.

This chapter offers insights on the significance of various types of herbal treatments, with special emphasis on the herbs found in India. The potential of herbal programs and spaces to promote herbal tourism is discussed using a variety of examples. A synopsis is offered of the history and progress in the development of herbal products to promote herbal tourism. This is followed by an overview of various frameworks and models employed in some countries to plan and develop a sustainable form of herbal tourism. India's progress in herbal tourism is specifically discussed, followed by suggestions for a forward-looking inclusive path to develop and promote sustainable herbal tourism.

Literature Review

Healing systems (Ravishankar & Shukla 2007) that have evolved over ages include Indian ayurvedic medicine and traditional Chinese medicine. The healing systems found in India include allopathy, Ayurveda, Siddha, Unani systems, yoga, naturopathy, homeopathy, acupressure, and acupuncture. Some people may associate healing with meditation and chanting. Although these systems are seen all over India, the Sowa Rigpa (Amchi) system is followed in Ladakh. The Sowa Rigpa system is rooted in the principles of Jung-wa-nga and Ngepa-Sum (panchamahabhutas and Tridosa) which advocate that all living beings and nonliving objects in the universe are composed of cosmic elements "Jung-wa-nga," viz., Sa, Chu, Me, Lung, and Nam-kha (Prithvi (earth), Jal (water), Agni (fire), Vayu (air), and Akash (sky)). According to this system, any imbalance in the five cosmic elements can lead to disease, and this fundamental principle guides the treatment plan. The Siddha system of medicine originated in South India and has become prevalent in the states of Tamil Nadu and Kerala. The Siddha system views disease as a condition caused due to a disturbance in three humors: *vaadham* (airy), *pittham* (fiery), and *kapam* (watery). According to this system, disequilibrium is the result of factors such as malformed environment, climatic conditions, diet, physical activities, and stress.

The ayurvedic system (Sarma 1931) includes treatment with herbal medicines, diet, exercise, meditation, breathing, physical therapy, and other methods. In India, the use of herbal treatment is mentioned in ancient texts. Dhanvantari (Murthy 1997) is considered the guru of medicine and is mentioned in ancient texts (the Puranas). He is regarded as the physician of gods (Devas) and described as an avatar of Vishnu (deity worshipped by Hindus). Charaka is famously known as the father of Ayurveda. In his book called *Charaka Samhita*, Charaka has discussed a variety of plants and animals. In the epic Ramayana, written by sage Valmiki dating back to 500 BCE to 100 BCE, the plant sanjeevani is attributed with life-saving healing powers. The sanjeevani plant was flown all the way by Hanuman, (Rongmei 2022) traveling from the Dronagiri Himalayan mountains across the Indian subcontinent to the island kingdom of Sri Lanka, to save the life of Laxmana (the exiled king's brother), who was grievously wounded during the epic battle with Ravana. In the Hindu epic Ramayana, Ravana is described as the king of the island of Lanka (Cartwright, 2016). He was a multiheaded rakshasa (demon having supernatural powers who harms others). He was a great scholar blessed with special invincible powers and boons given by lord Shiva and Brahma. Ravana was the prime antagonist in Ramayana, who became a powerful and arrogant ruler (Cartwright, 2016). He abducted Prince Rama's wife (Sita) and held her captive. Prince Rama, with the support of King Sugriva and his army, launched an invasion against Ravana in Lanka. Ravana was subsequently slain by Prince Rama.

Since ancient times, besides the cultural exchanges, there flourished across continents, a trading business of numerous items including traditional medicine knowledge and herbal remedies. These were carried out through the network of ancient silk routes (Kurin 2002) traversed by traders and healthcare dealers. UNESCO has recorded that as early as 2000 BC, trade involving herbal spice cinnamon (Cassia) was carried out via silk routes from Sri Lanka and China to the Arabian Peninsula and Iranian Plateau. The silk route lasted for about 2000 years and connected different countries and cultures.

Besides traders, physicians traveled to disperse medicine and knowledge to other countries (Singh & jaunjale 2022). As an example, Honigberger, (Honigberger n.d.) a successful practitioner and homeopathic physician, visited Lahore between 1829 and 1830 traveling through Levant, Egypt, Arabia, and Persia. He successfully treated Maharaja Ranjit Singh (the king of Punjab) for paralysis of vocal cords and swollen feet. Homeopathy therefore entered India in the eighteenth century and was completely assimilated here. Besides India and Pakistan, the physician also introduced homeopathy to Afghanistan, Egypt, Persia, Romania, Russia, and Turkey.

As mentioned in the previous chapters, health tourism implies traveling to distant places to seek treatment. In fact, herbal tourism has existed since ages and was driven by the desire to stay healthy. When traveling was difficult and primitive, sick people traveled to nearby areas for treatment. As per Dewan Pushkar Nath Madan, a prominent Kashmiri Pandit from Kashmir, sometime around 1880, a sick and desperate ancestor traveled from Kashmir on Ekka (traveling cart) to seek better treatment in Lahore. She had to undergo an agonizing travel through Kashmir forests on crude stone laden tracks (C.K. Atal, personal communication, 1978). Because of sickness and bumpy ride, she fell from her cart and ultimately died. This incident highlighted the arduous travel routes. Many years later, around 1945, one of the first hospitals of Kashmir, based on modern treatment, namely, Rattan Rani hospital, was established in Barbar Shah, Srinagar, Kashmir. It was managed by the husband and wife team (Dr. Jagat Mohini Thussu, a renowned social worker and gynecologist, graduate from King Edward Medical School, Lahore, and her husband Dr. Omkar Nath Thussu, a renowned pathologist) (Atal & Goyal, 2009).

Prior to this, there were only prayers and traditional remedies in Kashmir as difficult terrains made it almost impossible to travel far for treatment. With the passage of time, as mules, horses, and carts were replaced by sophisticated means of transportation, travel became easier, faster, and more comfortable. People with poor health who were compelled to travel far away were able to do it quickly and efficiently to seek treatment. Today, traditional and herbal treatment as a tourism industry is becoming popular across the globe. Besides the developing countries, it is fast catching up in the developed countries also. New modalities of latest research like genomics and artificial intelligence have contributed to the mushrooming of herbal and traditional medicine, industries for herbal medicines, natural products, health, wellness, and health-related travel and tourism. Today, there is a surge in herbal tourism globally with many countries promoting herbal tourism.

Since herbal tourism is an emerging field of inquiry, literature is relatively sparse on this topic. Of the handful that have appeared, it is discussed as a subset of medicine or wellness tourism and even agro-tourism (Nolan & Schneider, 2011; Berdzenishvili et al., 2013), medical tourism (Chambers & McIntosh, 2008).Worthy of note is the study from Iran that illustrates the manner in which subtle plants, with medicinal qualities, are used in food and tea servings as part of the traditional food heritage (Torabi et al., 2018). For instance, the authors write that "borage is very commonly associated with refreshing drinks; lemon verbena (Aloysia citrodora) tea is useful for relaxation and decreases depression and nervousness; saffron and mint teas are relaxing drinks with healthful benefits. However, most customers, tourists, and visitors who drink these teas or beverages are unfamiliar with the medicinal plants and their therapeutic properties" (Torabi et al., 2018, p. 3). The authors also share that a traditional herbal medicine tourist park is being planned in Isfahan (Iran) to promote "traditional medicine, botany and medicinal plants and as outdoor strategy for knowledge transfer" (Torabi et al., 2018, p. 2).

Promoting herbs as part of the farm produce and tapping into herbal gardens as part of an agro-tourism portfolio have also been noted by various scholars. For instance, Nurzyńska-Dzida et al. (2014) identified stakeholders of agro-tourism, including the tourists, to obtain insights on the demand for herb gardens and use of herbal plants as part of the farm offerings. The authors reported that almost 90% of the stakeholders acknowledge the significance of herbal gardens in boosting the image of agritourism farms. Furthermore, demand for natural aromatherapy was noted, and herbs were considered useful as spice plants (by 43%), medicinal plants (by 33%), and ornamentals (by 24%) (Nurzyńska-Dzida et al., 2014).

Several studies that focus on places such as Solo and Catur have emerged from Indonesia. Solo has rich Javanese culture and has become the hub for developing Javanese traditions. The Indonesian Ministry of Tourism and Creative Economy launched a book on wellness tourism scenario which presents the theme of "Tracing the History of Jamu." This voyage to revive the past roots of herbal medicine and traditions in Yogyakarta, Solo, and Semarang regions is aimed to not only inform tourists of the historical roots of herbal medicine and traditions but also to share the manner in which their history and traditional use interact from the perspective of the modern society (Widarni & Wijaya, 2022). Widarni and Wijaya (2022) interviewed local tourism stakeholders in Solo (Indonesia). They share that traditional wellness treatment elements (in the form of herbs and spices) are inherited from the kings and queens of the Javanese palace.

The key feature of uniqueness is that the herbal programs are guided by culturally driven local sagacity and insights passed down from different generations. Some wellness tours include activities that promote physiological and mental well-being and range from "physical activities to consuming healthy foods and herbal drinks to increase body stamina or also include meditation to calm the mind" (Widarni & Wijaya, 2022, p. 4). Another study focuses on Surabaya (name of an innovative business model used to establish the herbal tourism village) in Solo. The model suggests offering value-added packages in an innovative manner. Another example is the Catur Kintamani village which is agrarian, and 80% of its local residents rely on agriculture for sustenance (Utama et al., 2021).

Herbal tourism has emerged in this village as a potential economic base. However, its success is impeded by barriers associated with entrepreneurship, poor accessibility, and lack of demand. The new strategies aim at encouraging overnight stays, producing local souvenirs, and augmenting local wisdom. The Catur Village has met the criteria of herbal-themed rural tourism development model based on components such as attractions, accessibility, amenities, ancillary, agents of change, empowerment of local villagers, and use of adaptable mechanisms for post-COVID times. The goal is to actualize the potential of herbal products to financially support the tourism village while emphasizing on CHSE (Clean, Health, Safety, and Environment) guidelines.

Herbal plants in Catur are unique and give it a competitive edge. As part of the tour package, tourists are offered herbal drinks and knowledge of various types of herbs and uses of herbal plants, imparted in local as well as English language. Community gardens contribute and lunch is based on traditional menus. Another article on Catur explores herbal tourism village concepts to promote alternative niche forms of tourism that resonate with the contemporary post-pandemic times (Utama et al., 2021).

The Ministry of Tourism and Creative Economy continues to prioritize the concept of Sapta Persona (the seven charms) and excellent service in the herbal tourism sector. The seven elements of Sapta Persona are safe, orderly, clean, cool, beautiful, friendly, and memorable. Catur is particularly popular for coffee agro-tourism. It has a community empowerment program which promotes health, equality, and prosperity with herbs. It fosters community empowerment, particularly for women, through the cultivation of herbal plants and processing herbal products that are ready to be used and marketed (Uttama et al., 2021, p. 8). Also notable is the TOT (Training of Trainers) program which focuses on training in making herbal oil products and herbal drinks such as Sakuntala (Utama et al., 2021). Sakuntala is a traditional herbal drink made from turmeric and betel. It is served to guests/tourists. Training also focuses on making moringa and anti-mosquito herbal oil using local plants. Financial management training is offered in the tourism village.

Several studies conducted in Indonesia have shared local sustainable herbal tourism development paradigms and initiatives include Sapta, TOT, and CHSE. Moreover, Widarni and Wijaya (2022) describe the modified version of Smith's communications framework SOSTAC which consists of six stages (situation, objectives, strategy, tactics, actions, and control). Additionally, the authors posit that three key operating sources (+3Ms, man/woman (human resources), money (budget), and minutes (time scale)) need to be considered in the model. SOSTAC and +3Ms offer a fundamental arrangement to develop a cohesive marketing communication agenda. Five interconnected components suggested by Middleton (1999) are market demand, travel organizers, destination organizers, product supply, and accessible physical traits. Another exemplary example is the book on wellness and herbal tourism which features the Jamu herb (legalized by the government) (Susanto et al., 2018; Widarini & Wijaya, 2022). The Jamu theme is selected and promoted as a wellness tourism attraction in Yogyakarta, Solo, and Semarang regions. The book describes the voyage to rediscover the history of herbal medicine to assemble archaeological traces, preserve traditions, and facilitate the scientific/modern growth of herbal medicine. As noted in the book, "tourists not only get information about the history of herbal medicine and herbal traditions in Java, but also experience impressive cultural interactions in modern society. The values of wisdom and modernity become a perspective to see how the Javanese people keep Jamu as a tradition. On the other hand, it also describes the efforts to make herbal medicine a part of a modern lifestyle" (Widarini & Wijaya, 2022, p. 309).

The development and preparation of the wellness and herbal tourism program is not at an optimal stage because the conditions are shaped by both internal and external stakeholders. Juneidi et al. (2021) stress on the importance of increasing entrepreneurship knowledge and identify key barriers as accessibility and poor visitations. Herbal tour package should be designed in coordination with the PKM—a community partnership program. Utama et al. utilize downstream innovation in the development of herbal tourism village "strengthening homestay management and community-based tourism by empowering tourism village managers and villagers in developing tourism products based on the principles of Sapta Persona, including developing and packaging Catur Village souvenir products as something to buy, so that tourists to get a thorough tourism experience" (2022, p. 40). Some herbal programs aim at uplifting the status of women and promote women empowerment through cultivation of herbal plants and processing herbal products that are ready to be used and marketed (example of a program is "Healthy, Equal, Prosperous with Herbs").

Local ownership by community empowerment has been suggested through a 7D approach: D1, developing relationship; D2, discovering capacities; D3, directions of community actions; D4, designing community actions; D5, delivering community actions; D6, delivering planned activities; and D7, documenting outputs, outcomes, and learning. Another tool is PESTLE analysis which focuses on political, economic, social, technological, legal, and ecological aspects (Susanto et al., 2018). A community empowerment concept matrix has also been suggested: somewhere to stay-somewhere to stay-CHSE homestay managers (Clean, Health, Safety, and Environment), something to see, something to buy-women empowerment program-GESI, and something to do-excellent service component.

Findings Based on Literature Review and Participant/Nonparticipant Observation

This section offers insights on herbs, from a physician's viewpoint. A mix of techniques are used to present this section. These include use of secondary sources and nonparticipant observation and participation observation techniques. The first author is a physician, whereas the second author (late Dr. Atal) pioneered herbal cure for tuberculosis using modern scientific tools of research. Historical records were used to report contributions of late Dr. Atal in addition to review of books, articles, signature websites, and nonparticipant observation techniques, by the first author, to accumulate information to present in this chapter. The first author was mindful of his positionality and used critical self-reflexivity to offer a filtered holistic view when he conducted participant observation and nonparticipant observations.

Herbal Wealth and Herbal Health

Nature has blessed mankind with a vast herbal wealth consisting of innumerable varieties of plants, shrubs, herbs, and trees. Herbs are natural products and used by man in every sphere of life, for meeting daily needs, and having endless uses. Examples include food, food additives, spices, coloring agents, seasoning and flavoring agents, perfumery industry, cosmetics, aromatic materials, household products like insect repellents, and insect antifertility agents (Saxena et al., 1986). Out of the countless herbal wealth that nature has bestowed mankind, one thing that is outstanding is herbs that heal. The healing herbs are nature's gift to the world and a precious asset for man. Herbal culture is as ancient as it is global, the faith in this system is deep-rooted, and the valuable knowledge has been passed down from one generation to the next (Whelan & Dvorkin, 2006). Since ancient times, human beings have used time-tested observation-based herbal treatments to heal the human body. Herbs have healing, curative, cosmetic, nutritional, and palliative properties. In health and nutrition, they are used as carminatives, digestive agents, nutritional supplements, herbal remedies, antifungal agents, anti-infectives, anti-parasitic, immunity builders, etc.

In India, the classical texts like *Ayurveda* and *Charaka Samhita* have served as materia medica (knowledge related to healing properties of medicinal plants) for generations of physicians in imparting treatment for the masses. Inspired by these ancient texts are the equally valuable contributions of our modern scientists and researchers who have played a major role in bringing the benefits of herbal treatment to the masses which is praiseworthy (Bhutani & Gohil, 2010). The next few paragraphs discuss the concept of herbal treatment, its different types, benefits and disadvantages, issues, the intricacies of herbal tourism, and a cautionary note. It is important to understand that herbal treatment, especially if supported by modern scientific research for safety and efficacy, will provide superior healing for anyone seeking herbal treatment. Without the support of modern science, herbal remedies may face the risk of becoming obsolete in the face of stiff competition by modern medicine.

The Systems of Medicines

The World Health Organization (WHO) defines traditional medicine as "the health practices, approaches, knowledge and beliefs incorporating plant, animal and mineral-based medicines, spiritual therapies, manual techniques and exercises, applied singularly or in combination to treat, diagnose and prevent illnesses or maintain well-being" (WHO, n.d.). An article by the World Health Organization (WHO) differentiates between traditional medicine, herbal medicine, and complementary (or alternative) medicine, all of which are used for the health (Curtis & Gaylord 2004) and well-being of human beings. While traditional medicine is somewhat broadly based on traditions, herbal medicines are based on plant-based sources. The standard medical treatment (SMT) is a system which is recognized as the proper treatment for a type of disease and accepted or used by treating healthcare professionals. The various other health systems for disease treatment are presented in the forthcoming paragraphs.

Herbal Treatment and the Herbal Industry

Bhatt's (2016) article "Phytopharmaceuticals: A new drug class regulated in India" describes the choices in the herbal industry based on herbal plants, herbal plant materials, and other herbal products:

1. *Plants*: their leaves, roots, fruits, bark, seeds, and stems are used to improve health, prevent illness, and cure disease constitutes herbal treatment.
2. *Herbs*: are used for treatment include unrefined crude plant material, like leaves, stems, flowers, fruit, seed, and even roots.
3. *Herbal materials*: include crude extracts derived from herbs, like fresh juices, dried powdered plants, gums and resins, fixed oils, essential oils, etc.
4. *Herbal preparations*: include finished herbal products like powdered herbal materials, extracts, tinctures, and fatty oils of herbal materials. Herbal preparations may include preparations made from one or more herbs.
5. *Herbal-derived medicines*: called phytopharmaceuticals include a purified and standardized fraction, with a defined minimum of four bioactive or phytochemical compounds of an extract of a medicinal plant or its part, for internal or external use of human beings or animals for diagnosis, cure, treatment, mitigation, control, or prevention of any disease or disorder.

Herbs in the Era of Modern Science

In the contemporary era, laboratory-tested evidence-based medicines are used for the purpose of healing. The herbal medicinal system, therefore, has to compete with modern medicine systems for its survival. Survival depends on constant upgrades using modern technology. Herbal culture has also undergone this metamorphosis with better evidence-based herbal research, proper standardization, modern packaging, and promotional campaigns. The preference between modern medicines and herbal treatment is a matter of personal choice, but many factors like cost, availability, safety, trust, belief, and application of modern science to the herbal system have changed the way people perceive herbs. Today, all things natural are generally perceived as trustworthy. However, from a safety standpoint, it is important to embrace a rational approach and scrutinize the pros and cons of various alternatives. As with any treatment modalities, it is equally important to take herbal treatment under expert supervision. Self-medication, based on the Internet or untrained advocates may prove to be harmful. Questions arise regarding the use of herbal medicines compared to modern medicine. Issues may arise such as scientific validation of safety and efficacy, ethical correctness of use, informed and authorized treatment provider, and informed treatment user.

In the WHO Traditional Global Summit, the WHO mentions that for centuries, mankind has depended upon traditional and complementary medicine for household use and for society's benefits (WHO Team, 2023). Traditional knowledge and herbal medicines have since ancient times been the first port of treatment for the global population, and this knowledge has also been the foundation on which modern medical literature and texts was established. The traditional systems were developed when science was not developed and the medicinal plants were selected on the basis of observation of effects and safety. Pharmacopeia is a book containing directions for the identification of compound medicines. It is published by an authorized entity such as the government or a medical/pharmaceutical society. Pharmacopoeia till the turn of last century primarily consisted of medicinal plants. With the discovery of morphine by Friedrich Sertürner (from an opium plant), the era of modern drugs commenced. When the development of synthetic techniques enabled scientists to develop synthetic drugs, the traditional medicines started fading away (Yuan et al., 2016). That is, there was a period of decline in use of natural products after the nineteenth century as modernism became fashionable, and there was a race to develop modern medicines. The big companies in the pharmaceutical industry were aggressively marketing new drugs, in trendy/eye-catching packaging backed by solid scientific evidence-based data. However, today the interest in medicinal plants has been regenerated. The modern era has revolutionized treatment with discoveries of modern-day drugs. Popularity and success of drugs depend upon scientific validation, effectiveness (efficacy), safety, availability, and cost. Modern researchers have taken keen interest in studying many ancient ayurvedic plants like Ashwagandha for their medicinal values (Atal & Schwarting, 1961).

In modern times, the "observational therapeutics" of traditional knowledge and "reverse pharmacology" of this knowledge have led to many discoveries in medicine. While the availability and economics of herbal plants are usually not a major issue, the effectiveness (efficacy) and safety of many herbal drugs have been established by modern research. Therefore, the cost, availability, efficacy, and safety make herbal medicines popular, bringing them within reach of the vast majority of the world population. Herbal plants have also contributed greatly toward new drug discoveries. Modern research has now isolated and extracted the important or pharmacologically active principle from the plant, discarded the rest of the plant as useless or having side effects, and chemically synthesized the main molecules to be marketed as a drug (Altemimi et al., 2017).

This has strengthened the belief in the usefulness of herbal medicines, scientifically supported by modern research. The use of herbal medicinal plants for various illnesses is an acceptable practice all over the world. In some countries like Ghana, where the rural population predominates, 60–70% of the population depends upon herbal remedies, as the first line of cure (Mintah et al., 2022). Easy availability of herbal medicines, cost-effectiveness, and accessibility to masses ensure its sustainability.

Evidence-Based Research of Medicinal Plants

The landmark discovery of salicin (Wang et al., 2021), an analgesic and antipyretic from *Salix alba*, paved the way for the global pharmaceutical industry to recognize and research medicinal plants as sources of bioactive phytochemicals. This paved the way for new drug discoveries. Advancement in modern research has not only led to discovery of new drugs researched from traditional or herbal remedies but also validated their mechanism of action, established their efficacy, and also established their safety. Research tools used, whether applied on traditional plants for establishing effectiveness in disease or used for bioassays in identifying the active principles in the plant, have been successful in validating the usefulness of herbal drugs. This idea of evidence-based research (EBM) was first conceived by Archibald Cochrane. Breakthrough in plant research has led to many exciting drug discoveries, contributing modern drugs like artemisinin, Boswellia, colchicine, opioids, and many more famous plant-based drugs. Around 40% of pharmaceutical products in the modern era are constituted by natural products and landmark drugs derived from traditional medicine, and examples include aspirin, artemisinin, and others.

Lessons from Ayurveda: Tuberculosis and the Landmark Discovery of Bioenhancers

An Indian breakthrough in medical science is the pioneering evidence-based discovery (Atal et al., 1985) of world's first bioenhancer piperine, discovered from a clue in Ayurveda (Atal et al., 1981). Piperine is derived from black pepper, an ancient spice used as a food additive and an ingredient of numerous ayurvedic formulations. Bioenhancer piperine was discovered in 1979 by Dr. Atal, who was researching ayurvedic and medicinal plants. He also coined the term bioavailability enhancers (bioenhancers). Bioenhancer piperine was first experimented on vasicine and sparteine, which became the world's first experimentally bioenhanced drugs. With this discovery, a new chapter of bioenhancers was scientifically validated in medical science. The term bioenhancers did not exist in any modern scientific literature prior to 1979. This outstanding contribution from India in medical science, which is based on ancient Ayurveda, made India a pioneer in bioenhancers research. It has opened new horizons in bioenhancers and bioenhanced drugs.

This discovery led to the development of the world's first bioenhanced antitubercular (Zutshi 1985) formulation (Atal & Bedi 2010) Risorine, which was officially released by the Government of India at Anusandhan Bhawan on World TB Day in 2011. On the same day, it was presented to the Philanthropist and Chairman of Microsoft, Mr. Bill Gates, in Delhi. Risorine has benefited innumerable patients suffering from tuberculosis in India. Piperine is a safe and effective bioenhancer if used in prescribed doses. Table 3.1 in Appendix chronologically traces the development, recognition, and advocacy of this breakthrough discovery by late Dr. Atal. This landmark discovery is worthy of recognition by societies at the local, national, and global levels. Tourists should also be informed so that they realize the medicinal potential of herbs in treating serious illnesses.

Rising Popularity of Herbal Drugs

The scientifically researched success stories and worldwide recognition of treatments such as the one by late Dr. Atal have boosted the confidence in the herbal system and brought many herbal drugs into limelight for effectiveness and safety. Popularity of herbal drugs, therefore, has grown by leaps and bounds within the scientific community as well as the public (Deshmukh 2023). A study from Germany assessed the reasons regarding the use of herbal treatment by the public (Welz et al., 2018). It found that the reason for the rising popularity of herbal treatment was not clearly known, but it highlighted some findings on this issue. Among these were the dissatisfaction with conventional medicines, adhering to family traditions and beliefs, positive experiences of herbal use in the past, and preference of the public for its use in mild to moderate illness. It is obvious that acute and serious illnesses are mostly treated by allopathic medicines.

As per Johns Hopkins (n.d.) comment on herbal medicine, the FDA (Food and Drug Administration, USA) considers herbal supplements as food, not drugs. It issues an advisory on the correct use of herbal medicines, like gaining self-knowledge on the medicine, using a reputed brand, following label instructions, using the prescribed dose, being treated under good supervised guidance of experienced professionals, being alert for allergic reactions or any adverse effects, and using standardized formulation. The manufacturing company can be contacted for further clarifications if needed. Some herbs commonly used in the USA include black cohosh, echinacea, evening primrose, feverfew, garlic, ginkgo biloba, ginseng, goldenseal, green tea, hawthorn, saw palmetto, and St. John's wort (Shabi & Abdullah, 2023). Chronic illnesses are generally better treated with herbal drugs rather than acute ones. Alongside the soaring popularity and support for herbal treatments and the surge in herbal tourism, it is important for both suppliers, regulators, and consumers (tourists) to exercise caution.

Cautionary Note

Herbal tourism needs planning, knowledge, and expenses. Because people travel to distant lands for treatment, it is important to properly understand the concept of herbal medicine before indulging in a herbal tourism experience for treatment or for preventive undertaking. These include gaining proper or correct tourist information and knowledge, choices of travel agency, choices of country, calculating treatment expenses wisely, the expenses for travels, the duration of stay as a herbal tourist, visa issues, additional expenses in case of unforeseen prolonged stay, and many travel related issues. All the relevant medical records and documentation need to be carried for fast assessment of medical problems. The issues related to herbal treatment include inadequate concept of herbal treatment, different choices of herbal treatment, the availability of various herbal types, their effectiveness and side effects, availability of good herbal treatment centers, good physician preferences, and many other aspects. Accurate knowledge is important, and lack of information can result in failure, economic loss, waste of time, and exposure to side effects. With the exponential growth (Zhang 2015) in demand for natural products across the globe, a modernized stance has become essential. Hence, it is important to undertake travel to the homeland of experienced and modern professional centers renowned for herbal medicines.

A tourist seeking good-quality herbal treatment may be denied (Ekor 2014) advantages of herbal treatment due to various factors. These include short shelf life, wrong timing of harvest of plant raw materials, lack of information on the composition of herbal drugs, poor and damaging storage conditions, like excessive sun, humidity, dust, and insect damage. Others include use of inappropriate anatomical parts of plants, like wrongly selected stems, leaves, roots, and bark, wrongly selected geographical area of harvest, non-sustainable supply, or inconsistent supply of plant raw materials. In spite of widespread use traditionally by people around the world, herbal medicines therefore may not have the desired effect or may be totally devoid of any effectiveness. Sometimes the active principle is present in very low concentrations, leading to poor efficacy. Furthermore, some herbal medicines are generally considered safe but may have serious side effects, particularly for the elderly. The herbal drugs is an important industry that is unfortunately poorly regulated in many countries, unregistered, unaccounted, and unsupervised by the health authorities (Bhatt, 2016).

Late Dr. Atal, (co-author) of this article highlighted some preventable errors observed in the manufacture of herbal products (Dr. C.K. Atal, Personal Communication, 1978). He pointed out that many herbal companies hired tribals to collect the plant raw material from jungles. It is observed that some tribals who are not professional botanists, sometimes collect an undesired misidentified plant for the company which may be devoid of activity and may be toxic. This results in a faulty herbal preparation devoid of effectiveness and also being contaminated with an undesired plant raw material, might even be toxic. Therefore, an important issue with herbal medicine can be non-standardized remedies. Late Dr. Atal further stressed that for herbal medicines to compete with modern medicines, it is essential to standardize herbal formulations so that formulation consistency is achieved in terms of efficacy and safety (Dr. C.K. Atal, Personal Communication, 1978). Modern scientific methods and tools for standardization used with traditional herbal products solve this problem. Some other issues are related to substandard companies manufacturing poor-quality or adulterated medicines. Heavy metal toxicity can become a dangerous issue. In some chronic diseases, the herbal drugs can become toxic for weak organs. Integration of expertise in natural products is an issue but can be resolved by establishing a platform and task force to integrate data. Next, initiatives from the Government of India are shared followed by examples of herbal attractions for herbal tourists.

Government Initiatives in India

Several initiatives by the Indian government can be noted. As an example, the President's promotion of the herbal industry can be evidenced in his patronage of Amrit Udyan, the herbal garden in the President's house in India. The President's palace has sprawling gardens including the East Lawn, Central Lawn, Long Garden, Circular Garden, Herbal-I, Herbal-II, Tactile Garden, Bonsai Garden, Arogya Vanam, vibrant Bal Vatika, and a Spiritual Garden. These showcase herbal knowledge with pride. Other examples are presented below.

The Indo-US Workshop In the first Indo-US workshop on traditional medicine in 2016, the use of analytical techniques, such as liquid chromatography-mass spectroscopy to help upgrade the Ayurveda, Siddha, and Unani pharmacopeia and bring it on a par with the US Pharmacopeia, was discussed. These tools have been in use in renowned Indian medicinal plants research centers like RRL (Regional Research Laboratory, Jammu), Jammu (now renamed IIIM (Indian Institute of Integrative Medicine, Jammu), since the early 1980s. In the USA, various aspects of herbal product quality are assessed, including standardization, selection of marker compounds, adulteration, and misidentification (White et al., 2017).

The Indo-US 2016 workshop concluded that organizational reforms are needed for improvements in traditional medicine. Support and strengthening by use of basic life sciences would bring a positive change in disharmony and provide a solid base in understanding and building the industry. Other suggestions included need for scholarships, establishing excellence centers, exchange programs, centralized laboratories, conferences, workshops and seminars, using multidisciplinary perspectives in research and development, reforms in regulations, advisory bodies, using state-of-the-art technologies, establishing schools for traditional medicine, certificate courses, producing traditional medicinal scientists and practitioners, research centers, biotechnology centers, involving the industry in manufacturing process, and market research and survey activities. Promotion campaigns can boost the participation of all sectors and support the industry. To top it off, participation of government resources can further uplift the quality of traditional medicine. Cooperation with the tourism industry and ministry of tourism will ensure sustainable development, promotion, and use of traditional medicine.

Mutual agreements In March 2022, recognizing the significance of traditional knowledge across the globe, the World Health Organization (WHO) and the Government of India signed an agreement (WHO 2022) to establish the WHO Global Centre for Traditional Medicine, supported by an investment of USD 250 million from India. It aims to exploit the potential of global traditional medicine through modern science and technology to improve the health of people on the planet.

National AYUSH Mission AYUSH mission society and Government of India have implemented the national AYUSH mission scheme (Ayush n.d.). AYUSH stands for Ayurveda, yoga, naturopathy, Unani, Siddha, and homeopathy treatments. It supports AYUSH wellness clinics for the health of the public.

Herbal Philately: The Indian postal department has issued stamps to create awareness of herbal plants. Figure 3.1 depicts flowering trees and herbal plants in India on the first day cover issued by the Philately department of India. Herbal plants depicted are Gloriosa (source of colchicine), Kadamb, rhododendron (used as

Fig. 3.1 Herbal themed stamps and first day cover issued by the Government of India. (Source: Personal collection of the authors)

a herbal tonic), lotus, and the tree *Cassia fistula* (golden rain, used for digestive issues).

Another example is the Janaki Ammal Herbarium (n.d.) in Jammu (J&K State in India) (Janki Ammal Herbarium n.d.). Herbariums offer authentic information on herbal plants. This herbarium is located at the Indian Institute of Integrative Medicine (managed by the Council of Scientific and industrial research), and various medicinal plants are digitized here. Digitization of herbs has enabled dissemination of herbal information to tourists, researchers, scientists, students, and the general public. Figure 3.2 shares a moment of celebration with the renowned scientists who spearheaded and inspired this initiative.

Last, natural herbal parks and gardens are a tourist attraction and are also patronized by scientists, researchers, and the medical practitioners in addition to the local residents. In fact, herbal parks can be found all over the country. For instance, in the State of Haryana, every district has a herbal garden which is also referred to as herbal gyan (knowledge) kendras. One such park spans 160 acres of land at Chuharpur Reserved Forest of Yamuna Nagar which is famous for its agroforestry practice. This park is a breeding ground for native and exotic species of medical plants and also offers training to farmers to grow species with medical/economic value. Another example of a world-renowned natural herbal garden is located in the Chamoli District of the State of Uttarakhand in India (Fig. 3.3) which is a world heritage site and called the valley of flowers. It is a natural habitat for many species of medicinal plants. The famous blue poppy as seen in the image is found in the rocky crevices high up in the valley.

Fig. 3.2 Celebrating the Janaki Ammal Herbarium. (Source: Personal Collection of the authors. Renowned Indian scientist, Dr. Janaki Ammal (second from right) at her farewell event at the Janaki Ammal Herbarium. Late Dr. Atal is standing next to Dr. Janaki Ammal (third from right))

Fig. 3.3 The valley of flowers, Chamoli, Uttrakhand, is an example of natural herbal garden. (Source: Personal collection of authors)

Conclusion

Medicinal plant species, used in different cultures across the world, have been estimated to be between 35,000 and 70,000 out of which at least 6500 species can be found in Asia (Mashayekhan et al., 2015). If tapped for their curing potential, these plant species hold tremendous potential for bringing relief to the sick and diseased populations across the globe. Herbal tourism (Tung 2021) villages and parks are becoming popular tourist attractions. Strategic community-empowered models are being developed which are guided by a bottom-up stakeholder collaboration approach (Wibawa et al., 2016; Widarini & Wijaya, 2022). Several herbal garden projects and herbal attractions have also mushroomed in India. The traditional use of herbal health remedies in India offers important health and nutrition benefits while generating income and environmental gains for rural communities (Deka et al., 2015). Since herbal tourism is an emerging field of inquiry, literature from a tourism standpoint is relatively sparse although numerous articles exist from a medicinal standpoint.

Aligned with the above statement, an article in the Tehran Times (Majlesi, 2021) describes the importance of traditional knowledge in contributing to the well-being of the country. The author emphasizes that the ancient remedies can lift herbal tourism in a big way by paving the path for alternative medicines. He also suggested that herbal tourism holds potential to expand the conventional tourism portfolio as well as extend the tourist season duration in destinations well-known for medicinal herbs. The article describes a couple of studies from Iran that have highlighted the significance of traditional and herbal tourism. In the studies, it has been estimated that about 30,000 people from various countries visited Iran in 2012 for medical treatment, mostly from nearby countries, bringing significant foreign exchange, and created jobs for millions directly or indirectly (Majlesi, 2021).

In the contemporary era, sustainable use of natural resources has become paramount. The tourists, the scientific community, the practitioners, and the hosts should be committed toward preservation of the history and use of herbs. The herbal pharmaceutical industry can play a major and ethical role in this aspect and provide economical and quality products in treatment modalities. Experienced and trained herbal health specialists can play a vital role in ensuring professional ethical care. The travel agencies must ensure best services at a pocket-friendly budget for the ill and suffering tourists.

Herbs provide healing and health benefits. As illustrated in this chapter, the healing herbs have traditionally provided safe relief since ancient times at a very economic cost. Modern (Hamet & Trembley 2017) technology has advanced the benefits and safety (Wakhloo et al., 1980) of many traditional herbal systems of medicine. It has also exposed the red flags and dangers of using traditional medicine. Therefore, the adoption of these ancient healing systems, as a safe and cheap alternative to modern medicine, has witnessed a renewed interest in the world communities globally. Consequently, the surging popularity of herbal treatment across the world has seen a corresponding surge in herbal tourism. The faith in herbs is strengthened by the readily accessible and authentic information on authorized digital platforms.

Appendix

The landmark milestones of the drug discovery of bioenhancers by late Dr. Atal which led to the discovery of tuberculosis drug (antitubercular formulation Risorine), and its subsequent advocacy at national and international level is traced in Table 3.1.

Table 3.1 Chronological order of the historical discovery of herbal bioenhancers (from discovery to market launch) by the late Dr. Atal

Year	Event
1979	Dr. Atal discovered and validated the world's first scientifically researched evidence-based bioenhancer piperine in 1979 which scientifically validated the concept of bioenhancers. Link: https://insa.nic.in/writereaddata/UpLoadedFiles/IJHS/Vol49_4_5_KNagarajan.pdf
1979	He published the world's first scientific literature on bioavailability enhancers Atal C.K. 1979, A breakthrough in drug bioavailability, a clue from age old wisdom of Ayurveda. I.D.M.A. Bulletin. 10, 483, 4

(continued)

Table 3.1 (continued)

Year	Event
1979	He first coined the term bioavailability enhancers (bioenhancers) Bioenhancers, a revolutionary concept to market, J.Aim, 2010, Apr to Jun, 1(2), 96 to 99. Link: https://www.ncbi.nlm.nih.gov/pmc/articles/PMC3151395/
1979	He researched bioenhancers on vasicine and sparteine in 1979, which became the world's first experimentally bioenhanced drugs An update on bio-potentiation of drugs using natural options. Link: https://journals.innovareacademics.in/index.php/ajpcr/article/download/38889/23967
1979	He discovered the world's first bioenhancer piperine in 1979 Risorine, a novel CSIR drug curtails TB treatment. Link: https://niscpr.res.in/includes/images/csirnews/csirnews_mar10.pdf
1985	He scientifically derived the mechanism of action of bioenhancers Biochemical basis of enhanced drug bioavailability by piperine. J Pharmacol Exp Ther. 1985,232, 258, 62. [https://jpet.aspetjournals.org/content/232/1/258.short]
1980–85	This discovery was patented by IIIM Jammu. https://niscpr.res.in/includes/images/csirnews/csirnews_mar10.pdf
1984-85	He initiated the world's first bioenhanced antitubercular drug project Influence of piperine on rifampicin blood levels in patients of pulmonary tuberculosis https://pubmed.ncbi.nlm.nih.gov/4044481/
2009	His discovery helped develop the world's first bioenhanced antitubercular formulation Risorine, launched in the market in 2009 at Hotel Lalit, Delhi. https://indianexpress.com/article/cities/ahmedabad/cadila
2011	Risorine was officially released on world TB day 24 March 2011 by Government of India at Anusandhan Bhawan, Delhi. (CSIR headquarters Rafi Marg Delhi) World Tuberculosis Day observed with release of TB drug Risorine, Press Information Bureau, Government of India. Link: https://pib.gov.in/newsite/PrintRelease.aspx?relid=71235
2011	Risorine was officially presented by the Government of India to Mr. Bill Gates, chairman Microsoft (Hotel Le Meridien, Delhi) on world TB day 24th March 2011 Press information bureau, World Tuberculosis day observed with release of TB drug Risorine, link provided in above row.
2010	He received the Indian Medical Association, Chikitsa Seva Rattan award, 2010, (IMA EDB award ceremony)IMA Bhawan, EDB, Delhi Links: https://www.flickr.com/photos/13059662@N06/4816530020 https://www.flickr.com/photos/13059662@N06/13401939445
2014	He received the EDPA letter of appreciation award 2014, (EDPA Midcon 2014) Hotel Lemon tree, Vaishali Link: https://www.flickr.com/photos/13059662@N06/14786913873 https://www.earlytimes.in/m/newsdet.aspx?q=128781
2010, 2014, 2016	Few examples of various medical conferences held on bioenhancers include, NITRD, EDPA, IMA, Vardhman Mahavir Medical College, Vamscon 2016 YouTube and Image links https://youtu.be/jncrDaMChIk?si=2yida8ItfYl3NYTt https://www.flickr.com/photos/13059662@N06/4816530020/ https://www.flickr.com/photos/13059662@N06/14989069496/ https://www.flickr.com/photos/13059662@N06/14786913873/.

(continued)

Table 3.1 (continued)

Year	Event
1979 to 2023	Bioenhancers have been researched globally since its discovery in 1979 and have facilitated a huge trajectory in drug research Bioenhancers from mother nature: an overview, SS Thorat et al. 2023, springeropen https://fjps.springeropen.com/articles/10.1186/s43094-023-00470-8

Source: Atal and Khurana (Guest lecture) (2016). *VAMSCON, YouTube link* https://youtu.be/jncrDaMChIk?si=2yida8ItfYI3NYTt

References

Altemimi, A., Lakhssassi, N., Baharlouei, A., Watson, D. G., & Lightfoot, D. A. (2017). Phytochemicals: Extraction, isolation, and identification of bioactive compounds from plant extracts. *Plants, 6*(4), 42.

Atal, C. K. (1979). A breakthrough in drug bioavailability-a clue from age-old wisdom of Ayurveda. *IDMA Bulletin, 10*, 483–484.

Atal, N., & Bedi, K. (2010). Bioenhancers: Revolutionary concept to market. *Journal of Ayurveda and Integrative Medicine, 1*(2), 96–99. https://doi.org/10.4103/0975-9476.65073. https://pubmed.ncbi.nlm.nih.gov/21836795/

Atal, N., & Goyal, N. (2009). *Dr. Jagat Mohini Thussu, a true Rattan of Rattan Rani hospital. KPN, Kashmiri Pandit Network, image gallery.* https://ikashmir.net/gallery/details.php?image_id=2475

Atal. N., & Khurana, P. (2016). Guest lecture at VAMSCON, Youtube video. Vardhman Mahavir Medical College, Safdarjung hospital. M M studio: https://youtu.be/jncrDaMChIk?si=2yida8ItfYI3NYTt

Atal, C. K., & Schwarting, A. E. (1961). Ashwagandha—An ancient Indian drug. *Economic Botany, 15*, 256–263. https://doi.org/10.1007/BF02862166

Atal, C. K., Zutshi, U., & Rao, P. G. (1981). Scientific evidence on the role of Ayurvedic herbals on bioavailability of drugs. *Journal of Ethnopharmacology, 4*(2), 229–232.

Atal, C. K., Dubey, R., & Jaswant Singh, J. (1985). Biochemical basis of enhanced drug bioavailability by piperine: Evidence that piperine is a potent inhibitor of drug metabolism. *Journal of Pharmacology and Experimental Therapeutics, 232*(1), 258–262.

Ayush. (n.d.) National Ayush mission (NAM), ministry of Ayush, Government of India. https://namayush.gov.in/

Berdzenishvili, T., Abesadze, A., Arabidze, M., & Cotoidze, P. (2013, November). Agro-tourism infrastructure development and personnel training project in Kakheti-Alvani-Mtatusheti region. In *8th Silk Road International Conference "Development of Tourism in Black and Caspian Seas Regions"*.

Bhatt, A. (2016). Phytopharmaceuticals: A new drug class regulated in India. *Perspectives in Clinical Research, 7*(2), 59–61. Retrieved from: https://www.ncbi.nlm.nih.gov/pmc/articles/PMC4840792/

Bhutani, K., & Gohil, M. (2010). *Natural products drug Discovery research in India: Status and appraisal*. https://pubmed.ncbi.nlm.nih.gov/21046972/

Butthongdee, U. (2015). Sustainable tourism development planning in Klong-Blab village, Surat Thani Province Thailand. *International Journal of Social Sciences, 4*(3), 20–29.

Cartwright, M. (2016). *Ravana*. World History Encylopedia. Retrieved in December 5 from: https://www.worldhistory.org/Ravana/; https://www.worldhistory.org/Ravana/#google_vignette

Chambers, D., & McIntosh, B. (2008). Using authenticity to achieve competitive advantage in medical tourism in the English-speaking Caribbean. *Third World Quarterly, 29*(5), 919–937.

Coghlan, A., & Gooch, M. (2011). Applying a transformative learning framework to volunteer tourism. *Journal of Sustainable Tourism, 19*(6), 713–728. https://www.researchgate.net/publication/254367299_Applying_a_transformative_learning_framework_to_volunteer_tourism

Curtis, P.. & Gaylord, S. (2004). *Concepts of healing and models of care*. https://www.med.unc.edu/phyrehab/pim/wp-content/uploads/sites/615/2018/03/Concepts-of-Healing.pdf.

Deka, B. P., Rupam, B., & Dibyajyoti, B. (2015). Prospects of herbal gardens as new vistas of medical tourism in North East India. *Annals of Pharmacy and Pharmaceutical Sciences, 6*(1/2), 46–49.

Deshmukh, A. (2023). *How Ayurveda is the new medical tourism attraction, by Aditi Deshmukh*, Updated Feb, 2023. https://travel.economictimes.indiatimes.com/news/tourism/wellness/how-ayurveda-is-the-new-medical-tourism-attraction/98043429

Ekor, M. (2014). The growing use of herbal medicines: Issues relating to adverse reactions and challenges in monitoring safety. *Frontiers in Pharmacology, 4*(177). https://doi.org/10.3389/fphar.2013.00177. PMCID: PMC3887317 PMID: 24454289.

Farasani NT, Zeinali, H., & Moaiednia, M. (2016). Food heritage and promoting herbal medicine-based niche tourism in Isfahan, Iran, *Journal of Heritage Tourism 13*(1):1–11. https://doi.org/10.1080/1743873X.2016.1263307; https://www.researchgate.net/publication/311237631_Food_heritage_and_promoting_herbal_medicine-based_niche_tourism_in_Isfahan_Iran

Firenzuoli, F., & Gori, L. (2007). Herbal medicine today: Clinical and research issues. *Evidence-based Complementary and Alternative Medicine*, 37–40. https://doi.org/10.1093/ecam/nem096. PMID: 18227931; PMCID: PMC2206236. https://www.ncbi.nlm.nih.gov/pmc/articles/PMC2206236/

Hamet, P., & Tremblay, J. (2017). Artificial intelligence in medicine. *Metabolism, 69*, S36–S40. https://builtin.com/artificial-intelligence

Honigberger, J. (n.d.). *Biographies 1795–1869*. https://www.hahnemannhouse.org/johann-martin-honigberger-1795-1869/#:~:text=Honigberger%20arrived%20at%20Lahore%20in,low%20potency%2C%20which%20cured%20him

Janaki Ammal herbarium. (n.d.). *CSIR- IIIM, Jammu*. Retrieved in November 2023 from: https://iiim.res.in/herbarium/introduction.htm

John Hopkins. (n.d.). *Herbal medicine*. https://www.hopkinsmedicine.org/health/wellness-and-prevention/herbal-medicine

Junaedi, I. W. R., Utama, I. G. B. R., & Waruwu, D. (2021). Increasing entrepreneurship knowledge in the development of Catur Kintamani Herbal Tourism Village Mengwi. *International Journal of Social Science, 1*(4), 539–548.

Kala, C. P. (2015). Medicinal and aromatic plants: Boon for enterprise development. *Journal of Applied Research on Medicinal and Aromatic Plants, 2*(4), 134–139.

Kurin, R. (2002). *The silk road, connecting peoples and cultures*. Smithsonian Folklife festival program book. https://festival.si.edu/2002/the-silk-road/the-silk-road-connecting-peoples-and-cultures/smithsonian

Laub, A. (2019). *Plant-based medicine: Blood glucose*. American college of healthcare sciences. https://www.researchgate.net/profile/alixandra-laub-2/publication/335157972_plant-based_medicine_blood_glucose_hypoglycemic_plant-based_medicine/links/5d5356a3a6fdcc85f890474b/plant-based-medicine-blood-glucose-hypoglycemic-plant-based-medicine.pdf

Majlesi, A. (2021). How could ancient remedies lift tourism? *Tehran times*, Sept 18. https://www.tehrantimes.com/news/465204/How-could-ancient-remedies-lift-tourism

Mashayekhan, A., Pourmajidian, M. R., Jalilvand, H., Gholami, M. R., & Teimouri, M. S. (2015). Ethno botanical urvey of herbal remedies traditionally used in North Khorasan province of Iran. *Medicinal & Aromatic Plants, 4*(3), 1–5.

Middleton, V. (1999). More Sustainable Tourism: A Marketing Perspective. *Insights, 10*(6), A165–173.

Minteh, S., Archer, M., Asafo-Agyei, T., Ayertey, F., Atta-Adjei, P., Boamah, K., & Barimah, B. (2022). Medicinal plant use in Ghana: Advancement and challenges. *American Journal of Plant Sciences, 13*(3), 2158–2750. https://www.scirp.org/journal/paperinforcitation.aspx?paperid=115859

Murthy, A. (1997). Dhanwantari: The God of Hindu medicine. *Bulletin of the Indian Institute of History of Medicine Hyderabad, 27*(1), 1–14. https://pubmed.ncbi.nlm.nih.gov/12572586/

National Thailand. (2024). *Thailand boosts wellness tourism with five-year old traditional Thai medicinal plan*. Retrieved on January 26, 2024: https://www.nationthailand.com/thailand/tourism/40028426

Nipo, D. T., Chan, J. K. L., & Saikim, F. H. (2021). Exploring the potential of herbal tourism in sabah: A case of dai-c-recreation centre, kiulu. *Bimp-eaga Journal for Sustainable Tourism Development, 10*(1), 1–10.

Nolan, J. M., & Schneider, M. J. (2011). Medical tourism in the backcountry: Alternative health and healing in the Arkansas Ozarks. *Signs: Journal of Women in Culture and Society, 36*(2), 319–326.

Nurzyńska-Wierdak, R., Dzida, K., Zawiślak, G., & Popławska, Ż. (2014). The importance of herbal gardens and the possibility of using herbal plants on a agro-tourist farm in the opinion of the potential recipients of agro-tourism. *Annales Universitatis Mariae Curie-Skłodowska. Sectio E, Agricultura, 69*(1), 34–43.

Pandey, P. K. (2007). Rare medicinal herbs of the Himalayas: An ardent need for preservation. *Chemistry & Biodiversity, 4*(7), 1605–1607.

Priadarshi, V. (2023). *Readers blog, Times of India, Medical tourism: Trends, drivers and growth potential in India*. Retrieved from: https://timesofindia.indiatimes.com/readersblog/out-loud/medical-tourism-trends-drivers-and-growth-potential-in-india-52005/

Rangari Vinod, D. (2012). *Pharmacognosy and Phytochemistry*, 3rd edition, Career Publication, Nashik, India.

Ravishankar, B., & Shukla, V. J. (2007). Indian systems of medicine: A brief profile. *African Journal of Traditional, Complementary and Alternative Medicines, 4*(3), 319–337.

Rongmei, P. (2022). Why are the locals of this village in Uttarakhand upset with Lord Hanuman? Precious Rongmei, Times of India.com/travel trends, times travel, Chamoli/Updated: Oct 14. https://timesofindia.indiatimes.com/travel/destinations/why-are-the-locals-of-this-village-in-uttarakhand-upset-with-lord-hanuman/articleshow/94855269.cms

Sarma, P. J. (1931). Hindu medicine and its antiquity. *Annals of Medical History, 3*(3), 318.

Saxena, B. P., et al. (1986). Insect antifertility and antifeedant allelochemics in Adhatoda vasica. *International Journal of Tropical Insect Science, 7*, 489–493.

Shabi, F., & Abdullah, A. (2023). *Herbal supplements, National library of medicine*. National Center for Biotechnology Information. https://www.ncbi.nlm.nih.gov/books/NBK536964/

Sheldon, P. J. (2020). Designing tourism experiences for inner transformation. *Annals of Tourism Research, 83*, 102935.

Singh, S., & Jaunjale, A. (2022). *History of homeopathy in India*. Madhav university. https://madhavuniversity.edu.in/history-of-homopathy.html#:~:text=John%20Martin%20Honigberger%20visited%20India,This%20medicine%20cured%20him

Susanto, P. C., Adiada, A. A. K., & Suyasa, N. L. C. P. (2018). Identification of tourism potentials in Catur Village Kintamani. *Seminar Ilmiah Nasional Teknologi, Sains, Dan Sosial Humaniora (SINTESA), 1*(1), Article 1. https://doi.org/10.36002/snts.v0i0.524

Torabi, N., Zeinali, H., & Moaiednia, M. (2018). Food heritage and promoting herbal medicine-based niche tourism in Isfahan, Iran. *Journal of Heritage Tourism, 13*(1), 77–87.

Tung, F. W. (2021). Rediscovering herb lane: Application of design thinking to enhance visitor experience in a traditional market. *Sustainability, 13*(7), 4033.

Utama, I. G. B. R., Erfiani, N. M. D., Wuruwu, D., Susanto, P. C., Turker, S. B., Suyasa, N. L. C. P., et al. (2021). *Innovation of herbal tourism village concepts as alternative for tourism in the post-pandemic new normal era*.

Wakhloo, R., Kaul, G., Atal, C., et al. (1980). Safety of vasicine hydrochloride in human volunteers. *Indian Journal of Pharmacology, 13*, 129.

Wang, Q., Zhao, L., Gao, C., Zhao, J., Ren, Z., Shen, Y., Yao, R., & Yin H. (2021). Ethnobotanical study on herbal market at the Dragon Boat Festival of Chuanqing people in China. *Journal of Ethnobiology Ethnomedicine 17*, 19. https://doi.org/10.1186/s13002-021-00447-y

Welz, A. N., Emberger-Klein, A., & Menrad, K. (2018). Why people use herbal medicine: Insights from a focus-group study in Germany. *BMC Complementary and Alternative Medicine, 18*, 1–9.

Wibawa, B. M., Baihaqi, I., Hakim, M. S., Kunaifi, A., & Anityasari, M. (2016). Business model and value proposition design for the establishment of the herbal tourism village in Surabaya. In International Conference on Innovation in Business and Strategy, Kuala Lumpur. Malaysia. https://www.researchgate.net/publication/312194067_Business_Model_and_Value_Proposition_Design_for_The_Establishment_of_The_Herbal_Tourism_Village_in_Surabaya

Whelan, J. S., & Dvorkin, L. (2006). Plants from many healing landscapes: Gathering information and teaching clinicians about the cultural use of medicinal herbs. *Journal of the Medical Library Association, 94*(2), 223.

White, J. D., O'Keefe, B. R., Sharma, J., Javed, G., Nukala, V., Ganguly, A., et al. (2017). India-United States dialogue on traditional medicine: Toward collaborative research and generation of an evidence base. *Journal of Global Oncology, 4*, 1–10.

WHO. (2022). *WHO establishes the Global Centre for Traditional Medicine in India, Maximizing potential of traditional medicines through modern science and technology*. New Release, Geneva, 25 March.

WHO. (n.d.). *Traditional, complementary and integrative medicine, an overview*. https://www.who.int/health-topics/traditional-complementary-and-integrative-medicine#tab=tab_1

WHO TEAM. (2023). WHO traditional medicine global summit 2023 meeting report: Gujarat declaration. *Journal of Ayurveda and Integrative Medicine, 14*(5).

Widarini, P. S. I., & Wijaya, M. (2022). Wellness and herbal tourism based on local wisdom as an alternative to 'new normal' tourism. *Jurnal Komunikasi, 14*(2), 299–321.

Yuan, H., Ma, Q., Ye, L., & Piao, G. (2016). *The Traditional Medicine and Modern Medicine from Natural Products, 21*(5), 559. https://doi.org/10.3390/molecules21050559; https://www.mdpi.com/1420-3049/21/5/559

Zhang, J., Onakpoya, I. J., Posadzki, P., & Eddouks, M. (2015). The safety of herbal medicine: From prejudice to evidence. *Evid-Based Complementary and Alternative Medicine, 2015*, 316706.

Zutshi, R. K., Singh, R., Zutshi, U., Johri, R. K., & Atal, C. K. (1985). Influence of piperine on rifampicin blood levels in patients of pulmonary tuberculosis. *The Journal of the Association of Physicians of India, 33*(3), 223–224.

Chapter 4
Wellness in the Spa Industry: An Investigation of Hot Spring Tourism in Australia

Gaelle Hugues Gregoire Joson, Joanne Pyke, Terry Delacy, and Charles Davidson

Abstract The evolution of wellness as a tourism segment has progressively grown as a field of academic and practical interest. This trend is a response to the growing awareness of the concept of wellbeing which is widely understood as a state of being or feeling which has an emphasis on making one's life physically and emotionally better. This can include happiness, financial stability, health and spirituality.

Spas or wellness hotels are an important element of wellness tourism and are often associated with nature-based activities. These activities are centred around some form of water treatments. Hence hot springs are a growing segment of the wellness tourism industry. Hot springs have long been known for their therapeutic benefits; however, there has been little academic investigation of the industry, its wellness impacts or its significance as a tourism phenomenon.

The chapter will explore Australian tourism and wellness tourism, the importance of Australian hot springs in regional destinations and the importance of developing indicators specific to the hot spring tourism industry. This will be achieved through an investigation of the emergence of the tourism and hot spring industry in Australia and the review of tourism indicators to help develop a set of indicators as a tool to measure the impacts of hot springs.

Keywords Wellness tourism · Spas · Hot springs · Healing · Sustainable tourism indicators

G. H. G. Joson (✉) · J. Pyke · T. Delacy
Victoria University, Melbourne, Australia
e-mail: gaellejoson@gmail.com

C. Davidson
Peninsula Hot Springs, Fingal, VIC, Australia

D. Chhabra et al. (eds.), *Sustainable Development and Resilience of Tourism*,
https://doi.org/10.1007/978-3-031-63145-0_4

Introduction

As a growing phenomenon, the wellness tourism industry has attracted the interests of researchers, students and professionals worldwide. Discussions around the wellness paradigm in the tourism industry has generated interest due to its holistic approach which aims to promote health and wellbeing of individuals and communities more widely. Wellness tourism is defined as travelling for the purpose of feeling physically and emotionally better. Current industry trends show that the consumer focus has shifted to a healthier lifestyle and are more prone to consume products from the wellness industry whether it is a stay at a wellness resort or a day at a hot spring facility (Bendegul & Heather, 2022). At its most simple, the wellness paradigm has been described as relaxation and proactive action towards a healthy lifestyle (Smith & Puczko, 2008). Considered as a niche tourism market, this industry became a trend where people would travel for the purpose of feeling physically and emotionally better.

The wellness tourism industry sits at the intersection of two large and growing industries, the tourism industry and the wellness industry. It is the largest sector of the wellness industry which was accounted for $5.6 trillion in 2022 and covers an array of different sectors that allow consumers to build wellness activities into their daily lives as illustrated in Table 4.1 (Global Wellness Institute, 2023). Wellness tourism encourages rejuvenation and rest while discovering healthier lifestyles, authentic experiences and preventing or managing diseases. Overall, travel for wellness is motivated by the search for meaning, connection and joy (Bendegul & Heather, 2022, p. 7).

To gain further understanding of the wellness tourism industry and its evolution, it is important to define the terms wellness and wellness tourism. Wellness is defined by the GWI as "the active pursuit of activities, choices, and lifestyles that lead to a

Table 4.1 Distribution of Global Wellness Economy (after Global Wellness Institute, 2023)

Global Wellness Economy: $5.6 trillion in 2022	
Sectors	Value ($)
Personal care and beauty	$1089b
Healthy eating, nutrition and weight loss	$1079b
Physical activity	$976b
Wellness tourism	$651b
Public health, prevention and personalised medicine	$611b
Traditional and complementary medicine	$519b
Wellness real estate	$398b
Mental wellness	$181b
Spas	$105b
Workplace wellness	$51b
Springs	$46b

state of holistic health" (Global Wellness Institute, 2021, p. 1). Wellness is also defined as being a complex and multilayered state of wellbeing (Dunn, 1959a). Wellness tourism is hence defined as "travel associated with the pursuit of maintaining or enhancing one's personal wellbeing" (Global Wellness Institute, 2021, p. 71). Both definitions foreground the importance of one's overall health which is simply identified as the absence of disease or illness. It is, however, argued that the approach to wellness is not only the absence of disease but a proactive approach in maintaining and enhancing health (Stará & Peterson, 2017).

The current literature reveals three main conceptualisation of wellness: "wellness is a state of being", "wellness as a process" and "wellness as an approach to professional care" (Stará & Peterson, 2017, p. 4). In research, the terms "wellness" and "illness" are often used interchangeably. Myers (1992) argues that even though these terms can be easily intertwined, wellness cannot be used as a synonym to health. The reason is that health consists of "social, spiritual, emotional and physical", aspects whereas wellness is the holistic aspect of a positive state of being where there is an optimal functioning of all these components (Stará & Peterson, 2017, p. 4). Wellness is viewed as a proactive path towards a state of holistic health. It is important to note that the concepts of wellness and health/medical work in harmony and complement each other. Broadly, if one's end goal is to be healthy, then wellness is the pathway to achieve it (Dillette et al., 2021). While the concept of wellness tourism is a relatively new term, wellness is referred by the GWI as a modern word with ancient roots (Global Wellness Institute, 2021).

The importance of wellness was prominent in 3000–1500 BC with the practice of Ayurveda and traditional Chinese medicine as portrayed below in Fig. 4.1 (Global Wellness Institute, 2021, n.d.). The practice of yoga, acupuncture and meditation are also examples of ancient wellness practices which have had sustained popularity. These practices emerged as a holistic perspective to health and wellness by fostering harmony between the mind, body, and spirit (Bendegul & Heather, 2022). In 1860s, hydrotherapy, also known as water cure, was introduced to cure any illnesses and purify the body using cold and hot baths. In the 1960s, Dunn introduced the concept of "high level wellness" which recognised health as a continuous lifestyle process (Dunn, 1959a, b).

While the concept of wellness was over-shadowed by medical science through much of the 1900s, the limitations of western medicine were increasingly recognised, and the wellness movement gained momentum from the 1980s to 2000s (Global Wellness Institute, 2010). The trend was signalled by a widespread growth in practices such as the adoption of workplace wellness programs, the growth of the fitness and spa industries and celebrity promotion of wellness activities. The COVID-19 pandemic arguably gave further momentum to the pursuit of wellness and is likely to be associated with the increasing popularity of spas and hot and mineral springs which have unique healing properties. The wellness economy is predicted to rapidly grow amidst and beyond the pandemic (Global Wellness Institute, 2021).

As illustrated in Table 4.1, springs and spas respectively accounted for \$46b and \$105b in 2022 (Global Wellness Institute, 2021). These figures demonstrate the

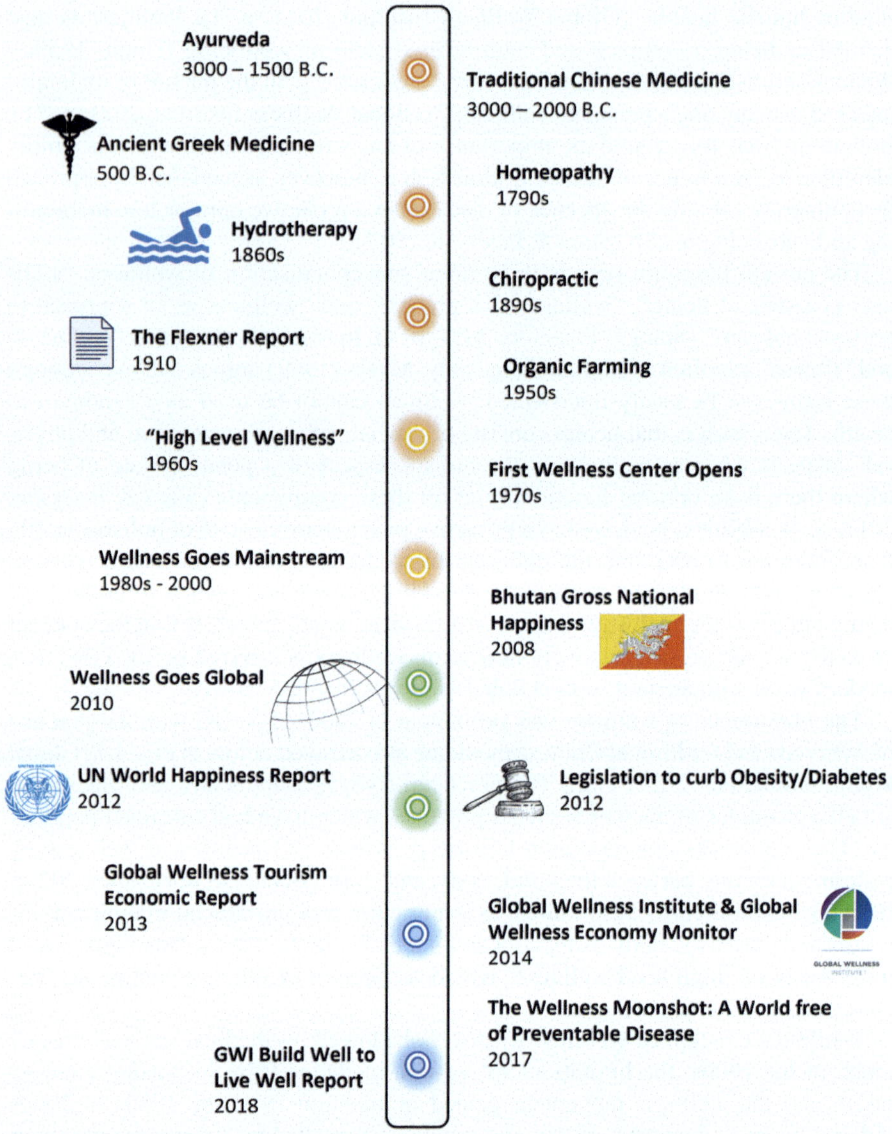

Fig. 4.1 Evolution of wellness tourism. (After Global Wellness Institute, n.d.)

importance and the economic impact of these industries in the global wellness economy. Spa tourism is associated with activities involving hydrotherapy or balneotherapy. The spa tourism industry promotes a place where overall wellbeing is prioritised. As a new niche market, there have been discussions around the meaning of the term spa. Simply defined as "health through water" or "hydrotherapy", spa relates to tourism as the participation of tourists in any spa relevant activity (Ramos

& Untong, 2016). This unique form of tourism has evolved from a place where natural hot and mineral hot springs were located to a place where physical, mental and emotional improvements are experienced through the use of water. This can be referred to as hot spring tourism. Hot spring tourism is an emerging form of spa tourism and has not yet been properly defined. Hot spring tourism can be considered as a unique and special type of tourism.

Bathing in hot healing waters is recognised as an ancient wellness practice. Hot baths gained popularity due to their natural healing properties which was proven to be an innovative treatment and therapy (Erfurt-Cooper & Cooper, 2009; Smith & Puczko, 2008). The bathing experience has been shaped by culture and embedded within local cultural practices across the world. Many countries in the Asia Pacific region and Europe have both established strong bathing cultures. Bathing in hot tubs was believed to be a form of spiritual and physical purification by many Europeans, Africans, Americans, Middle Easterners and Asians (Paige et al., 1988). Over time the bathing culture has progressively changed from being seen as purely medicinal to being an important form of relaxation and pleasure. The trend of bathing in a hot source of water has emerged with the appearance of the spa travel trend (Erfurt-Cooper & Cooper, 2009).

In Australia, the potential of this industry is increasingly attracting interest as a niche market with potential for growth (Sykes, 2019). Hot springs being an important element of the wellness industry have always been present in Australia. Most importantly, soaking in hot springs has been a well-established practice by indigenous Australians who have valued and protected hot spring sites (Sykes, 2019). With the emergence of hot spring centres in development in Regional Australia, the industry is poised to grow significantly over the next 10 years.

This chapter discusses the emergence of spa and hot spring tourism. It investigates the type of research that exists on spa tourism and hot spring tourism. The chapter provides discussions around the growing importance of developing specific indicators to measure and assess the sociocultural, economic and environmental impacts of hot springs. The Delphi technique was the method used to develop a set of indicators to measure the impacts of hot springs on the community and the environment. At this stage, preliminary data shows that hot springs have a positive impact on the community and the economy.

Australian Tourism

Tourism makes a significant contribution to Australia's economy. In 2022, tourism generated $96.8 billion in total tourism consumption and contributed approximately $36.5 billion to Australia's GDP in 2021–2022 (Tourism Research Australia, 2022a). Despite being impacted severely by the pandemic and the imposition of government restrictions on movement and business operation, the industry is steadily recovering with a robust national strategy that brings stakeholders together to support industry recovery (Tourism Australia, 2022). The tourism industry is crucial to critical infrastructure such as airports, roads and accommodations and plays a crucial role in regional development with 44% of tourism revenue flowing to regional Australia (Deloitte Access Economics, 2021). Tourism is the ninth largest employing industry and was ranked fourth biggest exporting industry in Australia (Deloitte Access Economics, 2021). This shows how the Australian tourism industry is an important driver of economic growth through the generation of jobs, investment, trade and the promotion of regional growth.

Australia has a wealth of assets that places it competitively in terms of international tourism. This advantage stems from its spectacular and natural attractions, pristine beaches and rainforest as well as a unique opportunity to experience the oldest living culture on Earth, Australia's First Nations peoples. First Nations peoples offer visitors an exclusive insight into their culture, relationship with the land and sea and their connection to country helping visitors create a deeper connection to Australia. In 2019, 17% of international visitors and 1% of domestic visitors (representing an increase of 12% since 2014) participated in an indigenous tourism experience (Tourism Australia, 2021a, b). As travellers seek out unique and authentic experiences as well as a cultural connection, the interest in indigenous tourism experiences continues to grow (Tourism Australia, 2021a).

Australia is a popular destination for international visitors, and prior to COVID-19, Australia welcomed nearly ten million visitors in 2019 (Tourism Australia, 2022) and is poised to grow at levels exceeding pre-COVID levels. Australia has a high-yield market where international visitors stay for longer periods of time and have a larger spend compared to visitation in other countries. Before the pandemic, the total worth of tourism was estimated at A$166 billion with A$45.4 billion from international tourism. Australia was ranked 7th out of 117 countries for the Travel and Tourism Development Index 2021 which measured the factors and policies that support sustainable and resilient development of the Travel and Tourism sector which hence contributes to the development of the country (World Economic Forum, 2022).

By 2027, total visitor expenditure is forecast to surpass pre-pandemic levels with a total of $227.7 billion (Tourism Research Australia, 2022b). Australia is also predicted to welcome 10.2 million international visitors by 2025 which will set a new record (Tourism Research Australia, 2023). The annual domestic overnight trip and day trip expenditures are also expected to increase. International visitor expenditure is expected to reach above the pre-pandemic levels by 2024 which is 1 year sooner that the predictions (Tourism Research Australia, 2023). Due to international border closures, operators benefited from a strong domestic tourism market which World Travel and Tourism Council (2018) accounts for 73% of the visitor economy. Tourism Research Australia (2022b) forecasts further growth domestic travel as restrictions are eased despite the increase in the price of fuel and the rising costs of travel.

Endowed with the raw ingredients of nature-based assets, premium-quality produce, cultural and spiritual attractions and wellness tourism are identified as an emerging industry in Australia (South Australian Tourism Commission, 2021). The Global Wellness Institute ranks Australia ninth in the wellness tourism markets in

2022 (Global Wellness Institute, 2023, p. 58). Governments recognise the high value of wellness travellers and are investing in new facilities, renovating and improving existing ones (Bendegul & Heather, 2022). As the world is in a phase of recovery from the COVID-19 pandemic, visitors/travellers are now prioritising their health and wellbeing evidenced by an expected annual growth rate of 21% for wellness tourism from 2020 to 2025 (Bendegul & Heather, 2022, p. 208; Global Wellness Institute, 2021).

Tourism Australia's strategy to develop *Signature Experiences of Australia* aligns with travellers seeking to discover destinations offering wellness experiences (Tourism Australia, 2023). Tourism Australia's strategic focus is to encourage regional dispersion while promoting unique experiences such as Aboriginal-guided experiences, wellness activities, cultural attractions and luxury lodges amongst others. With thermal/mineral springs establishments operating across 130 countries and an estimation of over 200 projects in the pipeline for future openings and developments, the sector is expected to grow amidst the pandemic (Global Wellness Institute, 2023, p. 77)

Hot Spring Tourism in Australia

Australian hot springs are often found in unexpected places. With no currently active volcano in Australia, many hot springs are fed by the Great Artesian Basin (GAB) which occupies quarter of the continent (Erfurt, 2021; Lambert & Jenni, 2011). The first bore was sunk into Great Artesian Basin was at Kallara Station in northwest New South Wales in 1878 (Brake, 2020). Where there are no active volcanoes, groundwater is heated by heat radiating from the Earth's molten core which then rises under pressure from an aquifer to the surface. Bores or wells are used to make water rise to the surface at a constant velocity (Erfurt, 2021, p. 22). The Great Artesian Basin is the largest artesian basins in the world. In Queensland, Australia, the *Cunnamulla Hot Springs* in the town of Cunnamulla is an example of a natural hot spring sourced from the Great Artesian Basin which attracts both local and international hot spring enthusiasts from all over the world.

The evolving Australian bathing culture is grounded in a rich and ancient culture. Australia has one of the strongest hot spring connections with Aboriginal people meaning that bathing in Australian hot healing waters is a unique and authentic experience (Sykes, 2019). Water has always played a critical role for First Nations people and has a deep cultural and spiritual value. Water courses define language boundaries and ceremonial places and are used for healing purposes as well as a food source (Erfurt, 2021; Waterwise Queensland, 2021). Dreamtime stories are stories that have existed for thousands of years which refer to the worldview stemming from Australian Aboriginal beliefs. These ancient cultural stories show the connection between the oldest community on earth and water sources (Waterwise Queensland, 2021). The state of Victoria, Australia, has a unique position in terms of the number of geothermal water aquifers along its coastline. The "Great Victorian Bathing Trail" project helped establish 900 km iconic tourism route connecting hot springs, mineral springs and sea baths across the coastline Victoria, Australia (Bathing Australia, 2022; Sykes, 2019). The bathing trail has help to boost tourism in one of the coldest areas of Australia during winter. With the number of existing and emerging hot and mineral springs in the state, there is the opportunity to investigate their impacts on the destinations they are located in and on the community.

It remains the case that few destinations within Australia are marketed as a wellness tourism destination. This gap is being rectified by Tourism Australia as wellness has been named as one of five niches for driving regional tourism development (Global Wellness Institute, 2018). The number of visitors seeking wellness tourism experiences in Regional Australia has seen a growth of 9.6% in visitation per year from 2015 to 2019 (South Australian Tourism Commission, 2021). These figures are expected to increase with the growing interest of people in health and wellness. With a large number of geothermal and mineral springs, this trend is anticipated to contribute to Australia's competitive advantage as a wellness destination (Tourism Australia, 2019; Voigt & Pforr, 2013). The growing popularity of hot spring and the wellness industry in Australia provide an opportunity to drive the economy of regional destinations as well as future developments. With little academic investigation on the impacts of hot springs or their significance as a tourism phenomenon, there is an opportunity to measure and assess this trend.

Literature Review

Existing literature indicates that spa tourism has been primarily researched to understand consumer behaviour, spa tourist motivations and tourist satisfaction and loyalty towards the spa destination. As a relatively new phenomenon, most research in this field provides an assessment of the role of the spa and wellness tourism industry in different destinations (Erfurt-Cooper & Cooper, 2009). To gain a deeper understanding of this niche market, most researchers explored tourist profiles and their key motivations to travel to a wellness destination (Azman & Chan, 2010; Clark-Kennedy & Cohen, 2017; Mi et al., 2019). Exploring tourist motivation seemed important for most researchers as it would help to understand the wellness travellers and would establish a benchmark for further studies. Azman and Chan (2010) analysis supported the claims made by Smith and Puczko (2008) who attributed spa tourist motivation to personal needs and lifestyles. Smith and Puczko (2008) acknowledged the increasing tourism interest in restoring and maintaining health and wellbeing (Moscardo, 2011). This research was done to mainly develop appropriate marketing strategies to enhance tourist satisfaction. Wellness tourism was found to be beneficial for baby boomers from a medical and wellness perspective. As an ageing cohort, baby boomers seek alternative forms of treatments for maintaining wellbeing (Patterson & Balderas-Cejudo, 2022).

The wellness industry has become more competitive with a number of existing and emerging wellness destinations around the world. Achieving destination competitiveness is hence the key focus for most destinations (Lee & King, 2006, 2009).

With the unique characteristics and attributes of hot springs, it can be a destination competitive advantage, leading to repeated visitation, tourist satisfaction and loyalty. Tourist satisfaction amongst spa visitors is a topic frequently researched (Han et al., 2017; Shavanddasht & Allan, 2019; Wisitnorapatt & Sirirat, 2023). Tourist satisfaction and loyalty were found to be important to focus on to gain competitive advantage over other destinations (Backman et al., 2023). With tourist loyalty brings economic benefits and local developments to the destination. Some authors investigated the factors of tourist loyalty to determine its relevance towards tourist behaviours (Han et al., 2017). Shavanddasht and Allan (2019) investigated the difference between first time and repeat tourists visiting spas. The authors found that repeat visitors were more satisfied and loyal compared to first timers. Most of the studies done in this field aimed at investigating customer loyalty towards the wellness destination.

Other studies addressed the contribution of spa tourism to the regional economy of the destination. The aim of most of these studies was to explore wellness tourism and spa tourism in their respective destinations (Mueller & Kaufmann, 2001; Smith & Puczko, 2008; Vystoupil et al., 2017). Spa tourism can act as a catalyst for strategic opportunities to help destinations recover from the recent pandemic. While investigating this niche market, it was observed that most hot springs/spas facilities were focused on aesthetic and recreational use. With the pandemic, Pinos Navarrete and Shaw (2021) outlined the potential opportunities for hot springs and spas to regain their traditional wellness and medical functions. The contribution of hot springs to regional development represents an opportunity to further investigation.

Importance of Hot Springs to Regional Development

Can thermal/hot spring tourism be a driver of regional development? With an array of assets, hot springs can be a catalyst for bringing positive changes to the destination in which they are located. Some of the numerous benefits of having a hot spring in regional destinations are the creation of jobs for local communities, infrastructure/roads/facilities development, increased income, increased visitor's economy, increased visitation numbers to the destination, sustainable development in the destination, environmental protection, cultural development and creation of diversity and inclusion initiatives with the local community (Clark-Kennedy & Cohen, 2017; Erfurt-Cooper & Cooper, 2009; Wang & Lin, 2021). With the emergence of health-conscious travellers, hot springs can bring sociocultural change to the destination. While all these benefits are crucial in regional development, hot springs are commonly known for its curative/therapeutic purpose, may it be physical health such as treating chronic diseases or mental health. It is assumed that having a hot spring in a destination can be beneficial to the community's health and wellbeing and support community development (Global Wellness Institute, 2018). However, research on how and to what extent to which these benefits accrue at a regional level is limited, despite much speculation that hot springs bring a competitive advantage to destination development (Erfurt, 2021).

The Need to Develop Indicators Specific to the Hot Spring Industry

The need for sustainable tourism development is an increasing priority in the context of contemporary issues such as climate emergencies, over tourism and rising inequality. It is an assumption of this research that hot spring developments can be an important agent for change and to transform destinations into an emerging green economy path (Law et al., 2016). The concept "regenerative" has recently emerged in the tourism literature as promoting a meaningful and positive transformation of the planet and the tourism industry (Ateljevic, 2020). The concept of "regenerative" can be related to the way hot springs can assist in this transformation (Ateljevic, 2020; Cave & Dredge, 2020; Pollock, 2019; Zari, 2009). The regenerative process involves a fundamental paradigm shift in relation to the recreation of social and ecosystems' capacity (Cave & Dredge, 2020; Pollock, 2019; Zari, 2009). This ongoing process requires an understanding of the uniqueness of a destination's cultural, social, environmental and economic components to seek regenerative development and design (Sykes, 2019; Zari, 2009). The connections between various components, such as energy, ecosystems, building, people, water, waste and so on (Zari, 2009), are a crucial consideration for regenerative development to be viable. Hot spring development can be an important element of the shift from a transformative perspective to a regenerative one (Cave & Dredge, 2020).

Indicators are a way of measuring a change or progress towards achieving a specific outcome (World Tourism Organization, 2004, p. 8). Tourism indicators are purposefully chosen from a range of indicator sets to identify and measure the results of our actions with regard to key issues faced by tourism destinations (Whittlesea et al., 2019; World Tourism Organization, 2004).

Given the economic significance of the tourism industry, there has been a growing interest on the impacts of the tourism sector and the industry's ability to implement sustainability practices. The development of appropriate tourism indicators was crucial to measure and monitor strategies and to assess the progress of the industry in the future. Several indicator schemes have been developed in the tourism field including sustainable tourism indicators, green economy indicators and European Tourism Indicators scheme (ETIS) amongst others (European Union, 2016; Law et al., 2017; World Tourism Organization, 1996). These schemes were developed to measure and monitor the impact of tourism at a global, national and destination levels.

Sustainable tourism indicators (STIs) have been used for general tourism destinations and can be assessed using sustainable development strategies in destinations (Rasoolimanesh et al., 2020). It has an extensive indicator reference with over hundred indicators which allows destinations managers to select issues that are the most pertinent to the destination and apply them (Kristjánsdóttir et al., 2018). Different methods were used to develop sustainable tourism indicators. Some researchers suggested that there were two ways of developing sustainable indicators for tourism which are the scientific or academic approach and the policymaker approach (Rasoolimanesh et al., 2020). The scientific or academic approach describes sustainability as a complex subject which requires technical methods of assessment, whereas the policymaker approach requires consensus amongst stakeholders and is easily understood by locals. Measuring and assessing the impacts of tourism is quite complex and mostly requires a scientific or academic approach. This approach requires an assessment of the phenomenon over time and its impacts. This has led to the development of various sets of indicators attempting to quantify sustainability, especially at a destination level (Rasoolimanesh et al., 2020).

The green economy indicator framework aims to improve production processes and consumption practices by measuring green economy policies. The green economy indicators attempt to understand the extent to which tourism activities are being "greened" (Law et al., 2017, p. 1436). With 119 indicators across 52 components, it may be time-consuming to measure and assess. While this approach provides a deep understanding of the environmental issues persisting in the industry, it is deemed hard to assess and measure with the extended value chain of the tourism industry and constant changes in tourism flows.

ETIS was designed specifically to manage, inform and monitor sustainability on a destination level. With 43 indicators, the aim of ETIS was to provide a framework to reflect on the impact of tourism on the local economy, environment and community (European Union, 2016). Destinations may share common core indicators, but the differentiation between their sustainability issues and their governance policy may require a different approach with unique and specific objective (Rasoolimanesh et al., 2020). It is hence crucial to have a valid, reliable and unique set of indicators specific to each industry to achieve specific objectives.

Current performance measures of tourism are mainly focused on economic benefits rather than understanding and preventing wider impacts on the society and the environment. From a green economy perspective, tourism indicator frameworks need to be inclusive of three main pillars of a sustainable approach which are the people, planet and profit (Whittlesea et al., 2019). The performance measurement framework developed for Australia's visitor economy encompasses economic, environmental and sociocultural indicators. Whittlesea et al. (2019) recognise the need to incorporate indicators which can assess past performance in relation to a strategy and indicators and can provide guidance on future trends and expectations. Whittlesea et al. (2019) also emphasise the importance of stakeholder engagement in developing and designing an indicator framework. Stakeholders' engagement can provide an understanding of any limitations and opportunities faced by the industry.

With the number of studies on the development of indicators in the wellness tourism field, some economic and environmental indicators can easily be translated to the hot spring industry. However, the aim of developing economic and environmental indicators specific to the hot spring industry is to make it easy to use and easy to understand for stakeholders. On the other hand, the challenge is to develop a set of indicators capable of assessing the impacts of hot springs on its user's perception of health and wellbeing and on the community. The research literature around sociocultural issues has a strong focus on visitors' experiences which combines satisfaction, the reason for visiting (Tonge et al., 2005). Wellness indicators can be subjective. With the uniqueness of the hot spring industry, there is a need a specific yet measurable set of sociocultural indicators.

After carefully reviewing indicator schemes developed for the tourism industry (ETIS, STIs and green economy indicators), it was noted that they were used and adapted as a general framework for tourism at a national or regional scale. Some indicators could not potentially be adapted to hot springs' destination as such. With the unique characteristics of hot springs, there is a need for their own set of indicators. It is explained by the fact that the indicator schemes developed by larger associations do not always highlight problems and trends. It does not always evaluate and monitor progress in the hot spring industry at both destination and individual levels. Hence, developing indicators specific to the hot springs will be a valuable tool to measure and assess its environmental, economic and sociocultural impacts. It will help monitor and assess the progress of the industry at a larger scale and help make changes if needed. At a regional level, there is a need for specific indicators measuring the wellness impact of hot springs at an individual and community level.

Methodology

This chapter aims to understand the impacts of hot springs in regional Australia and reviews indicator framework used in the tourism field. This will help identify an indicator set specific to the hot springs. Hence, a qualitative method can be used to develop a set of indicators to measure and assess the impacts of hot springs on the economy, environment and community. The Delphi technique was used to develop an indicator set specific to hot springs. This technique involved an in-depth set of questionnaires sent to a panel of experts to gather trustworthy and reliable information (Dalkey & Helmer, 1963; Garrod & Fyall, 2005). 32 experts were purposefully gathered with global expertise hot spring and tourism development. The experts were identified from advice from key people in the community, industry and government.

The aim of the first round of questionnaire was to gain deeper understanding of the potential sociocultural, environmental and economic impacts of hot springs on its destination. It presented a series of open-ended questions to the panel of experts. The responses gathered in the first-round questionnaires was used to create the second-round questionnaires and similarly the third-round questionnaires (Hsu & Sandford, 2007). The second round of questionnaires presented a list of proposed indicators identified from the responses received from the first questionnaire. The experts were asked to rate the proposed indicators on a scale of importance. A comment box was included after each proposed indicators to give an opportunity to the experts to comment freely on their ratings. Consensus was reached based on the ratings given to each proposed indicators and whether it will be feasible and easily measured.

This technique was widely used in the tourism and hospitality literature to forecast future developments (Kaynak et al., 1994; Miller, 2001). In this context, the technique was used to gather information on a central problem and encourage independent analysis by the experts. This technique has been proven to be useful to build a framework for a specific area of research and to gather trustworthy information (Miller, 2001).

Preliminary Results

At the time of writing, the final indicators are being refined and unavailable for publication. However, the following discussion highlights themes encompassed by the framework (as shown in Fig. 4.2), plans for data generation and intended applications of the data.

The framework encompasses three categories: economic, environmental and sociocultural under which themes were identified. The themes were identified from the responses received from the panel of experts on two rounds of the Delphi technique. Under the economy category, direct economic impacts of hot spring operations in the destination and indirect economic impacts were identified as themes.

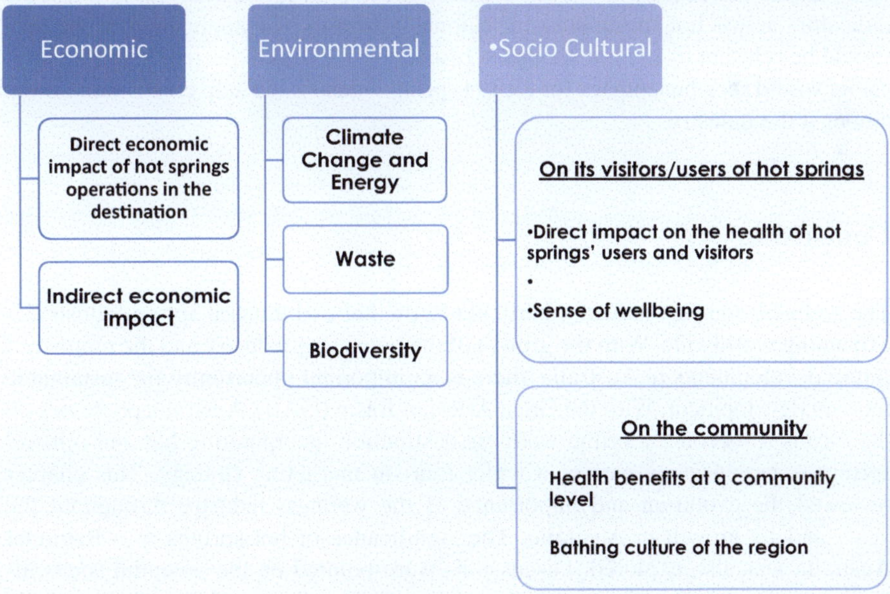

Fig. 4.2 Indicator framework developed from the responses of the experts

The economic indicators were identified easily by the panel of experts. It was noted that the experts identified the need for a tool to measure hot spring businesses' economic performance. Quantifying the traffic of visitors to the hot springs, employment status and total expenditure were important aspects mentioned by the panel of experts.

Under the environmental category, climate change and energy, waste and biodiversity were identified as emerging themes. With climate change being a hot topic worldwide, it was important to recognise the potential impacts of hot springs on the environment by quantifying the amount of energy and waste produced by the facility. The environmental impacts of hot springs will be subjective to where hot springs are located and how big the facility is.

The sociocultural category was the most challenging to refine as no prior research has been done specifically on the impacts of hot springs on both its visitors/users and the community. On an individual level, direct impact on the health of hot spring users and visitors and sense of wellbeing were recognised as key themes. The literature mostly covers positive impacts of hot springs on its user's physical health. Sense of wellbeing has been merely discussed as a potential indicator to measure the impacts of hot springs individually. On the community level, health benefits at a community level and bathing culture of the region were recognised as themes. In the destination in which the hot springs are located, the experts deemed important to assess the quality of life of its residents. The level of engagement with the community was also one key aspect that the panel of experts discussed. With this specific set of sociocultural indicators, it is an opportunity to quantify the impacts of hot springs at both an individual and community level.

To assess the validity of the developed indicators, a future study should test the indicators across hot spring sites in regional Australia. The application of the data should be relevant to any hot springs across the world. It is expected that these indicators would set a benchmark for the hot spring industry and help create momentum amongst the industry.

Conclusion

The wellness tourism industry, particularly the hot and mineral spring industry, is expanding worldwide. With the growth of the hot spring industry and the number of future developments in Australia, there is an important opportunity for sustainable tourism development. With the Great Artesian Basin (GAB), Australia possesses all the means to become a prime wellness destination incorporating hot and mineral springs experiences as part of a wider tourism marketing strategy. This chapter reviewed the evolution and importance of the wellness industry throughout the years and its growth in Australia. The significance of hot springs in a Regional Australia was also explored. Discussions were pointed on the potential sociocultural, environmental and economic impacts of hot springs and the importance of developing specific indicators to measure and assess the impact of hot springs on its community and on the destination. This has led to the development of a set of indicators specific to the hot spring industry. It is assumed that the results from the study will acknowledge that hot springs can be a powerful catalyst in enhancing the quality of life for both its visitors and the community. It is expected that the hot spring industry will play a proactive role in promoting sustainable and regenerative growth leading the way towards a greener and more sustainable future.

References

Ateljevic, I. (2020). Transforming the (tourism) world for good and (re) generating the potential 'new normal'. *Tourism Geographies, 22*(3), 467–475.

Azman, I., & Chan, K. (2010). Health and spa tourism business: Tourists' profiles and motivational factors. *Health, Wellness and Tourism: Healthy Tourists, Healthy Business, 9*, 24.

Backman, S. J., Huang, Y.-C., Chen, C.-C., Lee, H.-Y., & Cheng, J.-S. (2023). Engaging with restorative environments in wellness tourism. *Current Issues in Tourism, 26*(5), 789–806.

Bathing Australia. (2022). *Baths.* https://www.bathing.org/baths

Bendegul, O., & Heather, L.-K. (2022). *Wellness management in hospitality and tourism.* Goodfellow Publishers.

Brake, L. (2020). Development, management and rehabilitation of water bores in the Great Artesian Basin, 1878-2020. *Proceedings of the Royal Society of Queensland, The, 126*, 153–175.

Cave, J., & Dredge, D. (2020). Regenerative tourism needs diverse economic practices. *Tourism Geographies, 22*(3), 503–513.

Clark-Kennedy, J., & Cohen, M. (2017). Indulgence or therapy? Exploring the characteristics, motivations and experiences of hot springs bathers in Victoria, Australia. *Asia Pacific Journal of Tourism Research, 22*(5), 501–511.

Dalkey, N., & Helmer, O. (1963). An experimental application of the Delphi method to the use of experts. *Management Science, 9*(3), 458–467.

Deloitte Access Economics. (2021). *The value of tourism.*

Dillette, A. K., Douglas, A. C., & Andrzejewski, C. (2021). Dimensions of holistic wellness as a result of international wellness tourism experiences. *Current Issues in Tourism, 24*(6), 794–810.

Dunn, H. L. (1959a). High-level wellness for man and society. *American Journal of Public Health and the Nations Health, 49*(6), 786–792.

Dunn, H. L. (1959b). What high-level wellness means. *Canadian Journal of Public Health/Revue Canadienne de Sante'e Publique, 50*(11), 447–457.

Erfurt, P. (2021). *The Geoheritage of Hot Springs*. Springer Nature.

Erfurt-Cooper, P., & Cooper, M. (2009). *Health and wellness tourism: Spas and hot springs.* Channel View Publications.

European Union. (2016). *The European Tourism Indicator System: ETIS toolkit for sustainable destination management, Luxembourg*.

Garrod, B., & Fyall, A. (2005). Revisiting Delphi: The Delphi technique in tourism research. In *Tourism research methods: Integrating theory with practice* (Vol. 85). CABI.

Global Wellness Institute. (2010). *Spas and the global wellness market: Synergies and opportunities.* https://globalwellnessinstitute.org/wp-content/uploads/2018/06/Spas-and-the-Global-Wellness-Market-Final-4.25.2010.pdf

Global Wellness Institute. (2018). *Global wellness economy monitor.*

Global Wellness Institute. (2021) *The global wellness economy: Looking beyond COVID*, viewed 05/08/23. https://globalwellnessinstitute.org/wp-content/uploads/2021/11/GWI-WE-Monitor-2021_final-digital.pdf

Global Wellness Institute. (2023). *Global wellness economy monitor 2023*, viewed 17/01/2023. https://globalwellnessinstitute.org/the-2023-global-wellness-economy-monitor/

Global Wellness Institute. (n.d.). *History of wellness*, viewed 11/06/23. https://globalwellnessinstitute.org/what-is-wellness/history-of-wellness/

Han, H., Kiatkawsin, K., Kim, W., & Lee, S. (2017). Investigating customer loyalty formation for wellness spa: Individualism vs. collectivism. *International Journal of Hospitality Management, 67*, 11–23.

Hsu, C.-C., & Sandford, B. A. (2007). The Delphi technique: Making sense of consensus. *Practical Assessment, Research, and Evaluation, 12*(1), 10.

Kaynak, E., Bloom, J., & Leibold, M. (1994). Using the Delphi technique to predict future tourism potential. *Marketing Intelligence & Planning, 12*, 18–29.

Kristjánsdóttir, K. R., Ólafsdóttir, R., & Ragnarsdóttir, K. V. (2018). Reviewing integrated sustainability indicators for tourism. *Journal of Sustainable Tourism, 26*(4), 583–599.

Lambert, S., & Jenni. (2011). *Australia's great thermal way*. Australia's Great Thermal Way.

Law, A., De Lacy, T., Lipman, G., & Jiang, M. (2016). Transitioning to a green economy: The case of tourism in Bali, Indonesia. *Journal of Cleaner Production, 111*, 295–305.

Law, A., DeLacy, T., & McGrath, G. M. (2017). A green economy indicator framework for tourism destinations. *Journal of Sustainable Tourism, 25*(10), 1434–1455.

Lee, C.-F., & King, B. (2006). Assessing destination competitiveness: An application to the hot springs tourism sector. *Tourism and Hospitality Planning & Development, 3*(3), 179–197.

Lee, C.-F., & King, B. (2009). A determination of destination competitiveness for Taiwan's hot springs tourism sector using the Delphi technique. *Journal of Vacation Marketing, 15*(3), 243–257.

Mi, C., Chen, Y., Cheng, C.-S., Uwanyirigira, J. L., & Lin, C.-T. (2019). Exploring the determinants of hot spring tourism customer satisfaction: Causal relationships analysis using ISM. *Sustainability, 11*(9), 2613.

Miller, G. (2001). The development of indicators for sustainable tourism: Results of a Delphi survey of tourism researchers. *Tourism Management, 22*(4), 351–362.

Moscardo, G. (2011). Searching for well-being: Exploring change in tourist motivation. *Tourism Recreation Research, 36*(1), 15–26.

Mueller, H., & Kaufmann, E. L. (2001). Wellness tourism: Market analysis of a special health tourism segment and implications for the hotel industry. *Journal of Vacation Marketing, 7*(1), 5–17.

Myers, J. E. (1992). Wellness, prevention, development: The cornerstone of the profession. *Journal of Counseling & Development, 71*(2), 136–139.

Paige, J. C., Soullière, L. E., & Harrison, L. S. (1988). *Out of the vapors: A social and architectural history of Bathhouse Row: Hot Springs National Park, Arkansas*. US Department of the Interior, National Park Service.

Patterson, I., & Balderas-Cejudo, A. (2022). Baby boomers and their growing interest in spa and wellness tourism. *International Journal of Spa and Wellness, 5*(3), 237–249.

Pinos Navarrete, A., & Shaw, G. (2021). Spa tourism opportunities as strategic sector in aiding recovery from Covid-19: The Spanish model. *Tourism and Hospitality Research, 21*(2), 245–250.

Pollock, A. (2019). Flourishing beyond sustainability. In *ETC Workshop in Krakow* (pp. 1–10).

Ramos, V., & Untong, A. (2016). Spa tourism. In J. Jafari & H. Xiao (Eds.), *Encyclopedia of tourism* (pp. 886–888). Springer International Publishing. https://doi.org/10.1007/978-3-319-01384-8_185

Rasoolimanesh, S. M., Ramakrishna, S., Hall, C. M., Esfandiar, K., & Seyfi, S. (2020). A systematic scoping review of sustainable tourism indicators in relation to the sustainable development goals. *Journal of Sustainable Tourism*, 1–21.

Shavanddasht, M., & Allan, M. (2019). First-time versus repeat tourists: Level of satisfaction, emotional involvement, and loyalty at hot spring. *Anatolia, 30*(1), 61–74.

Smith, M., & Puczko, L. (2008). *Health and wellness tourism*. Taylor & Francis Group.

South Australian Tourism Commission. (2021). *Wellness tourism: Fact sheet*, viewed 15/02/23. https://tourism.sa.gov.au/media/ogjn21uk/satc-wellness-tourism-fact-sheet-final.pdf
Stará, J., & Peterson, C. (2017). Understanding the concept of wellness for the future of the tourism industry: A literature review. *Journal of Tourism & Services, 8*(14), 18–29.
Sykes, M. (2019). *A great Victorian bathing trail*, Australasian Leisure Management, no. 136.
Tonge, J., Moore, S., Hockings, M., Worboys, G., & Bridle, K. (2005). *Developing indicators for the sustainable management of visitor use of protected areas in Australia*. Cooperative Research Centre for Sustainable Tourism.
Tourism Australia. (2021a). *Innovate reconciliation action plan*, viewed 15/02/2023. https://www.tourism.australia.com/content/dam/digital/corporate/documents/tourism-australia-innovate-rap-document-03082022-v1.pdf
Tourism Australia. (2021b). *Tourism Australia annual report 2020–2021*, viewed 15/02/23. https://www.tourism.australia.com/content/dam/digital/corporate/documents/ta-annual-report-fy21.pdf
Tourism Australia. (2022). *THRIVE 2030: The re-imagined visitor economy*, viewed 14/02/2023. https://www.austrade.gov.au/news/publications/thrive-2030-strategy
Tourism Australia. (2023). *Signature experiences of Australia*, viewed 15/02/23 2023. https://www.tourism.australia.com/en/resources/industry-resources/industry-programs/signature-experiences-of-australia.html
Tourism Research Australia. (2022a). *National tourism satellite account 2021-22*, viewed 15/02/2023. https://www.tra.gov.au/data-and-research/reports/national-tourism-satellite-account-2021-22/national-tourism-satellite-account-2021-22
Tourism Research Australia. (2022b). *Tourism forecasts for Australia: 2022 to 2027, viewed 19/01/2023.* https://www.tra.gov.au/en/economic-analysis/tourism-forecasts/tourism-forecasts-for-australia-2022-2027
Tourism Research Australia. (2023). *Tourism forecasts for Australia*, viewed 19/01/2023. https://www.tra.gov.au/en/economic-analysis/tourism-forecasts
Voigt, C., & Pforr, C. (2013). *Wellness tourism: A destination perspective*. Routledge.
Vystoupil, J., Šauer, M., & Bobková, M. (2017). Spa, spa tourism and wellness tourism in The Czech Republic. *Czech Journal of Tourism, 6*(1), 5–26.
Wang, W.-C., & Lin, C.-H. (2021). A model for sustainable tourism development of hot spring destinations following poverty alleviation: Understanding the tourists' perspective. *Sustainability, 13*(17), 9856.
Waterwise Queensland. (2021). How did Aboriginal peoples manage their water resources. *Waterwise: Qld, Australia.* Available at https://www.resources.qld.gov.au/__data/assets/pdf_file/0007/1408282/aboriginal-peoples-manage-water-resources.pdf [Verified 31 May 2021].
Whittlesea, E., Becken, S., Jago, L., & Pham., P.T. (2019). *A new indicator framework for Australia'S visitor economy*, viewed 11/11/2022.
Wisitnorapatt, S., & Sirirat, P. (2023). Measuring customers' satisfaction and loyalty in Thai wellness spas. *International Journal of Spa and Wellness, 6*, 1–18.
World Economic Forum. (2022). *Travel & tourism development index 2021: Rebuilding for a sustainable and resilient future*, viewed 14/02/23. https://www3.weforum.org/docs/WEF_Travel_Tourism_Development_2021.pdf
World Tourism Organization. (1996). *What tourism managers need to know: A practical guide to the development and use of indicators of sustainable tourism*. WTO.
World Tourism Organization. (2004). *Indicators of sustainable development for tourism destinations: A guidebook*. World Tourism Organization.
World Travel & Tourism Council. (2018). *Domestic tourism: Importamce and economic impact*, viewed 14/02/23. https://wttc.org/Portals/0/Documents/Reports/2018/Domestic%20Tourism-Importance%20Economic%20Impact-Dec%2018.pdf?ver=2021-02-25-182514-683
Zari, M. P. (2009). An architectural love of the living: Bio-inspired design in the pursuit of ecological regeneration and psychological wellbeing. *WIT Transactions on Ecology and the Environment, 120*, 293–302.

Chapter 5
The Importance of Mindfulness for Tourism Employees

Reyhan Arslan Ayazlar and Gökhan Ayazlar

Abstract Mindfulness can be defined as awareness and focus on the present. General well-being is important for all individuals including employees of the tourism industry. It is a point of view that has recently been accepted by both researchers and managers that optimal employee performance in the tourism industry is possible not by overloading employees with work but by improving employee motivation and well-being. The fact that uncertainty cannot be avoided in the industry and the majority of the employees are required to provide effective solutions, to customer concerns and needs in a short time, can put a mental strain on tourism employees. The aim of this study is to reveal the importance of mindfulness in tourism employees. The concept of mindfulness and its importance for tourism employees are scrutinized and insights are offered on how to reduce negative emotions and increase positive emotions in tourism employees.

Keywords Mindfulness · Tourism employees · Employee well-being

Introduction

Today, the ideal management approach has evolved from encouraging employees to produce the most output to actively inviting interaction in the workplace (Feng, 2022). In this context, an increase in the importance of employee health and happiness and academic attention on the subject are worthy of note. It is possible to say that within the scope of the concept of happiness, which is at the center of attention of positive psychology, concepts such as emotional intelligence, flow, stress management, psychological capital, and mindfulness stand out. Mindfulness is defined by Kabat-Zinn (1994: 4) as "paying attention in a particular way: on purpose, in the

R. A. Ayazlar (✉) · G. Ayazlar
Mugla Sitki Kocman University, Tourism Faculty, Tourism Management Department, Kötekli, Türkiye
e-mail: reyhanayazlar@mu.edu.tr

D. Chhabra et al. (eds.), *Sustainable Development and Resilience of Tourism*,
https://doi.org/10.1007/978-3-031-63145-0_5

present moment, and non-judgmentally." Mindfulness is also expressed as focusing on the present (Schutte & Malouff, 2011).

Mindfulness was previously applied in the field of psychology for its therapeutic benefits such as anxiety, stress, substance use, insomnia, depression, and post-traumatic disorders. The problems that employees may encounter in the workplace have brought the concept of mindfulness to the agenda in the organizational context and increased interest in the subject (Johnson & Park, 2020). Mindfulness does not make employees encounter fewer negative situations in the workplace. However, according to Said and Tanova (2021), employees with a high level of mindfulness can buffer themselves from getting overwhelmed by negative situations. This kind of disposition can also help them keep control of an adverse situation. Since most employees are focused on the past or the future, they do not realize that such dispositions can affect their current performance (McMahon & Collard, 2009). For instance, preoccupation due to lack of awareness can prevent individuals from noticing new information about the job (Han & Zhang, 2011). Mindfulness enables employees to see and evaluate different alternatives by taking a step back to observe the present moment instead of reacting instantly (Hülsheger et al., 2013). On the other hand, it is criticized that employees should use their minds to earn a living, but they are not trained to do so (Barbezat & Bush, 2013).

Mindfulness strategies can offer a range of benefits for the employees such as (1) improving employees' emotional regulation abilities and coping skills, enabling them to control negative situations and stand safely in these environments, (2) reducing reactions to negative situations (Parsons et al., 2017), (3) enabling them to look at tasks more objectively and concentrate fully by focusing on the present (Carciofo & Jiang, 2021; Feng, 2022), (4) experience of improved neuroplasticity and reduced age-related brain decline due to reduced stressful experiences (Good et al., 2016), (5) a high level of mindfulness in the individual that can become a powerful strategy for achieving individual career goals in the workplace (Reb et al., 2013).

Mindfulness training for employees has emerged as one of the priority topics in agendas of leading businesses. Businesses such as Aetna, Dow Chemical, General Mills, Goldman Sachs, Google, Intel, Nike, SAP, Target, and the United States Marine Corps have implemented mindfulness-based training to reduce employee stress and to increase their well-being and productivity (Jha et al., 2010; Wolever et al., 2012; Aikens et al., 2014). Mindfulness as a practice has become widespread in tourism destinations. Some tourism regions in southeastern Asia such as Thailand and Sri Lanka use the concept of mindfulness in destination branding (Kang & Gretzel, 2012). Other regions consider mindfulness as a strategy to develop a more authentic destination perception and place attachment (Lengyel, 2022; Stankov et al., 2021; Wang et al., 2021b). Mindfulness practices have started to be on the agenda of hotel businesses. For example, the health program called "Westin Well-Being Movement" created by Westin hotels is implemented for both employees and customers. The brand bases this health program on six pillars: feel well, work well, move well, eat well, sleep well, and play well. It is also possible to benefit from free online meditations by visiting the Westin Well-Being site (Puddicome, 2014).

Recently, the concept of mindfulness has started being discussed in the context of eliminating the negative effects on tourism employees in the tourism workplace. Tourism is a labor-intensive industry. Therefore, employee-consumer interface plays a pivotal role in ensuring successful management and operation of tourism and hospitality businesses (Johnson & Park, 2020). Tourism employees can have high level of stress, and there may be a need to regulate their moods during work time. Mindfulness can be seen as a feasible and low-cost solution to reduce negative psychological effects on employees in many tourism settings (Stankov & Filimonau, 2023).

Negative impacts faced by tourism employees can result in high turnover in the tourism industry. For instance, in the USA, the highest turnover is seen in the accommodation, food service, leisure, and hospitality industries (Izell, 2022). In addition, the fact that the tourism industry was one of the most affected industries during the pandemic has worsened the mental health of tourism employees who were trying to survive financially. There are limited studies that investigate the effect of mindfulness on tourism employees. This chapter examines the benefits of mindfulness on creativity (Kalyar et al., 2021) and its potential to reduce emotional exhaustion (Mekawy et al., 2022; Said & Tanova, 2021), negative effects of stay-at-home policies during the pandemic (Shi & Wang, 2022), prolonged sustainability during and after pandemic (Stankov & Filimonau, 2023), and collective thriving (Wu & Chen, 2019). For example, mindfully green frontline managers are sensitive to environmental problems and are more connected to nature (Kalyar et al., 2021). Tourism employees have lower level of cynicism, workplace incivility, and work alienation which have effects on emotional exhaustion with high level of mindfulness (Mekawy et al., 2022). Frontline service employees with a high level of mindfulness are less likely to be influenced by a negative situation at home which can lead to emotional exhaustion. For example, the emotional exhaustion at work of a frontline service employee can increase if he or she experiences poor sleep quality at night. Such negative emotional experiences at home can trigger a tendency to neglect customers or mistreat them. Employees with high workplace mindfulness tend to be less affected by these negative situations (Shi & Wang, 2022). More studies are, therefore, needed to scrutinize the significance and positive effects of mindful dispositions in the hospitality sector in terms of the ability to cope with stressful situations in the workplace and impact on well-being.

In this study, the importance of mindfulness for tourism employees is examined. Key research questions guiding the study are the following:

1. What is the importance of mindfulness for employees working in the hospitality and tourism industry?
2. What are the effects of mindfulness on these employees?

Literature Review

In this section, the concepts of mindfulness and mindfulness in the workplace are examined. At the same time, insights are offered on models developed for mindfulness, and their operationalizations are covered. The chapter concludes with the importance of mindfulness for emotion regulation in the workplace.

Jon Kabat-Zinn states that the words "mind" and "heart" mean the same thing in the Asian language and that mindfulness and compassion are interconnected with each other (as cited in Kaoun, 2019). Mindfulness refers to a nonjudgmental awareness of the present moment without obsessive thoughts about the past and the future (Kabat-Zinn, 1994). The non-mindful individuals are likely to act in "automatic mode." These individuals may not reflectively engage before taking action. Because they rely on their preexisting beliefs to make decisions (Kaoun, 2019). However, mindfulness is an important tool to practice by staying in the present moment and not entertaining thoughts about the past or the future. Aligned with this stance, Chesley and Wylson write that in a mindful state of mind, we "disengage from our habitual patterns of reactivity and come closer to clear comprehension of any situation" (2016: 319).

Mindfulness in the workplace means paying attention to what is happening in the workplace and being aware of both the individual's internal state and the reactions to the events around him/her (Dane, 2011). Similarly, Kahn (1990: 164) introduces the concept of engaged employees in the workplace and defines committed employees as those who "express themselves physically, cognitively, and emotionally during role performance." Extant literature reports that mindful employees are more aware of their environment and more responsive and nonjudgmental. Mindful tourism employees have a high awareness of their feelings and thoughts (inner) and are reflexive in their verbal communications and response to visual symbols (external) (Brown & Ryan, 2003). Mindfulness and mindful employees may be more ready to cope with the current working environment of the tourism industry. The intense employee-customer interaction in the tourism industry and the expected customer satisfaction and loyalty by tourism businesses reveal the necessity of employees who can manage their emotions and have appropriate tone of voice and facial expression (Wang et al., 2021a).

Two delineations associated with mindfulness can be noted: mindfulness as a trait or as a situation. Mindfulness can be considered as a state that is stable over time. However, this state of mindfulness, considered to be a constant, can be influenced by different states of awareness over a period of time (Kiken et al., 2015). Therefore, different levels of mindfulness can exist during different periods of time in an individual. Thus, it should not be considered as a homogeneous, that is, this state of mindfulness tends to differ from one person to the other (Kabat-Zinn, 1994).

Mindfulness can be a fixed feature that an individual has a high tendency to have, or it can be evaluated as a state of mind that can be activated through some interventions or tools (Brown & Ryan, 2003). Khoury et al. (2017) write that mindfulness has been operationalized with two approaches in documented literature. One of these approaches is Buddhist-based; the other one is based on a Western ideology. The Buddhism stance is based on different foundations of thought in different schools such as Theravada, Mahayana, and Vajrayana. According to Theravada Buddhism, mindfulness has two meanings: "to remember" and "to understand what is in the moment." This conceptualization of mindfulness is operationalized through daily and deep meditation practices. The Western-based approach is guided by two sub-approaches: the conceptualization practices put forward by Jon Kabat-Zinn and Ellen Langer. Although both mindfulness approaches have their roots in Buddhism, they differ from each other in terms of conceptualization and operationalization. Jon Kabat-Zinn's definition of mindfulness focuses on attention and awareness. In this context, the Stress Reduction and Relaxation Program (SPR), which was developed to reduce patients' medical pain, was later developed and evolved into the Mindfulness-Based Stress Reduction (MBSR) program. The MBSR program includes practices rooted in Buddhism such as breathing, sitting, walking, eating meditations, body scanning and stretching, as well as psychological techniques such as psychoeducation, group discussion, and individual support. In addition, we see that Kabat-Zinn and his colleagues operationalize mindfulness with different programs. These include mindfulness-based cognitive therapy (MBCT) (combines cognitive therapy and mindfulness meditations), mindfulness-based relapse prevention (MBRP) (aims to prevent substance abuse), and Mindfulness-Based Eating Awareness Training (MB-EAT) (which prevents overeating or programs that aim to regulate eating habits). Ellen Langer refers to mindfulness as a sociocognitive trait. In this context, she briefly defines mindfulness as being open to innovation. Based on this theory, Langer (1989) has included mindfulness practices with techniques such as noticing distinctions, multiple perspectives, and producing novelty. Thus, this school of thought aims to make mindfulness easy even for those who do not want to meditate. Different methods can assist in promoting emotional regulation of individuals.

Methodology

Secondary data was used as the main source of information for this chapter. Secondary data tends to be reliable and readily available (Dolata et al., 2015). This type of information is usually gathered from published materials such as journals, websites, statistics, and reports (Church, 2002). Secondary data is considered suitable data in the field of tourism for meta-analysis, and it can assist in identifying research trends (Saunders et al., 2000).

Secondary data was mainly designed and developed around the subject of mindfulness for tourism employees. In order to create a consistent resource pool of peer-reviewed materials, academic journal articles in the academic database were evaluated. In addition to basic studies on the subject, articles on mindfulness published from 2010 and onward were scanned through the scientific database and used in the study.

Findings

In line with the findings obtained from secondary data, it has been determined that mindfulness has some effects on the positive and negative emotions of tourism employees. It has also been found that mindfulness is an effective tool for employees' learning. The findings also show that mindfulness can be a tool for dealing with problems that may arise in the tourism industry after the pandemic.

The Effects of Mindfulness to Reduce Negative Emotions

Today, mindfulness is considered one of the important tools in terms of protecting and improving the mental health and well-being of employees. Mindfulness is seen as one of the concepts that supports the process in eliminating some of the negativities brought by working life. Emotional dissonance, emotional exhaustion and reduced job performance, burnout, depression, psychological contract breach, and rumination can be counted among these negative factors.

The concept of emotional labor is one of the important issues for the tourism industry, due to the intensity and significance of interaction between employees and customers (Zapf & Holz, 2006). Emotional labor is defined as "the effort, planning and control needed to express organizationally desired emotion during interpersonal transactions" (Morris & Feldman, 1996: 987). Emotional labor may "involve enhancing, faking or suppressing emotions to modify the emotional expression" (Grandey, 2000: 95). Employees are expected to have a public vocal, facial expression, and body language in accordance with the service standards of the business while meeting customer needs and demands. However, emotional dissonance may occur if employee feelings conflict with organizational goals (Morris & Feldman, 1996). Mindfulness practices may help tourism employees to manage their emotions by keeping focused in the present moment during interaction with customers (Johnson & Park, 2020).

Emotional exhaustion can result in poor job performance. Decreasing job performance in employees may lead to consequences such as decreased organizational commitment and increased employees' intention to leave work. A low level of job performance in the workplace can compromise service quality and negatively impact customer assessment of service quality (Said & Tanova, 2021). Mindfulness also assists employees in coping with customer incivility (Jang et al., 2020). Mindful employees also display less incivility toward their colleagues (Hülsheger et al., 2021).

Burnout is one of the elements that employees frequently encounter in industries where customer-employee interaction is intense (Maslach & Jackson, 1981). It occurs when employees cannot regulate their emotions. Maslach (1982) defined job burnout as "a syndrome of emotional exhaustion, depersonalization of others and a feeling of reduced personal accomplishment." Although burnout is one of the important causes that leads to employee turnover, managers seldom show interest in improving the situation (Johnson & Park, 2020). There are several studies that draw attention to the relationship between mindfulness and employee burnout. Burnout can develop in individuals due to long-term exposure to interpersonal stress factors (Garcia-Izquierdo & Rios-Risquez, 2012). Burnout not only affects the health of the individual negatively but also affects the ability to empathize (Neff, 2003). Mindfulness strategies can assist in reducing situations leading to burnout (Kemper et al., 2015).

Psychologically, overthinking, suppressing, and avoiding emotions bring depression. Mindfulness practices can help regulate thoughts (Beevers et al., 1999). The conservation of resources theory suggests that "strive to obtain, retain and protect that which they value" (Hobfoll, 1998: 55). From this point of view, it is possible to say that employees tend to protect what is valuable to them. At this point, mindfulness is considered a personal resource that helps to reduce stress in the workplace and mitigates emotional exhaustion (Chen & Eyoun, 2021). Work-related demands can cause employees to perceive resource depletion. In this case, mindfulness positively affects employees' attitudes toward acceptance and supports a self-regulation process with less rumination (Allen & Kiburz, 2012).

Rumination is shown as one of the most important causes of long-term stress. Rumination permanently negatively affects the individual's level of consciousness because it is affected by external factors and therefore interrupts the flow (Feng, 2022). In terms of employees, rumination is expressed as continuing to think about work at nonworking hours. It is limiting to consider rumination only for negative situations. While there may be intrusive, negative, and recurrent thoughts about the work, employees may find a solution to a work-related problem or develop an idea during nonworking hours. Positive work-related rumination can help employees develop constructive feelings toward work and overall well-being. Rumination, which develops in a negative direction, continues to recur as long as the individual does not move away from work. On the other hand, positive rumination improves problem-solving skills of employees (Cropley & Zijlstra, 2011). Martin and Tesser (1996) mention that rumination can benefit a person by facilitating his or her progress toward accomplishing goals. Therefore, rumination can be also considered a double-edged sword.

Situations such as seasonality in the tourism industry, vulnerability toward unpredictable macro-environment factors, and a fragile structure can shake the trust of the employees toward the industry. One of the most common situations at such times is the psychological contract breach. It has been contended that since all obligations toward employees cannot be specified in a formal contract (Shaffakat et al., 2022), a psychological contract can be divided into positive and negative reciprocity (Eisenberger et al., 2004). Negative reciprocity, which is the subject of mindfulness in this study, is a situation where employees believe that the business is abusing their duties (Chiu & Peng, 2008). This situation may be perceived by employees as a breach of duty, or it may arouse feelings that may trigger hostility toward the company (Shaffakat et al., 2022).

As a result of a psychological contract breach, hostility toward the business may occur among the employees. Mindfulness can be considered an important tool in

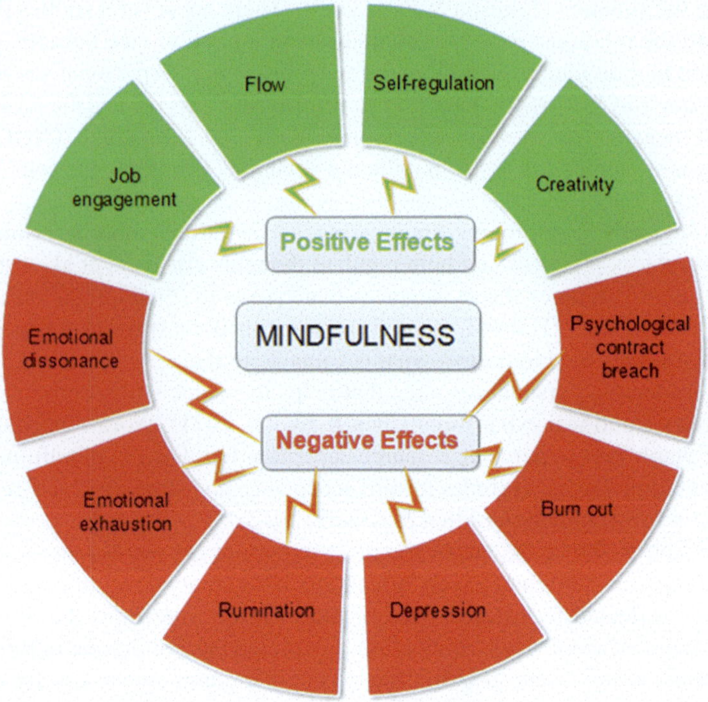

Fig. 5.1 Effects of mindfulness

regulating this feeling of hostility and psychologically distancing the individual from the experience (Feldman et al., 2010). Mindfulness helps the individual to see alternative perspectives of an experience. Therefore, employees can develop a more flexible attitude toward psychological contract breach in the workplace (Brown & Ryan, 2003). In other words, mindful employees can internally buffer themselves more efficiently against hostile feelings (Shaffakat et al., 2022) (Fig. 5.1).

The Effects of Mindfulness to Improve Positive Emotions

It has been stated in documented literature that mindfulness not only helps in reducing negative emotions in employees but also stimulates some positive emotions. These positive emotions include creativity, self-regulation, and flow. Creativity is a factor that can facilitate efficiency in employees, for instance, in terms of responding to customer expectations and needs. However, it is stated that creativity can flourish in a working environment where innovation or trying new things is encouraged. Similarly, mindfulness practices have a positive and supportive effect on a work environment (Wang et al., 2021a, b).

Moving forward, mindfulness is today considered a powerful self-regulation mechanism by several researchers (Chambers et al., 2009; Vago & Silbersweig, 2012; Shaffakat et al., 2022). Mindfulness provides two key functions for self-regulation. One of them is decoupling. Accordingly, when an individual can decouple his/her experience, feelings, and thoughts from the self, ego involvement decreases thereby allowing for more detached experiences. The second function is reduced automatically, that is, an individual can improve the practice of pausing and stepping back when facing negative situations. Working of both functions together can enhance both emotional and behavioral flexibility (Glomb et al., 2011). In terms of employees, "conscious and deliberate forms of self-regulation" (Schalk & Roe, 2007: 173) may be effective in regulating reactions against negative situations in the workplace. Mindfulness can also play an active role in an individual's experience of flow at work.

Mindfulness facilitates an individual to let himself or herself go with the flow of the lived experience (Epstein, 1990). The flow experience consists of clear goals, immediate feedback, a sense of control, and the integration of action and consciousness (Csikszentmihalyi & Rathunde, 1993). In this sense, flow in the workplace can be described as employees' interest in challenging tasks clear understanding of the goals. With the mental clarity achieved through mindfulness, employee ability to cope with challenging tasks increases. Therefore, it is possible to say that employees experience flow in the workplace if they are able to meet challenges that are compatible with their capabilities (Feng, 2022). Flow in the workplace is characterized by the fact that employees are immersed in their work, they find it enjoyable, and they are intrinsically motivated (Bakker et al., 2011). Employees who have flow experience in the workplace report that they are satisfied with their jobs because it enhances their performance and overall well-being (Sheldon et al., 2015). Therefore, improving the flow experience of employees at work through mindfulness is important for organizational development and personal growth of employees. Next section discusses the contributions of mindfulness to learning and the tourism industry after the pandemic.

Mindfulness in Learning: The Importance of Mindfulness for In-Service Training

Employee learning is one of the important factors affecting workplace success. However, unless there is effective learning, the effectiveness of training provided in the workplace becomes a debatable issue. It is possible to talk about the importance of mindfulness in effective learning of employees. In this section, mindfulness-based learning is mentioned, and the effectiveness of this type of learning is discussed.

Based on a review of documented literature, it is noted that university students are a population that is frequently used in studies associated with mindfulness education. Most studies are carried out on nursing and medical students. Such studies show that mindfulness has a positive effect on students' mental health and academic success (Kerrigan et al., 2017). However, studies on mindfulness employee training are limited, and this line of inquiry makes an interesting area of exploration, and the objective of this chapter is to add to sparse literature by applying learning models which are appropriate for tourism industry.

Kaoun (2019) made suggestions for the training of employees in the workplace. According to the author, prejudice and assumption are among the most important obstacles in the development of emotional intelligence and learning. For example, an individual will not need to learn anything he thinks he already knows. These concepts impede an individual's capability to understand himself/herself and realize his/her full potential. If prejudices and assumptions are mitigated through mindfulness, barriers to learning can be removed.

Kaoun (2019) proposes various models for the training of employees in mindfulness. In one of these models, he calls for the 70-20-10 notion. In other words, 70% of learning happens in the workplace, 20% can happen through informal coaching or mentoring, and 10% can be imparted through structured training. In this context, it is possible to say that mindfulness training given in the workplace is the most effective way of learning. However, after the training is completed, a suitable environment to practice mindfulness at work is important for the sustainability of the training. There are several factors that can prevent the implementation and adoption of mindfulness in the workplace. Examples include issues associated with loud noise and other distracting stimuli in the workplace, high stress due to multiple tasks, insufficient time to perform mindfulness practices, and failure to comply with mindfulness practices by other employees (Lyddy et al., 2016). In summary, the leaders in the workplace should not only give importance to mindfulness education but also focus on post-training in order to sustain positive effects of mindfulness practices.

The second model proposed by Kaoun (2019) is based on the principle that adults learn best through experiential exercises. However, having theoretical knowledge about mindfulness is not enough. If enough time is not devoted to mindfulness practices, learning has no value. Employees are less likely to take interest in training if they are not familiar of its outcomes and benefits. Therefore, educational material informing of benefits should be shared beforehand with the employees (Kaoun, 2019).

Mindfulness as a Coping Strategy After Pandemic

Mindfulness is considered an effective, promising, and low-cost practice in preventing or mitigating uncertainty and turbulent situations that occur during crisis periods (Zhu et al., 2021). The pandemic is one such example. The tourism industry has been one of the most negatively affected sectors, especially with the travel restrictions experienced around the world during the COVID-19 period (Gossling et al., 2020). Examples include closed tourism businesses, downsizing, and lay-off situations (Hassan & Soliman, 2021). According to Chen and Eyoun, "when employees develop fear of COVID-19 and worry about the possibility of job loss, they become overwhelmed and consumed by those thoughts. Those who can stay in the present and experience the emotions as they occur are more capable to cope with the strain" (2021: 3).

Jahel et al. (2021) visualize three post-COVID futures for countries around the world: the first is the return to normal as before (return to normal); the second is the creation of new rules in the functioning of the economy and society (new normal) and finally a radical transformation (post-normal). Various researchers have also made predictions about the future of the tourism industry if COVID-19 continues to loom in the background (Ateljevic, 2020; Lew et al., 2020). The proposed changes prioritize strengthening the resilience of tourism stakeholders (Sharma et al., 2021). In addition, uncertainty might continue to loom over tourists and businesses. In order to remove uncertainty, both parties should focus on safety procedures. In other words, while tourists make their future holiday plans, tourism businesses and destinations should prioritize safety measures in their future strategies and goals (Stankov et al., 2021). This is where mindfulness can play an important role, and as the mindfulness tourist segment grows, tourism suppliers will also embrace mindful products and services.

Stankov et al. (2020a) share the importance of mindfulness in contributing toward long-term transformation in the tourism industry in the post-COVID times. Stankov et al. (2020b) reveal the existence of mindfulness-oriented tourism as an alternative tourism market. In addition, there are other studies that use mindfulness as a useful tool to increase the well-being of employees and increase their resilience in the workplace during the lockdown period (Zhu et al., 2021). The concept of mindfulness is also evaluated as a way to reduce the negative effects of COVID-19 on tourism workers and tourists (Chen & Eyoun, 2021). It is predicted that the concept of mindfulness can be used in the general recovery of the tourism industry (Ma et al., 2021; Wen et al., 2021). At this point, it is important to mention an emerging form of alternative tourism called healing tourism which focuses on mindfulness. Unlike psychology tourism, healing tourism focuses on promoting "wellness over the lifespan" of a tourist's travel. At this point, authors argue that healing tourism should be part of the "new normal" (Ma et al., 2021). In psychology tourism, psychological treatment methods are applied to treat mental problems and/or diseases (Chen et al., 2013). An individual's mental health depends on how he perceives the universe, the world, society, and life. If a person individual can attain and maintain a balanced/harmonious mood or disposition through mindfulness, he or she is likely to be generally healthy, happy, and free from the symptoms of illness (Wen et al., 2021).

Stankov and Filimonau (2023) share the five most common skills of mindfulness and their relevance to the tourism industry. Please see Table 5.1. As the table illustrates, the skills of observing, not judging, not reacting, acting with awareness, and describing are considered important practices. These practices can be embraced both by tourists and tourism workers in the post-pandemic times.

Table 5.1 Basic mindfulness skills and examples of their relevance for post-pandemic tourism

Mindfulness skills	Description (based on Antonova et al., 2021)	An example for post-pandemic tourism
Observing	Deliberately turning toward and noticing present-moment experiences during daily activities (e.g., body sensations when walking, eating, and similar or sensory stimuli, such as sounds or smells)	Provides time and inner space for self-reflection during unpleasant COVID-19-related safety measures whilst traveling, such as mask-wearing on public transport
Non-judging of inner experiences	Reduces the tendency to judge one's experiences, emotions, feelings, and thoughts and minimizes the tendency to be critical of oneself more generally	Adaptability to new tourism offers that were previously unappealing (e.g., staycation or domestic trips in place of overseas travel)
Non-reacting to inner experiences	Ability to be aware of distressing feelings, thoughts, or images without getting caught up in them, noticing them in a decentered way, letting go of them, and returning to feeling calm soon after they have passed	Getting over anxiety or unpleasant associations with traveling in post- COVID-19 period or with traveling to certain locations, such as the past hotspots of the pandemic
Acting with awareness	Measures the propensity to get distracted from the present-moment experience by mind-wandering (daydreaming, worrying), running on automatic pilot, and rushing through activities without giving them attention	Increased tourist participation in preventative and protective behavior when on holidays, such as mask-wearing and social distancing
Describing	Ability to put one's sensations, feelings, thoughts, beliefs, opinions into words, or the tendency to do so	Provision of better quality and less biased tourist feedback on tourism products and services

Source: Stankov and Filimanou (2023)

Stankov and Filimanou (2023) talk about three different scenarios for post-pandemic mindfulness. Expressing the first situation as "sticking to the status quo," the authors create a scenario that is not very different from the pre-pandemic period. Accordingly, very few tourists will practice mindfulness when travelling, the concept of mindfulness will be mostly reserved for niche or luxury tourism products, and mindfulness will have a microscale effect. In other words, generating awareness is an initiative that is likely to remain at the discretion of destination managers and tourism providers in this scenario. For example, Hyatt's luxury wellness brand Miraval has offered awareness and digital awareness programs to build resilience.

The second scenario, called "the utilitarian approach," mentions the meso-level effects of mindfulness practices on the tourism industry. Mindfulness is considered to be the most effective and cost-effective coping method after the pandemic, and it is predicted that it will gain more acceptance in the tourism industry in this scenario. Therefore, mindfulness is expected to be supported by more private and public inputs and policies in the tourism industry. In this scenario, digital mindfulness support systems are more involved. For example, mindful.org, which is a public benefit corporation, has been one of the prominent applications throughout the pandemic as an online platform dedicated to awareness (Bayes-Fleming, 2020).

The third scenario is called "ultimate transformation." As this approach is garnering attention, progress toward transformation can be noted in the tourism industry, and mindfulness is becoming a benchmark both in the travel preferences of tourists and in the working environment of tourism industry employees (Stankov & Filimanou, 2023).

Conclusion and Discussion

This chapter shows the importance of mindfulness among tourism workers and presents the positive and negative effects of it in the tourism industry. The tourism industry is a labor-intensive industry, and in recent times, it has recognized the importance of mindfulness strategies. Long working hours, taking care of the customers one by one, and the concern of losing work because of seasonality issues in tourism make it important to make focusing on employee mental health more valuable. Mindfulness strategies show that they can be effective in reducing negative situations that may occur in employees and increase positive emotions in the studies carried out in the tourism industry. After devastating occurrences such as the pandemic, we find mindfulness practices can be valuable in terms of fostering the mental health and overall well-being of employees so that their productivity can be enhanced and sustained.

The well-being of employees through mindfulness in the tourism industry also positively affects the tourist experience. An employee who develops the ability to think about the current situation, as a result of interaction with the tourist, is more likely to behave in a calm manner and diffuse a volatile situation. Effective strategies in creating the tourist experience (Stankov et al., 2020a) can be formulated "when employees in the hospitality industry are aware of their surroundings and can pay attention to the present moment, observe their feelings and responses they would be able to have more control and be in a better position to focus on what they need to do in their job" (Said & Tanova, 2021: 4).

Mindfulness is considered a multifaceted structure (Salovey & Mayer, 1990). It is possible to consider the connection between mindfulness and EI through the five factors that Goleman (2004) stated for EI. According to the author, the five factors of EI are motivation, self-awareness, self-regulation, empathy, and social skills. When motivation is considered first, it is possible to say that mindful individuals can be more motivated internally. Similarly, this chapter also illustrates that mindfulness contributes positively to self-awareness (Brown & Ryan, 2003). Good et al. (2016) talk about the effects of mindfulness on self-control, which is necessary for self-regulation. One of the most distinctive features of mindfulness is non-judgment. Therefore, it is possible to talk about the relationship between non-judgment and empathy. Frizzell et al. (2016) also found that meditation affects the ability to communicate effectively among individuals and draws attention to the relationship between social skill and mindfulness, which is the last factor of EI. This chapter shows that mindfulness and EI are similar in many ways. While an individual can develop a higher level of awareness through mindfulness, EI goes one step further and emphasizes the importance of perception and suggests that high awareness can assist in interpreting perceptions. Mindfulness focuses on living in the moment and EI stresses on assessing emotions correctly. Mindfulness focuses on rigorous mental development. EI, on the other hand, is in an effort to grow both emotionally and intellectually. Therefore, it has been postulated that mental development can also contribute to emotional growth (Sethi, 2009; Kroon et al., 2017; Kaoun, 2019). There are various studies that indicate that an individual's attainment of greater mindfulness is associated with a high level of emotional intelligence (Chu, 2010). It seems that studies on this subject in the field of tourism are limited. The relationship between mindfulness and EI is important in terms of diversifying research on the mindfulness of employees in the tourism industry.

In summary, this chapter argues that although mindfulness is a new concept, its application has been noted in ancient times. Findings from studies on individuals' mindfulness in daily life have shown that it can also be applied in business life. This chapter offers several examples that reveal that employees' participation in mindfulness practices is mostly realized through training provided by the companies. However, if employees do not adopt mindfulness practices, it is not possible to sustain them. This is where intention comes into play. In our opinion, unless the intentions of the employees and individuals in general are committed, it is not likely for them to comprehend, embrace, and apply the concept of mindfulness, adopt it, and apply it in their lives in a sustained manner. Having a clear intention to start mindfulness is also effective in eliminating problems related to its sustainability. However, it should not be forgotten that although mindfulness is an important tool, it is not central to the well-being in an individual's life. In other words, we think that mindfulness is one part that of a whole wellness equation that includes other factors that affect the individual's well-being.

References

Aikens, K. A., Astin, J., Pelletier, K. R., Levanovich, K., Baase, C. M., Park, Y. Y., & Bodnar, C. M. (2014). Mindfulness goes to work: Impact of an online workplace intervention. *Journal of Occupational and Environmental Medicine, 56*(7), 721–731. https://doi.org/10.1097/JOM.0000000000000209

Allen, T. D., & Kiburz, K. M. (2012). Trait mindfulness and work–family balance among working parents: The mediating effects of vitality and sleep quality. *Journal of Vocational Behavior, 80*(2), 372–379. https://doi.org/10.1016/j.jvb.2011.09.002

Antonova, E., Schlosser, K., Pandey, R., & Kumari, V. (2021). Coping with COVID-19: Mindfulness-based approaches for mitigating mental health crisis. *Frontiers in Psychiatry, 12*, 563417. https://doi.org/10.3389/fpsyt.2021.563417

Ateljevic, I. (2020). Transforming the (tourism) world for good and (re)generating the potential 'new normal. *Tourism Geographies, 22*(3), 467–475. https://doi.org/10.1080/14616688.2020.1759134

Bakker, A. B., Oerlemans, W., Demerouti, E., Slot, B. B., & Ali, D. K. (2011). Flow and performance: A study among talented Dutch soccer players. *Psychology of Sport and Exercise, 12*(4), 442–450. https://doi.org/10.1016/j.psychsport.2011.02.003
Barbezat, D. P., & Bush, M. (2013). *Contemplative practices in higher education: Powerful methods to transform teaching and learning*. Wiley.
Bayes-Fleming, N. (2020). Five mindfulness lessons I learned while traveling during a pandemic. *Mindful*. Retrieved from https://www.mindful.org/five-mindfulness-lessons-i-learned-while-traveling-during-a-pandemic/
Beevers, C. G., Wenzlaff, R. M., Hayes, A. M., & Scott, W. D. (1999). Depression and the ironic effects of thought suppression: Therapeutic strategies for improving mental control. *Clinical Psychology: Science and Practice, 6*(2), 133–148. https://doi.org/10.1093/clipsy.6.2.133
Brown, K. W., & Ryan, R. M. (2003). The benefits of being present: Mindfulness and its role psychological well-being. *Journal of Personality and Social Psychology, 84*(4), 822–848. https://doi.org/10.1037/0022-3514.84.4.822
Carciofo, R., & Jiang, P. (2021). Deliberate and spontaneous mind wandering in Chinese students: Associations with mindfulness, affect, personality, and life satisfaction. *Personality and Individual Differences, 180*. https://doi.org/10.1016/j.paid.2021.110982
Chambers, R., Gullone, E., & Allen, N. B. (2009). Mindful emotion regulation: An integrative review. *Clinical Psychology Review, 29*(6), 560–572. https://doi.org/10.1016/j.cpr.2009.06.005
Chen, H., & Eyoun, K. (2021). Do mindfulness and perceived organizational support work? Fear of COVID-19 on restaurant frontline employees' job insecurity and emotional exhaustion. *International Journal of Hospitality Management, 94*, 1–10. https://doi.org/10.1016/j.ijhm.2020.102850
Chen, Q., Ding, M., & Feng, G. (2013). Preliminary discussion about psychological tourism products. *Tourism Research, 5*(1), 24–29. http://kns.cnki.net/kcms/detail/detail.aspx?FileName=KMDX201301006&DbName=CJFQ201
Chesley, J., & Wylson, A. (2016). Ambiguity: The emerging impact of mindfulness for change leaders. *Journal of Change Management, 16*(4), 317–336. https://doi.org/10.1080/14697017.2016.1230334
Chiu, S. F., & Peng, J. C. (2008). The relationship between psychological contract breach and employee deviance: The moderating role of hostile attributional style. *Journal of Vocational Behavior, 73*(3), 426–433. https://doi.org/10.1016/j.jvb.2008.08.006
Chu, L. C. (2010). The benefits of meditation vis-à-vis emotional intelligence, perceived stress and negative mental health. *Stress & Health, 26*(2), 169–180. https://doi.org/10.1002/smi.1289
Church, R. M. (2002). The effective use of secondary data. *Learning and Motivation, 33*, 32–45. https://doi.org/10.1006/lmot.2001.1098
Cropley, M., & Zijlstra, F. (2011). Work and rumination. In J. Langan-Fox & C. L. Cooper (Eds.), *Handbook of stress in the occupations* (pp. 487–502). Edward Elgar Publishing Ltd.
Csikszentmihalyi, M., & Rathunde, K. (1993). The measurement of flow in everyday life: Toward a theory of emergent motivation. In J. E. Jacobs (Ed.), *Nebraska symposium on Motivation, 1992: Developmental perspectives on motivation* (pp. 57–97). University of Nebraska Press.
Dane, E. (2011). Paying attention to mindfulness and its effects on task performance in the workplace. *Journal of Management, 37*(4), 997–1018. https://doi.org/10.1177/0149206310367948
Dolata, M., Kilic, M., & Schwabe, G. (2015). Unpacking the artifact knowledge: Secondary data analysis in design science research new horizons in design science. In *International conference on design science research in information systems* (pp. 327–342). Springer.
Eisenberger, R., Lynch, P., Aselage, J., & Rohdieck, S. (2004). Who takes the most revenge? Individual differences in negative reciprocity norm endorsement. *Personality and Social Psychology Bulletin, 30*(6), 787–799. https://doi.org/10.1177/0146167204264047
Epstein, M. (1990). Psychodynamics of meditation: Pitfalls on the spiritual path. *The Journal of Transpersonal Psychology, 22*(1), 17–34.
Feldman, G., Greeson, J., & Senville, J. (2010). Differential effects of mindful breathing, progressive muscle relaxation, and loving-kindness meditation on decentering and negative reactions

to repetitive thoughts. *Behaviour Research and Therapy, 48*(10), 1002–1011. https://doi.org/10.1016/j.brat.2010.06.006

Feng, X. (2022). Calm down and enjoy it: Influence of leader-employee mindfulness on flow experience. *Psychology Research and Behavior Management, 15*, 839–854. https://doi.org/10.2147/PRBM.S360880

Frizzell, D. A., Hoon, S., & Banner, D. K. (2016). A phenomenological investigation of leader development and mindfulness meditation. *Journal of Sustainable Social Change, 8*(1), 14–25. https://doi.org/10.5590/JOSC.2016.08.1.02

Garcia-İzquierdo, M., & Rios-Risquez, M. I. (2012). The relationship between psychosocial job stress and burnout in emergency departments: An exploratory study. *Nursing Outlook, 60*(5), 322–329. https://doi.org/10.1016/j.outlook.2012.02.002

Glomb, T. M., Duffy, M. K., Bono, J. E., & Yang, T. (2011). Mindfulness at work. In A. Joshi, H. Liao, & J. J. Martocchio (Eds.), *Research in personnel and human resources management* (pp. 115–157). Emerald.

Goleman, D. (2004). What makes a leader? *Harvard Business Review, 82*(1), 82–91.

Good, D. J., Lyddy, C. J., Glomb, T. M., Bono, J. E., Brown, K. W., Duffy, M. K., & Lazar, S. W. (2016). Contemplating mindfulness at work: An integrative review. *Journal of Management, 42*(1), 114–142. https://doi.org/10.1177/0149206315617003

Gossling, S., Scott, D., & Hall, C. M. (2020). Pandemics, tourism and global change: A rapid assessment of COVID-19. *Journal of Sustainable Tourism, 29*(1), 1–20. https://doi.org/10.1080/09669582.2020.1758708

Grandey, A. A. (2000). Emotional regulation in the workplace: A new way to conceptualize emotional labor. *Journal of Occupational Health Psychology, 5*(1), 95–110. https://doi.org/10.1037/1076-8998.5.1.95

Han, Y., & Zhang, Z. (2011). Enhancing managerial mindfulness: A way for middle managers to handle the uncertain situations. In *IACM 24th Annual Conference Paper* (pp. 1–44). https://doi.org/10.2139/ssrn.1872142

Hassan, S. B., & Soliman, M. (2021). COVID-19 and repeat visitation: Assessing the role of destination social responsibility, destination reputation, holidaymakers' trust and fear arousal. *Journal of Destination Marketing & Management, 19*(100), 495. https://doi.org/10.1016/j.jdmm.2020.100495

Hobfoll, S. E. (1998). *Stress, culture, and community: The psychology and philosophy of stress*. Plenum Press.

Hülsheger, U. R., Alberts, H. J. E. M., Feinholdt, A., & Lang, J. W. B. (2013). Benefits of mindfulness at work: The role of mindfulness in emotion regulation, emotional exhaustion, and job satisfaction. *Journal of Applied Psychology, 98*(2), 310–325. https://doi.org/10.1037/a0031313

Hülsheger, U. R., van Gils, S., & Walkowiak, A. (2021). The regulating role of mindfulness in enacted workplace incivility: An experience sampling study. *Journal of Applied Psychology, 106*(8), 1250–1265. https://doi.org/10.1037/apl0000824

Izell, C. (2022). Addressing mental health of hospitality employees: An exploratory study. *Management Undergraduate Honors Theses*. Retrieved from https://scholarworks.uark.edu/mgmtuht/19

Jahel, C., Bourgeois, R., Pesche, D., Lattre-Gasquet, M. D., & Delay, E. (2021). Has the COVID-19 crisis changed our relationship to the future? *Futures & Foresight Science, 3*(2), e75. https://doi.org/10.1002/ffo2.75

Jang, J., Jo, W. M., & Kim, J. S. (2020). Can employee workplace mindfulness counteract the indirect effects of customer incivility on proactive service performance through work engagement? A moderated mediation model. *Journal of Hospitality Marketing and Management, 29*(7), 812–829. https://doi.org/10.1080/19368623.2020.1725954

Jha, A. P., Stanley, E. A., Kiyonaga, A., Wong, L., & Gelfand, L. (2010). Examining the protective effects of mindfulness training on working memory capacity and affective experience. *Emotion, 10*(1), 54–64. https://doi.org/10.1037/a0018438

Johnson, K. R., & Park, S. (2020). Mindfulness training for tourism and hospitality frontline employees. *Industrial and Commercial Training, 52*(4), 185–193. https://doi.org/10.1108/ICT-10-2019-0095

Kabat-Zinn, J. (1994). *Whereever you go, there you are*. Hyperion.

Kahn, W. A. (1990). Pyschological conditions of personal engagement and disengagement at work. *Academy of Management Journal, 33*(4), 692–724. https://doi.org/10.5465/256287

Kalyar, M. N., Ali, F., & Shafique, I. (2021). Green mindfulness and green creativity nexus in hospitality industry: Examining the effects of green process engagement and CSR. *International Journal of Contemporary Hospitality Management, 33*(8), 2653–2675. https://doi.org/10.1108/IJCHM-09-2020-1079

Kang, M., & Gretzel, U. (2012). Effects of podcast tours on tourist experiences in a national park. *Tourism Management, 33*(2), 440–455. https://doi.org/10.1016/j.tourman.2011.05.005

Kaoun, T. M. (2019). Enhancing leaders' emotional intelligence: Why mindfulness? *Journal of Leadership Education, 18*(1), 200–212. https://doi.org/10.12806/V18/I1/T2

Kemper, K., Mo, X., & Khayat, R. (2015). Are mindfulness and self-compassion associated with sleep and resilience in health professionals? *The Journal of Alternative and Complementary Medicine, 21*(8), 496–503. https://doi.org/10.1089/acm.2014.0281

Kerrigan, D., Chau, V., King, M., Aff, M. P., Holman, E., Joffe, A., & Sibinga, E. (2017). There is no performance, there is just this moment: The role of mindfulness instruction in promoting health and well-being among students at a highly-ranked university in the United States. *Journal of Evidence-Based Complementary & Alternative Medicine, 22*(4), 909–918. https://doi.org/10.1177/2156587217719787

Khoury, B., Knäuper, B., Pagnini, F., Trent, N., Chiesa, A., & Carrière, K. (2017). Embodied mindfulness. *Mindfulness, 8*(5), 1160–1171. https://doi.org/10.1007/s12671-017-0700-7

Kiken, L. G., Garland, E. L., Bluth, K., Palsson, O. S., & Gaylord, S. A. (2015). From a state to a trait: trajectories of state mindfulness in meditation during intervention predict changes in trait mindfulness. *Personality and Individual Differences, 81*, 41–46. https://doi.org/10.1016/j.paid.2014.12.044

Kroon, B., van Woerkom, M., & Menting, C. (2017). Mindfulness as substitute for transformational leadership. *Journal of Managerial Psychology, 32*(4), 284–297. https://doi.org/10.1108/JMP-07-2016-0223

Langer, E. (1989). *Mindfullness*. Addison Wesley Longman.

Lengyel, A. (2022). Authenticity, mindfulness and destination liminoidity: A multi-level model. *Tourism Recreation Research, 47*(1), 31–46. https://doi.org/10.1080/02508281.2020.1815412

Lew, A. A., Cheer, J. M., Haywood, M., Brouder, P., & Salazar, N. B. (2020). Visions of travel and tourism after the global CO VID-19 transformation of 2020. *Tourism Geographies, 22*(3), 455–466. https://doi.org/10.1080/14616688.2020.1770326

Lyddy, C. J., Schachter, Y., Reyer, A., & Julliard, K. (2016). Transfer of mindfulness training to the work setting: A qualitative study in a health care system. *Journal of Continuing Education in the Health Professions, 36*(4), 240–248. https://doi.org/10.1097/CEH.0000000000000120

Ma, S., Zhao, X., Gong, Y., & Wengel, Y. (2021). Proposing "healing tourism" as a post-COVID-19 tourism product. *Anatolia, 32*(1), 136–139. https://doi.org/10.1080/13032917.2020.1808490

Martin, L. L., & Tesser, A. (1996). Some ruminative thoughts. In R. S. Wyer (Ed.), *Advances in social cognition*. Lawrence Erlbaum Associates.

Maslach, C. (1982). *Burnout. The cost of caring*. Prentice-Hall.

Maslach, C., & Jackson, S. E. (1981). The measurement of experienced burnout. *Journal of Organizational Behavior, 2*(2), 99–113. https://doi.org/10.1002/job.4030020205

McMahon, G., & Collard, P. (2009). Mindful feelings at work. *Training Journal*, 64–68.

Mekawy, M., Elbaz, A. M., Shabana, M. M., & Soliman, M. (2022). Breaking the psychological contract of travel agency employees during the COVID-19 pandemic: The moderating role of mindfulness. *Tourism and Hospitality Research, 22*(4), 387–402. https://doi.org/10.1177/14673584211054602

Morris, J. A., & Feldman, D. C. (1996). The dimensions, antecedents, and consequences of emotional labor. *Academy of Management Review, 21*(4), 986–1010. https://doi.org/10.5465/amr.1996.9704071861

Neff, K. (2003). Self-compassion: An alternative conceptualization of a healthy attitude toward oneself. *Self and Identity, 2*(2), 85–101. https://doi.org/10.1080/15298860309032

Parsons, E. M., Luebbe, A. M., & Clerkin, E. M. (2017). Testing the relationship between social anxiety schemas, mindfulness facets, and state and trait social anxiety symptoms. *Mindfulness, 8*, 1634–1643. https://doi.org/10.1007/s12671-017-0738-6

Puddicome, A. (2014). *Westin launches meditation program for guests, employees*. Retrieved from https://www.mindful.org/westin-launches-meditation-program-for-guests-employees/

Reb, J., Narayanan, J., & Ho, Z. W. (2013). Mindfulness at work: Antecedents and consequences of employee awareness and absent-mindedness. *Mindfulness, 6*, 111–122. https://doi.org/10.1007/s12671-013-0236-4

Said, H., & Tanova, C. (2021). Workplace bullying in the hospitality industry: A hindrance to the employee mindfulness state and a source of emotional exhaustion. *International Journal of Hospitality Management, 96*, 1–15. https://doi.org/10.1016/j.ijhm.2021.102961

Salovey, P., & Mayer, J. D. (1990). Emotional intelligence. *Imagination, Cognition and Personality, 9*(3), 185–211. https://doi.org/10.2190/DUGG-P24E-52WK-6C

Saunders, M., Lewis, P., & Thornhill, A. (2000). Collecting primary data through observation. In M. N. K. Saunders, P. Lewis, & A. Thornhill (Eds.), *Research methods for business students* (pp. 218–236). Financial Times/Prentice Hall.

Schalk, R., & Roe, R. E. (2007). Towards a dynamic model of the psychological contract. *Journal for the Theory of Social Behaviour, 37*(2), 167–182. https://doi.org/10.1111/j.1468-5914.2007.00330.x

Schutte, N. S., & Malouff, J. M. (2011). Emotional intelligence mediates the relationship between mindfulness and subjective well-being. *Personality and Individual Differences, 50*(7), 116–119. https://doi.org/10.1016/j.paid.2011.01.037

Sethi, D. (2009). Mindful leadership. *Leader to Leader, 2009*(51), 7–11. https://doi.org/10.1002/ltl.311

Shaffakat, S., Otaye-Ebede, L., Reb, J., Chandwani, R., & Vongswasdi, P. (2022). Mindfulness attenuates both emotional and behavioral reactions following psychological contract breach: A two-stage moderated mediation model. *Journal of Applied Psychology, 107*(3), 425–443. https://doi.org/10.1037/apl0000878

Sharma, G. D., Thomas, A., & Paul, J. (2021). Reviving tourism industry post-COVID-19: A resilience-based framework. *Tourism Management Perspectives, 37*(100), 786. https://doi.org/10.1016/j.tmp.2020.100786

Sheldon, K. M., Prentice, M., & Halusic, M. (2015). The experiential incompatibility of mindfulness and flow absorption. *Social Psychological and Personality Science, 6*(3), 276–283. https://doi.org/10.1177/1948550614555028

Shi, X., & Wang, X. (2022). Daily spillover from home to work: The role of workplace mindfulness and daily customer mistreatment. *International Journal of Contemporary Hospitality Management, 34*(8), 3008–30028. https://doi.org/10.1108/IJCHM-10-2021-1281

Stankov, U., & Filimonau, V. (2023). Here and now—The role of mindfulness in post-pandemic tourism. *Tourism Geographies, 25*(1), 374–389. https://doi.org/10.1080/14616688.2021.2021978

Stankov, U., Filimonau, V., & Vujičić, M. D. (2020a). A mindful shift: An opportunity for mindfulness-driven tourism in a post-pandemic world. *Tourism Geographies, 22*(3), 703–712. https://doi.org/10.1080/14616688.2020.1768432

Stankov, U., Filimonau, V., Gretzel, U., & Vujičić, M. D. (2020b). E-mindfulness—The growing importance of facilitating tourists' connections to the present moment. *Journal of Tourism Futures, 6*(3), 239–245. https://doi.org/10.1108/JTF-11-2019-0135

Stankov, U., Čikić, J., & Armenski, T. (2021). New age tourism—Challenges and positioning in a changing society. *Tourism Culture & Communication, 21*(2), 69–80. https://doi.org/10.3727/109830421X16191799471944

Vago, D. R., & Silbersweig, D. A. (2012). Self-awareness, self-regulation, and self-transcendence (S-ART): A framework for understanding the neurobiological mechanisms of mindfulness. *Frontiers in Human Neuroscience, 6*, 1–30. https://doi.org/10.3389/fnhum.2012.00296

Wang, X., Wen, X., Paşamehmetoğlu, A., & Guchait, P. (2021a). Hospitality employee's mindfulness and its impact on creativity and customer satisfaction: The moderating role of organizational error tolerance. *International Journal of Hospitality Management, 94*, 102–846. https://doi.org/10.1016/j.ijhm.2020.102846

Wang, Y.-C., Chen, P.-J., Shi, H., & Shi, W. (2021b). Travel for mindfulness through Zen retreat experience: A case study at Donghua Zen Temple. *Tourism Management, 83*, 104–211. https://doi.org/10.1016/j.tourman.2020.104211

Wen, J., Wang, C. C., & Kozak, M. (2021). Post-CO VID-19 Chinese domestic tourism market-recovery: Potential influence of traditional Chinese medicine on tourist behaviour. *Anatolia, 32*(1), 121–125. https://doi.org/10.1080/13032917.2020.1768335

Wolever, R. Q., Bobinet, K. J., McCabe, K., Mackenzie, E. R., Fekete, E., Kusnick, C. A., & Baime, M. (2012). Effective and viable mind-body stress reduction in the workplace: A randomized controlled trial. *Journal of Occupational Health Psychology, 17*(2), 246–258. https://doi.org/10.1037/a0027278

Wu, C.-M., & Chen, T.-J. (2019). Inspiring prosociality in hotel workplaces: Roles of authentic leadership, collective mindfulness, and collective thriving. *Tourism Management Perspectives, 31*, 123–135. https://doi.org/10.1016/j.tmp.2019.04.002

Zapf, D., & Holz, M. (2006). On the positive and negative effects of emotion work in organizations. *European Journal of Work and Organizational Psychology, 15*(1), 1–28. https://doi.org/10.1080/13594320500412199

Zhu, J. L., Schülke, R., Vatansever, D., Xi, D., Yan, J., Zhao, H., Xie, X., Feng, J., Chen, M. Y., Sahakian, B. J., & Wang, S. (2021). Mindfulness practice for protecting mental health during the COVID-19 pandemic. *Translational Psychiatry, 11*, 1–11. https://doi.org/10.1038/s41398-021-01459-8

Chapter 6
Mapping the Impact of Motorcycle Tourism in Trans-Himalayas: A Study on Alternative Livelihood Creation in Rural Ladakh, India

Parishmita Gogoi and Alka Maheshwari

Theme:

Developing quality of life paradigms for niche destinations

Abstract Motorcycle tourism has become one of the emerging tourism trends in India. This unique form of tourism requires professional expertise in addition to technical skilled guidance and specialized equipment. It is particularly popular in rural and remote regions of the Himalayas. However, it is important to promote it in a sustainable manner, and very few studies have scrutinized it from a sustainability standpoint. To fill this gap, this study aims to answer two research questions: (1) how can the current contribution of the motorcycle tourism market be mapped in a manner that it meaningfully contributes to the economy of rural communities located in the remote parts of Ladakh? and (2) what is the best technique to conceptualize innovative initiatives that support sustainable tourism underpinned on inclusivity and pro-poor guidelines? A conceptual sustainable livelihood model is proposed.

Keywords Motorcycle tourism · Rural tourism · Community livelihood · Niche tourism · Sustainable development · Ladakh

---

P. Gogoi (✉)
LEHO (Ladakh Environment & Health Organization), Leh, Ladakh, India

A. Maheshwari
Amity Institute of Travel & Tourism, Amity University, Noida, Uttar Pradesh, India

D. Chhabra et al. (eds.), *Sustainable Development and Resilience of Tourism*,
https://doi.org/10.1007/978-3-031-63145-0_6

Introduction

This section highlights the history and popularity of motorcycle tourism in India. Later it explores the geographical terrain of Ladakh, the scope of organizing motorcycle tourism in the region and the history and current trends in its tourism industry.

In today's world motorcycle touring in India has become one of the emerging tourism trends which professionally requires expertise along with the technical skilled guidance and equipment mandatory for the tour. Motorcycle tourism was initiated in 1903 spanning a distance of 6100 km (non-stop) from San Francisco to New York (Peek, 2022). This adventure journey was indigenously undertaken by Mr. George A. Wyman who became the first transcontinental rider to cross-travel, without any halt, from one continent to another (Peek, 2022). Comparatively, in India, motorcycle touring captured attention in 1955 with the manufacturing of the 350 cc Bullet by the Royal Enfield Company from England (The Royal Enfield Story, n.d.). Soon, the two-wheeler started being assembled in Chennai and its manufacturing followed soon after. Today, it has become a popular pursuit for many adventure travellers in many parts of India. Worthy of note is its popularity in the Ladakh region of the trans-Himalayan range. This region lies in the northernmost region of India and is host to adventure motorcycling thereby contributing to the surge in tourism. Sykes and Kelly (2016) note that this unique form of adventure tourism is more popular in rural destinations and is pursued by travellers with a high disposable income and propensity to spend. As described by Sykes and Kelly (2014, p. 03), "Motorcycle Tourism involves trips away from home whose main purpose is vacation, leisure, entertainment, or recreation. The trip may include participation in sporting events or attendance at events organized for commercial gain and/or charity, or independently organized motorcycling".

Also as described by Scol (2016), p. 01:

> Motorcycle tourism is often considered a marginal practice with special habits where taking the road, riding styles and camaraderie play essential roles. These rallies are a good example of the unusual and marginal nature of this tourism. Moreover, motorbike tourism has for a long time been ignored or rejected by regions and tourism professionals who had a negative image of motorcycles and motorcyclists. However, bikers have changed. They are on average older than in the past, well-integrated within civil society and often have good incomes. They are therefore high potential tourists who authorities should attract by offering products and services tailored to their needs and specialities. These services are driven by the community, professionals or associations, especially in the form of dedicated routes or accommodations identified or labelled for bikers. Those new customers have also caused a proliferation of specialized tour operators.

Ladakh is a high-altitude niche tourism destination. As stated in *Ladakh*, Wikipedia. (n.d.), Ladakh was established as a union territory of India on 31 October 2019, following the passage of the Jammu and Kashmir Reorganisation Act. Prior to that, it was part of the Jammu and Kashmir state, India. Ladakh is both the largest and the second least populous union territory of India. The largest town in Ladakh is Leh, followed by Kargil, each of which headquarters a district. This vast expanse of land is a mesmerizing confluence of mountains, deserts, lakes, rivers and sparse grasslands. It is equally complemented by a diverse amount of flora and fauna, despite the extreme climatic conditions. Ladakh is primarily spread across three major mountain ranges, namely, Ladakh, Zanskar and Himalayan ranges, littered with valleys like Lahaul-Spiti and Suru Valley in the south and Nubra, Kargil and Gilgit-Baltistan (now under POK) in the North, shown in Fig. 6.1.

The vast majority of its people live as farmers in small villages. Rainfall in the region is very low, and local people have to rely on glacial melt water, which is brought to the field through an elaborate channel system. The region's unique terrain attracts people from many walks of life, such as leisure tourists, adventure junkies and researchers alike. With improved infrastructure and on-going development of road networks, Ladakh has become increasingly accessible than ever before. It presents a rare and life-altering opportunity to motorists, seeking thrill experiences, to ride across a vast variety of nearly all kinds of terrains, altitudes and road conditions for the thrill-seeking motorist.

However, most tour operators, especially marketing motorcycle tourism, indulge in directional selling. A majority of the tour operators, mainly domestic, who are marketing Ladakh, are only focused on sales maximization targeting the mass market at cheaper package costs every season. These tourists visit only the major hotspots, namely, Leh, Nubra Valley, Pangong Tso and at most, some of the popular monasteries around Leh city, as shown in Fig. 6.2. These visitors are not aware of its fragile ecosystem, local culture and tradition, and they benefit only a minority of service providers in Leh and the mentioned hotspots, Fig. 6.2, at the cost of the rest 90% of the population living outside of the destination hotspots. A diagram showing the conventional trend of tourism in Ladakh is shown in Fig. 6.3.

Fig. 6.1 Map of Union Territory Ladakh with Leh city connection to Manali in Himachal Pradesh and Srinagar in Jammu & Kashmir (source: author)

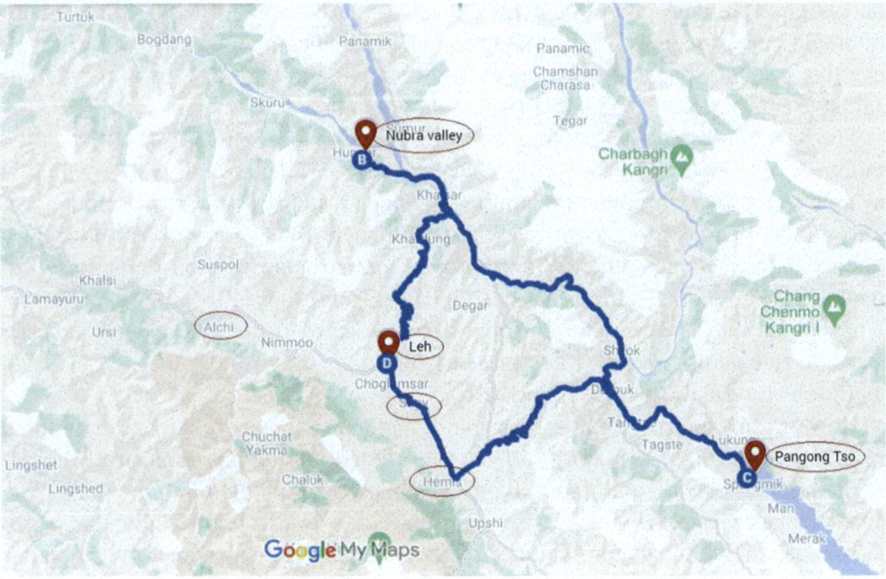

Fig. 6.2 The most visited tourist destinations of Ladakh. (Source: LEHO (Ladakh Environment & Health Organization), Leh, Ladakh, India)

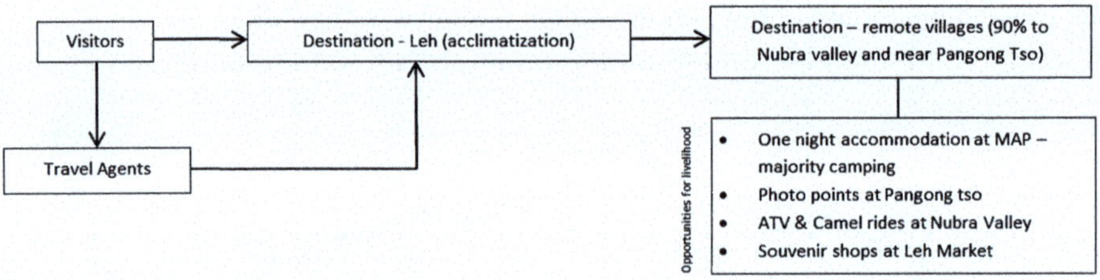

Fig. 6.3 Conventional tourism trend in Ladakh. (Source: LEHO (Ladakh Environment & Health Organization), Leh, Ladakh, India)

"When Ladakh was first opened for tourism in 1974, as per the government records, there were 527 tourists—500 foreign visitors and 27 domestic visitors" (Gupta, 2022, para. 08). Since then, the profile of Ladakh's tourism has been reflected by the dominance of foreign tourist arrivals. According to Smanla (2022), p. 2099, "The tourist season starting mid-April and lasting till to mid-September has become a period of intense economic activity. It has also brought significant changes in the traditional economy that were largely land and agriculture-centred". Gupta (2022), para. 09–10, also adds that:

> Almost 50 years later, the number of tourists has grown to 4.5 lakhs (450,000) this year, in eight months from January 1 to August 31 2022. The number of tourists in Leh this summer broke all previous records, according to official data. As many as 2.5 lakhs (250,000) people visited Leh for tourism in just two months, June and July. That's eight times more visitors

than Leh City's local population of around 30,870 people (2011 census) and around twice the population of Leh district which is approximately 1.33 lakh (133,000).

As a result, there can be several challenges if this influx is not managed well and brought a structured solution to mitigate the issues that result from mass tourism leading to overconsumption of resources. To deal with over-tourism and/or the impact of mass tourism, a strategic conceptual approach is required to inform/guide strategies that hold the potential to generate a balanced relationship between the motorcyclists and the numerous stakeholders from the local community. Based on the aforementioned, this study intends to map the scope of motorcycle tourism redefined with an immersive definition of responsible tourism. This model is cocreated with diverse yet region-specific creative elements to develop the life paradigms of the local communities living in the remote corners of Ladakh.

The effort is also to identify principles of sustainable development using a long-term and more cautious approach to ensure that future generations can enjoy at least the same quality of life as the present generation (Dar et al., 2022). Because motorcycle tourism is growing, it is attracting new riders who have high disposable income and leisure time at their disposal, both of which are key facets of successful tourism (Sykes et al., 2012). This study will therefore offer important implications by helping to map the niche motorcycle tourism market that holds the potential to bring significant and sustainable economic benefits to local residents of Ladakh.

The research aims to focus on three objectives: (1) map the economic contribution of the current form of motorcycle tourism to the rural economy of Ladakh, (2) study the scope of a reformed motorcycle tourism with more responsible and immersive experiences and (3) comprehend the possibility of the positive impact to be generated by the reformed motorcycle tourism on the quality.

Literature Review

This section examines existing research on motorcycle tourism and its impact on sustainable community development, exploring key themes, methodologies and gaps in knowledge.

Ladakh's motorcycle tourism thrives on the thrilling combination of challenging terrain and breathtaking natural beauty, making it a sought-after destination among adventure enthusiasts. Some of the motorcyclists travel to remote villages to explore lesser-known routes and scenic views, and they benefit the community by booking a room and staying at a house of a villager. However, the local homestay tourism model was impacted due to COVID-19, and the villagers lost the scope of income from tourism. Therefore, an alternative plan is needed so that tourism provides enhanced economic and social support to the local hosts. Richards (2020) argues that the notion of resilience, in the context of returning to the old normal, needs to be revisited.

Furthermore, several studies postulate that the concept of "responsible tourism" needs to be understood fully by all stakeholders (Changa & McCreeshc, 2022; Goodwin, 2017; Manente et al., 2014; McCombes & Vanclay, 2022). According to Mondal and Samaddar (2021), "Responsible Tourism (RT) comprises all the stakeholders' action and consciousness towards sustainable travel that regards the three pillars of sustainability which are environment, local community and economy". The RT initiative can prove to be a powerful unifier among various tourism stakeholders (Burrai et al., 2019). It is regarded as one of the pathways for meeting the 2030 Sustainable Development Goals (SDG) of the United Nations. Responsible tourism practice bestows maximum benefits to the local community and helps the local people to conserve their culture and habitats (Caruana et al., 2014; Mathew & Sreejesh, 2017).

However, responsible tourism is incomplete without understanding the significance of responsible consumption, and in this case, this has to be guarded by all stakeholders, primarily by the marketers who promote motorcycle tourism, to foster sustainable behaviour. According to Sheth et al. (2011), customer-centric sustainability (CCS) and mindful consumption (MC) are the two important aspects that can play a vital role in cocreating responsible tourism. Sheth et al. (2011), p. 04, explain that:

> In the CCS perspective, the environmental dimension relates to the impact of consumption on *environmental well-being*, that is, health and human well-being consequences of environmental change ensuing from consumption; the social dimension relates to the impact of consumption on the *personal well-being* of the consumers, reflecting individual (and family) well-being or quality of life, and associated welfare of the community; and, the economic dimension relates to the impact of consumption on the *economic well-being* of consumers associated with financial aspects such as debt burden, earning pressures, and work-life balance.

Moreover, when studying motorcycle tourism, Sheth et al. (2011), p. 07, write that "Consumption has a tangible facet—the *behaviour* of engaging in consumption, and in practice that is what appears to matter. There is also an intangible facet of consumption—the *mindset* pertaining to attitudes, values and expectations surrounding the consumption behaviour". Based on Richards (2020)'s suggestions, on p. 03, "Creativity can be applied in a variety of ways: through creative people, creative processes, creative products, and creative places. It is therefore important to understand not just the impact on the tourist system, but also all the interconnected parts of the economic, social, and cultural systems of places". Similar views were suggested by Ollivaud and Haxton (2019), p. 22, in their study that examined Indonesia as a niche destination, "The creative economy can help offer new products and services for new target groups away from conventional models of environmental or heritage-based cultural tourism: that can be through unconventional media advertisements, art creation in a specific building, and sound-and-light shows. Increasingly visitors are looking for experience-based, instead of destination-based, tourism".

The first author conducted a survey and observed that the current trends of motorcycle tourism in Ladakh, it is noted that motorcyclists mostly book fixed departure tours through travel agents and these tours cover the routes to Nubra

Valley and Pangong Tso. Nubra Valley is visited to experience Khardung La (one of the world's highest motorable roads at 18,000 feet), whereas, Pangong Tso is visited because of its vast size which is only one-third that located in India and the remaining two-thirds located in China (Tibet). Also, its mesmerizing azure colour makes the lake one of the must-visit unique attractions in the region.

The majority of the visitors spend a night at guest houses, camps or homestays in Hunder village (Nubra Valley) and Lukung or Spangmik village (near Pangong Tso) and leave the next day. This trend supports only those from the local community who can afford to offer camping or homestay type of accommodations. Also, outsiders from Leh City flock to villages, setting up campsites and guesthouses for tourism ventures. Unfortunately, this hampers local engagement, leaving them uninformed and less involved in the growing tourism industry. The extent to which these promote environmental and economic sustainability is still under question. Furthermore, Milano et al. (2019), p. 03, contend that "a place that is attractive and alluring to tourists should still be a place that is habitable and secure for local residents, and not sacrificed for the economic imperative above all else". Therefore, this research tries to map sustainable opportunities through an immersive form of motorcycle tourism while aiming to improve the livelihood of the local communities.

Methodology

This research is a combination of primary and secondary research methodologies and employs a non-probability sampling technique. One of the authors (LEHO (Ladakh Environment & Health Organization), Leh, Ladakh, India) who is now working in Ladakh was a part of an adventure tour company organizing motorcycle tourism to the region from Delhi for 2 years in her past career. This helped her to tap into her network of motorcycle riders, clubs as well as travel agents and local stakeholders in the business.

The motorcycle tourists are sampled with the help of a convenient non-probability sampling method. The travel agents are sampled with the help of mixing judgemental and snowball non-probability sampling methods. The village stakeholders are sampled on the basis of a quota sampling method. The sample sites were selected based on popularity and tourism patronage during the tourist seasons. The sites included were Nubra Valley in the northwest and villages around Pangong Tso in the eastern region of Ladakh. Two villages in Nubra Valley (Diskit and Hundar) and three villages near Pangong Tso (viz. Lukung, Spangmik and Man) were visited. The author visited both Nubra Valley and Pangong Tso herself to collect this data.

The data was collected, during 4 months in the winter season (October, November, December and January):

(1) A structured in-depth questionnaire was used which focused on two targeted stakeholder groups: "motorcycle riders" who participated in bike tours and "travel agents/tour operators" who organized the bike tours. Two sets of questionnaires were prepared with the help of Google Forms and distributed

online among the target stakeholders. Each response was collected and saved in a Google spreadsheet linked to every form. The response was received from 76 travel agents and 103 motorcycle riders (visitors).

(2) A semi-structured focused group interview was conducted with the local people in the study region.

Content analysis has been used for objective, systematic and quantitative description of the manifested content. The analysis resulted in varied subthemes from the open-ended questions in the structured questionnaire. For the purpose of content analysis, responses from open-ended questions were read and reread several times to interpret variations and discern similarities and differences in the content. Descriptive analysis also helped in summarizing information from the data collected through structured questionnaire.

In order to understand the interest of the motorcycle riders and travel agents towards immersive experiences supporting responsible tourism, some of the options included in the questionnaire are:

(a) Adventure activities—hikes, ATV, boulder, etc.
(b) Monastery festivals
(c) Seasonal sports—archery, ice hockey, horse riding, etc.
(d) Local art and craft
(e) Traditional healing (with Amchi)
(f) Guided village walks
(g) Local folk culture (music/dance)
(h) Farm-to-table
(i) Authentic local food as a part of the meal plan

Also, some knowledge exchange programmes are added to the questionnaire to understand the participants' perspectives on the activities towards promoting a sustainable and responsible model of tourism. They are the following:

(a) Photography workshop
(b) Provide groceries to underprivileged families
(c) Provide stationery items to schoolchildren
(d) Slow travel based on seasonal activities
(e) Organize awareness on climate change and responsible tourism
(f) Ensure to pay community development fund
(g) Organize workshops on motorcycle maintenance for the youth
(h) Provide training on hospitality and hygiene
(i) Organize waste management and clean-up drive
(j) Provide financial support for child education
(k) Learn sustainable high-altitude farming techniques
(l) Provide cooking lessons to locals
(m) Learn local cuisines from locals
(n) Local architecture and sustainable design
(o) Engage in meditation practices
(p) Local art and craft

Results and Discussions

The Results section presents the empirical findings and analysis, while the Discussion section interprets the results, contextualizes them within existing knowledge and explores their implications for the research topic.

Assessment of the Economic Status of the Local Communities

In this first section, insights are offered based on the analysis of preferences of travel agents to offer and riders to purchase the type of package tours and accommodation. The analysis highlights that the majority of the bike-ride tour packages belong fell to the cost bracket of 300–420 USD per person, and the main villages marketed by travel agents and visited by riders are Nubra Valley and Pangong Tso (as shown in Fig. 6.4), with most preferring to spend one night only in these destinations. Homestay is preferred only by 31% of the travel agents and 41.3% of the motorcycle riders in both Nubra Valley and near Pangong Tso, as shown in Fig. 6.5. The maximum room tariff per night that contributed for accommodation units falls in the range of 12 to 24 USD, as shown in Fig. 6.6. However, this revenue is mostly pocketed by non-native business stakeholders (83%) who partner with or own and run camping sites for tourists. Therefore, it is noted that the local village households are disconnected from the tourism business and only a few benefit from homestay business or any form of local activities. This finding is also supported by the data collected from the focused group interview of the local village communities.

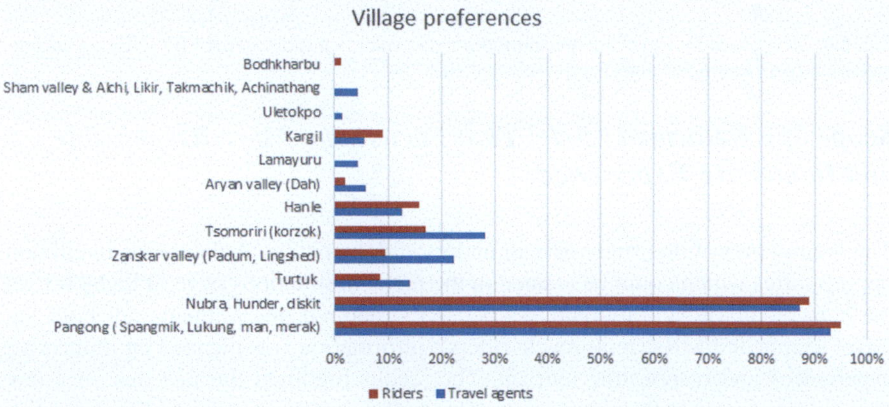

Fig. 6.4 Responses by both travel agents and riders on village (destination) preferences for night stay

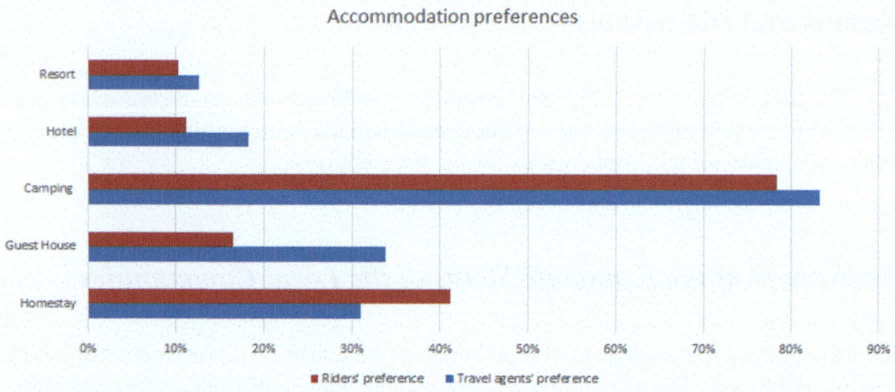

Fig. 6.5 Responses of travel agents and motorcycle riders (or MRs) on accommodation preferences at the selected sites

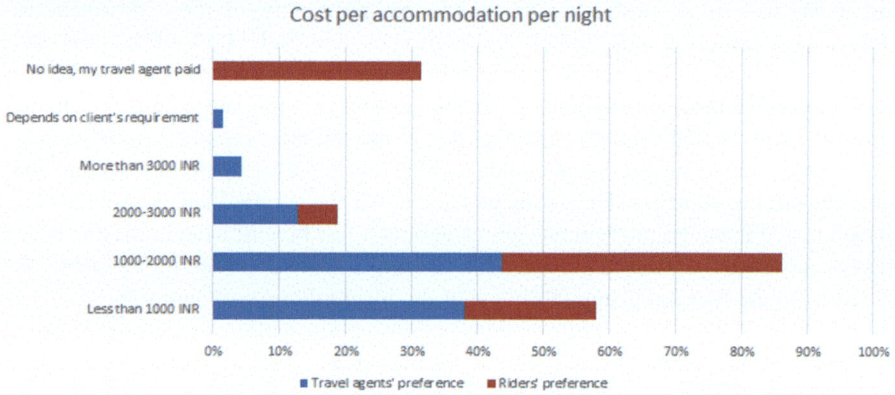

Fig. 6.6 Responses by travel agents and motorcycle riders on money spent (in INR) on accommodation (per person per night) at the selected sites

Scope of a Reformed Model That Focuses on More Responsible and Immersive Experiences

The second part of this study tries to understand the potential alternative livelihood opportunities which could be created in the niche destinations of rural Ladakh with the help of motorcycle tourism. As per the structured questionnaires of both the travel agents and MRs (motorcycle riders), information was gleaned on immersive experiences and responsible tourism. The results highlight the fact that very few respondents across both groups promote or experience the mentioned form of activities, as shown in Figs. 6.7 and 6.8, during their tours. According to the data analysed, the highest interest and involvement are only for ATV rides (56%) that are popular in the region, especially in the Nubra Valley. Activities like monastery

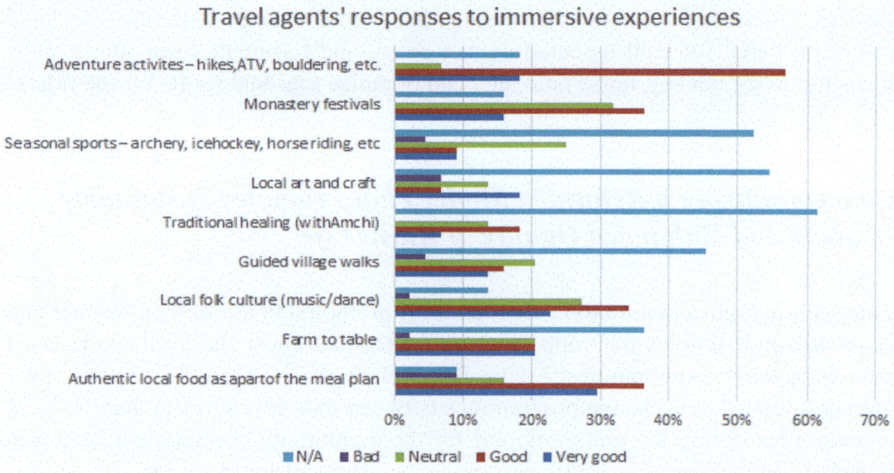

Fig. 6.7 Travel agents' perception on promoting/offering immersive experiences to their customers at the selected sites

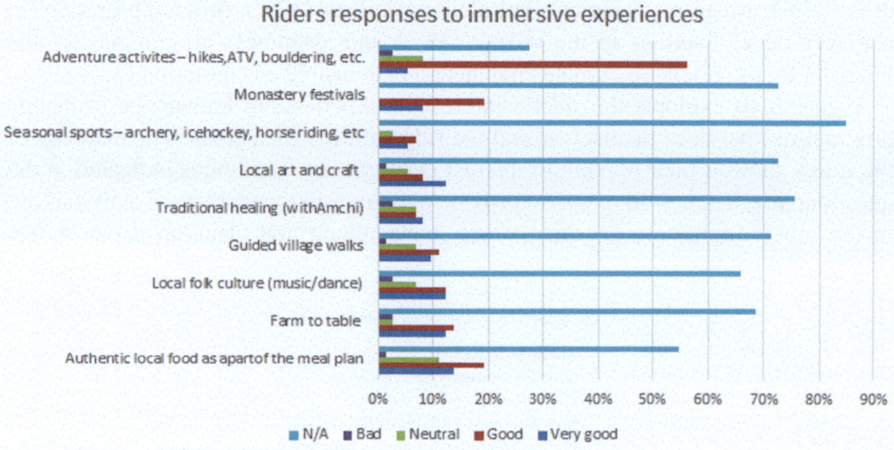

Fig. 6.8 Riders' perception of choosing immersive experiences during their tours at the preferred destination hotspots

festivals, seasonal sports, local art and craft, traditional healing and guided village walks are less promoted by the travel agents as the 46–60% response rates indicates "Not Applicable" to the tours. Among the riders' responses, except for ATV ride (56%), all other activity options with the highest response rate of 61% were unknown or unexplored by the riders.

Therefore, the local community can create immersive experiences for motorcycle riders by opening eateries which offer local food and an authentic experience. The farmers could also make extra income by giving the riders an opportunity for a farm-to-table experience, as they could visit the farms or greenhouse kitchen gardens, pluck vegetables and then indulge in the cooking experience too. The local youth, on behalf of their association, can create and formulate local sports clubs focusing on ice hockey, horse polo, etc. and organize seasonal sports for the riders.

Conceptualizing a Reformed Model That Promotes Sustainable Impacts and Enhanced Quality of Rural Life

This section argues that motorcycle tourism if promoted in a reformed manner can generate ample employment opportunities for the host (host and local can be used interchangeably) communities. Figure 6.9 explores travel agents' interest in promoting knowledge exchange programmes between motorcycle riders and the local communities during the tours. Support for the community development fund was found of the highest with a 54% response rate. Some of the other interesting programmes are in the order: photography workshops for local youth by the riders (48%), workshops for motorcycle maintenance for the youth by the riders with expertise on the subject (42%), providing stationery items for village school children (37%), waste management and clean-up drives (32%), providing grocery to underprivileged families in the village (32%) and awareness programme for the villagers by the riders on climate change and responsible consumption (32%).

Figure 6.10 explores the rider's interest in experiencing knowledge exchange programmes between themselves and the riders during their tours. The majority of the riders showed their maximum interest in all forms of activities indicated in the questionnaire. Their willingness to provide or engage in the mentioned activities are in the order: highest to organize waste management and clean-up drive (47%),

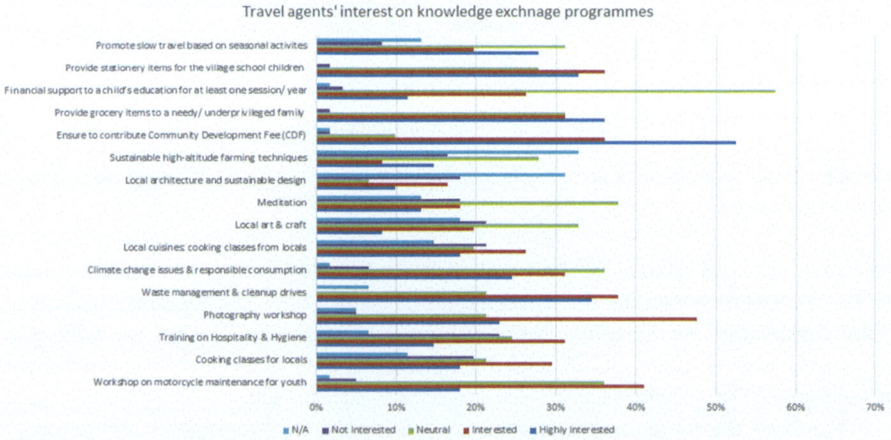

Fig. 6.9 Travel agents' responses on their future interest in promoting knowledge exchange programmes between the riders and the local communities during their tours at the niche destinations

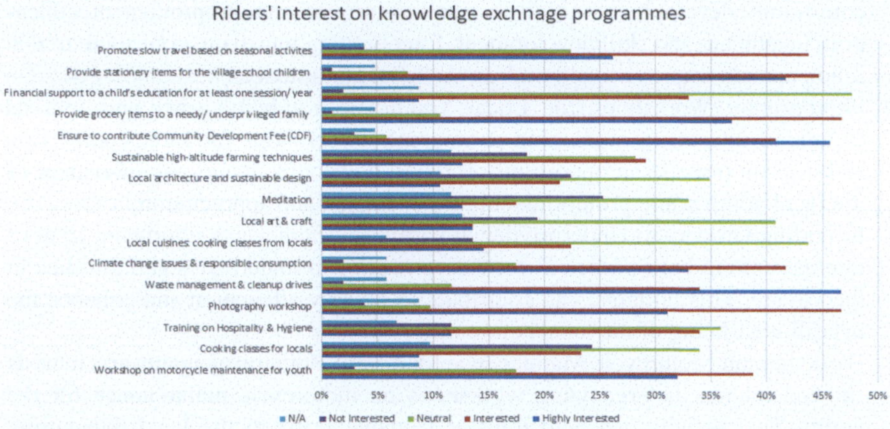

Fig. 6.10 Riders' responses on their future interest in experiencing knowledge exchange programmes between themselves and the local communities during their tours at the niche destinations

ensure to pay community development fund (46%), organize photography workshop (46%), provide grocery to underprivileged families (46%), provide stationery items to school children (44%), slow travel based on seasonal activities (43%), organize awareness on climate change and responsible tourism (42%) and organize workshop on motorcycle maintenance for the youth (39%). A little less interest was analysed in their willingness to provide training on hospitality and hygiene (34%), provide financial support to child education (33%), learn sustainable high-altitude farming techniques (28%), provide cooking lessons to locals (23%), learn local cuisines from locals (22%), local architecture and sustainable design (21%), engage in meditation practices (17%) and local art and craft (14%).

Therefore, this section comprehends the possibility of the positive impact that can be stimulated by a reformed motorcycle tourism plan that boosts the quality of life of the local communities in the niche destinations of rural Ladakh. Based on the collected data, Figs. 6.9 and 6.10 offer ideas on elements that can be included in the reformed model of motorcycle tourism to generate a positive impact. Stakeholders promoting the tourism business should be more mindful of the local environment and culture. Based on the common interests expressed by both travel agents and motorcycle riders, the authors identify several activities that can be added to a reformed model of motorcycle tourism to promote sustainable livelihood and an eco-friendly environment. Here are some recommendations:

- Waste Management and Clean-up Drives: Both travel agents and tourists showed a high level of interest in organizing waste management and clean-up drives. This activity should be prioritized as it helps in maintaining the cleanliness of the local environment and prevents pollution.
- Community Development Fund: Support for the community development fund was found to be high among travel agents. This fund can be utilized for various

community development initiatives, such as infrastructure improvement, education, healthcare and skill development. Encouraging tourists to contribute to this fund can help improve livelihoods in the local community.

- Photography Workshops for Local Youth: Both travel agents and tourists expressed interest in organizing photography workshops for local youth. This activity can provide an opportunity for skill development and empowerment of the local youth while promoting cultural exchange and appreciation.
- Providing Grocery to Underprivileged Families: There was a significant level of interest among both parties to provide groceries to underprivileged families in the village. This initiative can contribute to poverty alleviation and enhance the overall well-being of the local community.
- Workshop on Motorcycle Maintenance for Youth: Both travel agents and tourists showed interest in organizing workshops on motorcycle maintenance for the youth. This activity can help transfer valuable skills to the local population, opening up employment opportunities and promoting self-sufficiency.
- Awareness Programs on Climate Change and Responsible Tourism: Both parties expressed interest in organizing awareness programs on climate change and responsible tourism. This initiative can educate both tourists and local communities about sustainable practices, conservation and the importance of responsible consumption.

These recommended activities align with the interests of both travel agents and tourists, promoting sustainable livelihoods, community development, environmental conservation and cultural exchange. Implementing these activities in a reformed tourism model can contribute to a more eco-friendly and socially responsible approach to tourism.

The results indicate that the motorcycle riders share an interest in behaving in a responsible manner and participating in activities that can assist in supporting community life. Travel agents, motorcycle riders and locals are the key stakeholders and collectively they can plan/develop sustainable initiatives to support a holistic sustainable livelihood-generating and eco-friendly ecosystem while offering augmented immersive experiences.

Conceptual Model

This research study suggests that responsible tourism paradigms should feature sustainable livelihood strategies for niche tourism destinations. Drawn from the study results, several amendments are suggested for a responsible tourism model. As presented in the conceptual livelihood model, strategies can be categorized into three groups leading to three forms of benefits (direct, indirect and special), as shown in Fig. 6.11:

- *Direct Benefit*: The direct benefit is the monetary benefit that occurs directly whenever the local community residents:

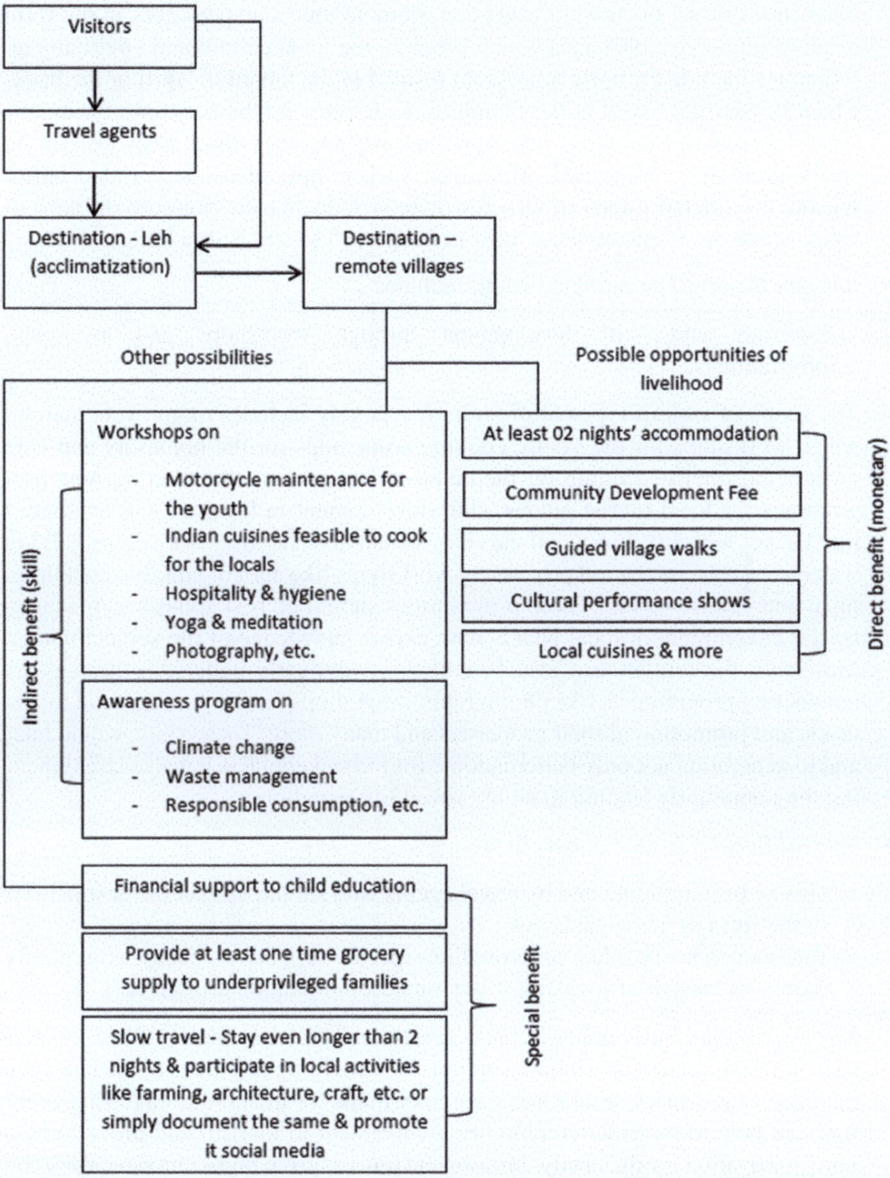

Fig. 6.11 A flowchart of the recommended model indicating three forms of benefits (direct, indirect and special). (Source: LEHO (Ladakh Environment & Health Organization), Leh, Ladakh, India)

- Host guests at their homestays.
- Sell local products in the form of souvenirs.
- Provide any other form of local services such as motorcycle repairs, café, guided village walks, cultural music, dance shows, etc.

- Apart from these, the travel agents can promote unique experiences in the form of more immersive packages which promote the local culture and environment. Examples include the apricot blossom festival in the month of April or the buckwheat flowering festival in the month of September. As the best time for motorcycle tourism in Ladakh is late April to early October, these festivals can be easily promoted and visited. Moreover, such unique festivals can also buffer seasonality-related issues in villages thereby reducing the pressure on the villages which are frequented the most in Pangong Tso and Nubra Valley.

• *Indirect Benefit*: The indirect benefit includes

  - Learning and skill development through workshops and awareness programmes.

- The scope of various types of programmes largely includes motorcycle maintenance workshops for the youth, cooking workshops for the homestay and café owners, hospitality training for the homestay owners, certification training programmes for local tourist guides, skill development in local art and handicraft like basket weaving and wood carving, sustainable farming workshops solving water scarcity issues, food processing workshops like apricot jam, sea buckthorn juice and similar others. These programmes should be best sponsored/managed by the government or local NGOs with expert knowledge of the subject matter. Moreover, the tourists can also benefit the community under this strategy by conducting programmes like photography workshops for the youth, social media usage and promotion of their properties and many more. These skills would later add to generating not only better income for the stakeholders but also strengthening the community leading to an improved life paradigm.

• *Special Benefit*: A special benefit strategy

  - Can be best implemented by travel agents who should market the destinations in the form of niche packages
  - Recommends the idea of providing special support to needy community members instead of focusing solely on accommodation providers

- There are many households in these rural areas that can barely afford to offer basic tourism-related amenities. Support like providing resources to the family in the form of groceries, books and stationery items for their children during every tour can help to motivate community involvement in tourism and prove to be a good motivation for the needy. Motorcycle tourists from higher income slabs can also offer financial support in the form of a donation to support a child's education. Special benefits can also accrue in the form of slow tourism offerings which means the riders can stay in the villages for longer duration and help the community in their day-to-day activities like during the farming season or the festive times.

Besides the aforementioned initiatives/strategies, there should be specific programmes that support the holistic development of the community and which support the conservation of the natural environment. This recommendation can be best implemented by introducing a community development fee (CDF) for every village visited by tourists, organizing cleanliness drives and plantation drives to mitigate environmental pollution, etc. These initiatives should also include tourists as part of the initiatives, for example, visitors participating in village clean-up drives in collaboration with the village youth association or any local self-help group. To ensure success, the proposed model stresses a participatory approach that is inclusive of the local community and suggests a purposeful co-learning and understanding process for developing a quality-of-life paradigm, for the village communities, through motorcycle tourism, as shown in Fig. 6.12. This will not only facilitate collaboration between the three stakeholders covered in this study but also include the government, local NGOs and other decision-makers and facilitate information seeking/sharing and views of a diverse group of stakeholders, integrating industry experience with local knowledge. The introduction of such a model not only motivates the entire village community to be interactive and more welcoming towards the tourists but also makes them feel proud of their native culture and tradition while earning their livelihood. Moreover, the proposed conceptual model suggests an open and transparent process to avoid cold clashes and potential conflicts between stakeholders with different agenda. Consequently, a huge sense of responsibility towards the entire ecosystem will be inculcated and later play a big role in the impact assessment of the model. The impact assessment strategy for a community livelihood development model involves collecting data on various indicators such as income levels, employment rates and overall quality of life to measure the model's effectiveness.

Conclusion

This research addressed the potential of motorcycle tourism as a responsible form of niche tourism in rural destinations of Ladakh in India. The analysis found that the local communities in Ladakh are meagrely involved in the development of tourism in their destinations because of the encroachment of facilities to entertain mass tourism which requires commercial accommodation units. The primary reason for the lack of progress and involvement of underprivileged communities at the niche destinations is due to the lack of awareness and participatory responsible tourism model, empowered by the community, which holds the potential to offer opportunities to offer economic benefits while at the same time, supporting the local ecosystems.

Therefore, a government-led development strategy is needed to promote cultural and environmental conservation and livelihood. To conclude, the study adopted a sustainable livelihood framework for motorcycle tourism as conservation as well as a development approach. The proposed conceptual model advocates that motorcycle tourism can be a great experiential and knowledge exchange platform for both

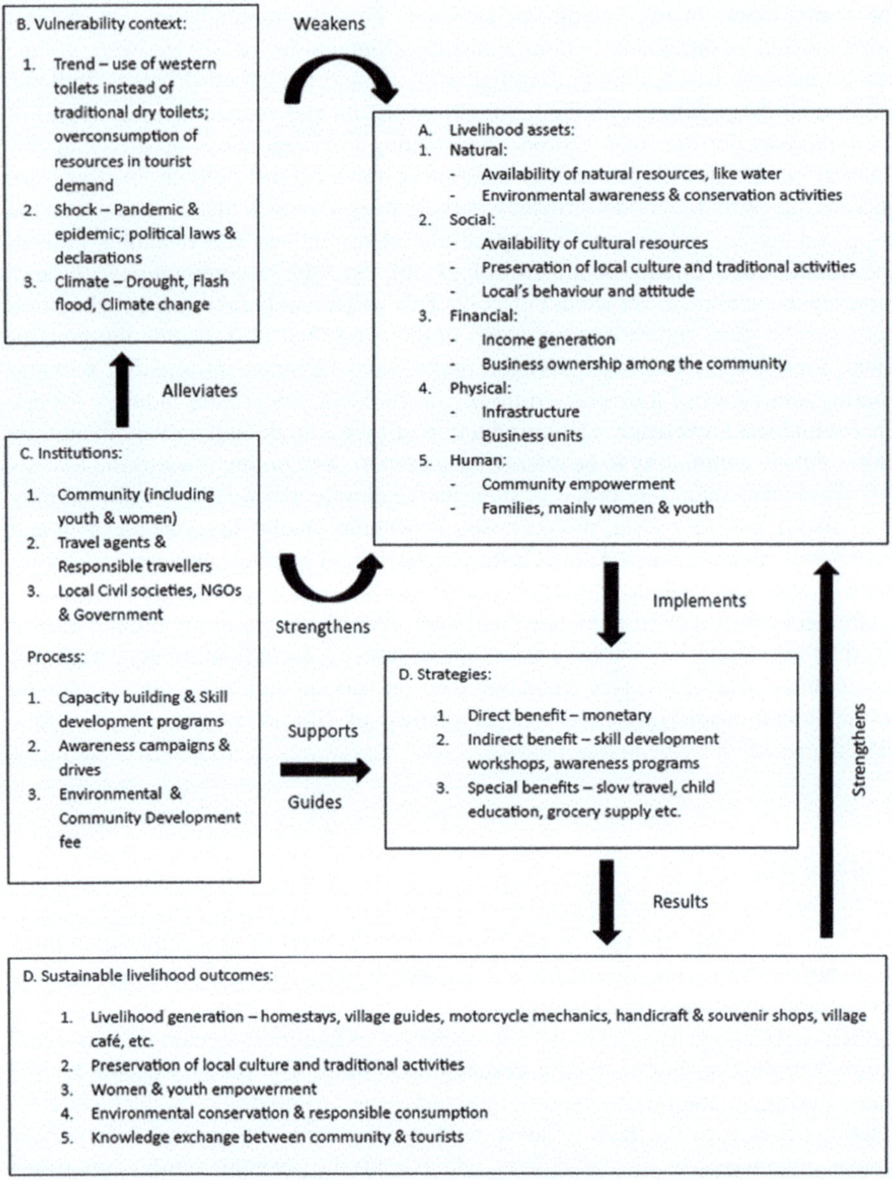

Fig. 6.12 A sustainable livelihood model diagram supporting the conceptual model described above. Idea adapted from Woyesa and Kumar (2021), Matiku et al. (2021). (Source: LEHO (Ladakh Environment & Health Organization), Leh, Ladakh, India)

the locals and tourists through an inclusive and mindful agenda. This responsible form of motorcycle tourism in rural Ladakh can impose a check on environmental degradation and mitigate current patterns of activities that encourage unsustainable livelihood.

References

Burrai, E., Buda, D. M., & Stanford, D. (2019). Rethinking the ideology of responsible tourism. *Journal of Sustainable Tourism, 27*(7), 992–1007.

Caruana, R., Glozer, S., Crane, A., & McCabe, S. (2014). Tourists' accounts of responsible tourism. *Annals of Tourism Research, 46*, 115–129.

Changa, H. H., & McCreeshc, M. (2022). The case for mountain biking in responsible tourism narratives. *Journal of Responsible Tourism Management, 2*.

Dar, S. N., Shah, S. A., & Wani, M. A. (2022). Geospatial tourist information system for promoting tourism in trans-himalayas: A study of leh ladakh India. *GeoJournal, 87*(4), 3249–3263.

Goodwin, H. (2017). The challenge of overtourism. *Responsible Tourism Partnership, 4*, 1–19.

Gupta, V. (2022, September 27). *As tourism grows in Leh, so does the concern about its environmental impact*. Mongabay-India. https://india.mongabay.com/2022/09/as-tourism-grows-in-leh-so-does-the-concern-about-its-environmental-impact

Ladakh. Wikipedia. (n.d.). https://en.wikipedia.org/wiki/Ladakh

Manente, M., Minghetti, V., & Mingotto, E. (2014). *Responsible tourism and CSR*. Springer.

Mathew, P. V., & Sreejesh, S. (2017). Impact of responsible tourism on destination sustainability and quality of life of community in tourism destinations. *Journal of Hospitality and Tourism Management, 31*, 83–89.

Matiku, S. M., Zuwarimwe, J., & Tshipala, N. (2021). Sustainable tourism planning and management for sustainable livelihoods. *Development Southern Africa, 38*(4), 524–538.

McCombes, L., & Vanclay, F. (2022). A responsible tourism perspective on tourist-host encounters in community-based tourism. In *Handbook of tourism impacts: Social and environmental perspectives* (pp. 36–59). Edward Elgar Publishing.

Milano, C., Novelli, M., & Cheer, J. M. (2019). Overtourism and tourismphobia: A journey through four decades of tourism development, planning and local concerns. *Tourism Planning & Development, 16*(4), 353–357.

Mondal, S., & Samaddar, K. (2021). Responsible tourism towards sustainable development: Literature review and research agenda. *Asia Pacific Business Review, 27*(2), 229–266.

Ollivaud, P., & Haxton, P. (2019). *Making the most of tourism in Indonesia to promote sustainable regional development*. OECD.

Peek, J. P. (2022, July 6). *The first motorized vehicle to cross the U.S. wasn't a car—it was a bike*. https://www.hagerty.com/media/automotive-history/the-first-motorized-vehicle-to-cross-the-u-s-wasnt-a-car-it-was-a-bike/

Richards, G. (2020). Tourism in challenging times: Resilience or creativity? *Tourism Today, 2020*(19), 8–15.

Scol, J. (2016). Motorcycle tourism: Renewed geographies of a marginal tourism practice. *Via Tourism Review*. https://doi.org/10.4000/viatourism.356

Sheth, J. N., Sethia, N. K., & Srinivas, S. (2011). Mindful consumption: A customer-centric approach to sustainability. *Journal of the Academy of Marketing Science, 39*(1), 21–39.

Smanla, T. (2022). Ecotourism in Ladakh: A critical aspect to contemplate. *International Journal of Research Publication and Reviews, 3*(12), 2096–2105

Sykes, D. M., & Kelly, K. G. (2014). Motorcycle tourism demand generators and dynamic interaction leisure. *International Journal of Culture, Tourism and Hospitality Research, 8*, 92–102.

Sykes, D., & Kelly, K. G. (2016). Motorcycle drive tourism leading to rural tourism opportunities. *Tourism Economics, 22*(3), 543–557.

Sykes, D. M., Kelly, K. G., & Ireland, B. N. (2012, June). Mapping the study of motorcycle tourism: Impacts and opportunities. In *Proceedings of international conference on tourism and events: Opportunities, impacts and change* (pp. 20–22). Ulster University.

The Royal Enfield Story | Since 1901 | Royal Enfield. (n.d.). https://www.royalenfield.com/in/en/our-world/since-1901/

Woyesa, T., & Kumar, S. (2021). Potential of coffee tourism for rural development in Ethiopia: A sustainable livelihood approach. *Environment, Development and Sustainability, 23*, 815–832.

Chapter 7
Transformative Potential of Sustainable Scuba Dive Tourism

Daniel Cohen

Abstract While recreational scuba diving is a relatively new phenomenon, human beings have been closely tied with the ocean and marine resources for thousands of years. This chapter presents an extensive literature review to scrutinize the sustainability of marine diving tourism and the extent to which it can create transformative experiences that may be beneficial for both the diver tourists and the host communities. It examines this based on the three pillars of sustainability including environmental, economic, and social outlook of dive tourism. Furthermore, viability of Sustainable Development Goals (SDGs) and the effectiveness of management goals are scrutinized based on a transformative paradigm. A case study is also used to discuss the transformative potential of the marine tourism industry on the Hawaiian Island of Maui following a devastating wildfire that occurred in August 2023.

Keywords Scuba diving · Sustainability · Marine ecotourism · Conservation biology · Human-wildlife coexistence · SDGs · Transformative paradigms · Outdoor recreation · Hawaii · Coral reef

Introduction

While recreational scuba diving is a relatively new phenomenon, human beings have been closely tied with the ocean and marine resources for thousands of years. Historically, motivation for participating in underwater diving has been to gather food, retrieve objects, or harvest resources of value (Ecott, 2002). Some of the earliest known examples of diving dates back roughly 6,500 years ago to the region of Mesopotamia, where pearls and sea sponges were harvested by free divers who

D. Cohen (✉)
School of Community Resources and Development, Arizona State University,
Phoenix, AZ, USA

D. Chhabra et al. (eds.), *Sustainable Development and Resilience of Tourism*,
https://doi.org/10.1007/978-3-031-63145-0_7

utilized rudimentary breath-hold techniques (Lahtinen et al., 2015). Examples of primitive diving goggles crafted out of wood and tortoise shells used by free-diving fishermen have been discovered on the Solomon Islands (Lahtinen et al., 2015). Elsewhere in the South Pacific Islands, diving without the use of respiratory equipment dates back centuries and has been documented as a universal practice to gather mollusks and other marine resources (Van Pel, 1961). Breath-hold diving has also been noted as a technique utilized by the Ama divers of Japan along with the Bajau divers of Indonesia to harvest marine resources for over 1,500 years (Dong et al., 2021). People all around the globe have been intrinsically linked with the ocean for millennia, and the modern scuba diving tourism industry is likely to continue to grow as technologies continue to improve. Each new advancement in the sustainable dive tourism industry offers the potential to provide opportunities for transformative experiences.

Until relatively recently, technology for underwater exploration and aquatic resource harvesting has remained unchanged from pre-modern times. Leonardo da Vinci, a Renaissance-era inventor and scientist, is credited with creating an early description of an underwater breathing device, while the English scientist Edmund Halley's diving bell represents the first patented example in 1691 (Chaline, 2017). Halley's diving bell supplied air from the surface and allowed users to record dives of 65 ft (20 m) lasting up to 90 minutes. Although it is far removed from the technologies of modern scuba diving, the primitive diving bell laid the groundwork for future technological advancement of recreational underwater exploration. In 1879, the first self-contained closed-circuit rebreather was introduced by British engineer Henry Fleuss and allowed divers to submerge in shallow water environments without a tether to the surface (Chaline, 2017). Based upon previous technological developments, early stages of recreational modern dive tourism as it is known today were pioneered primarily around coastal regions of the Mediterranean and other areas of Europe during the mid-twentieth century (Dimmock & Cummins, 2013). In the early 1940s, underwater exploration pioneers Jacques Cousteau and Émile Gagnan revolutionized the recreational diving industry with their invention of the Aqua-Lung, the first commercially successful self-contained open-circuit demand system (Matsen, 2009). This aquatic breathing apparatus facilitated the creation of underwater documentary films that helped catapult the wonders of recreational diving into the global mainstream consciousness (Cousteau & Dumas, 1953; Musa & Dimmock, 2013; Cousteau & Paisner, 2010). Further interest was generated in the 1950s with the release of *Skin Diver*, the first leisure magazine publication focused on recreational diving tourism (Musa & Dimmock, 2013).

Popular culture media that was released during the early days of modern recreational diving introduced the world to never-before-seen underwater landscapes and inspired interest in recreational diving across the globe. Films such as *A Silent World* and *World Without Sun*, released in 1956 and 1964, respectively, received critical acclaim for their introduction of inspiring underwater ecosystems to a global audience but have later received criticism for their irresponsible and unsustainable interaction with wildlife and the natural marine environment (Crylen, 2018). Cousteau's more recently produced films show improved interaction with the natural environment; however, his early documentaries which first captivated the global audience contain scenes that depict the divers engaging in irresponsible behaviors. This includes unfortunate examples such as "... the crew dynamiting [a] coral reef to collect dead specimens … and 'revenge' killing of a few sharks for feasting on a baby whale that the ship's propeller chopped up," (Crylen, 2018; Chris, 2006). Jacques Cousteau played a critical role in establishing global environmental awareness, and to date his legacy has had a lasting impact on the state of modern recreational diving.

Irresponsible interaction with the marine environment displayed in his early underwater films, however, foreshadow some of the sustainability concerns associated with the modern recreational dive industry.

In order for a recreational activity to be considered sustainable, its impact on all aspects of sustainability must be accounted for. Sustainable Development Goals (SDGs) must meet the needs of modern times without compromising the outlook of future generations. SDGs must also include measures that account for the longevity of the environment, economy, and social welfare (Brundtland, 1987). The interconectivity of the environemt and developmental challenges is key according to the Brundtland Commission, author of the highly influential report titled Our Common Future. In their words, "These are not separate crises: and environmental crisis, a development crisis, an energy crisis. They are all one" (Brundtland, 1987). It is implied that society, environment, and economy make up the three pillars of sustainability and each pillar is intrinsically connected to one another. This approach to breaking down sustainability into three pillars including social welfare, economics, and environmental health is known as the triple bottom line (TBL) model (Arowoshegbe et al., 2016). The interrelatedness of the three pillars that make up the TBL approach is visualized with Fig. 7.1.

Ensuring the long-term sustainability of the environment and natural biodiversity leads to economic and social benefits through various services that are provided by healthy ecosystems (Haines-Young & Potschin, 2010). However, for sustainable development to be effective, the goals must meet the current and future needs of the environment along with the socio-economics of host communities while simultaneously offering tourists and destinations an approach for transformative growth (Robert et al., 2005). This line of thinking is useful when considering the SDGs of recreation and tourism experiences. SDGs for marine recreation experiences such as scuba tourism should manage for the responsible use of the ocean and marine resources in addition to promoting opportunities for ethical consumption that provide benefit to the social welfare of locals and economic outlook of the host communities. To ensure long-term effectiveness for both the tourists and host communities, post-pandemic times are placing emphasis on a transformative stance for SDGs. Relevant transformative SDGs that focus on environmental sustainability may include examples such as advocating for policies around sustainable management and responsible use of marine resources in and around local dive sites, facilitating protection and restoration of vulnerable habitats to limit loss of biodiversity and promoting ethical consumption among tourists by eliminating greenwashing (Dwyer, 2022). Along with the environmental pillar of sustainability, marine

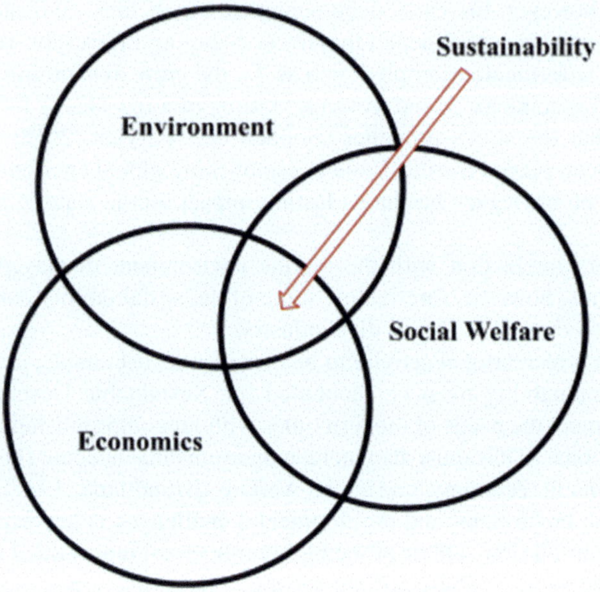

Fig. 7.1 Triple bottom line approach toward sustainability

tourism SDGs should also take into consideration social equity and economic viability of the operation and include goals that strive to reduce employment inequalities and promote regional economic growth by employing local residents (Buhalis et al., 2023).

This chapter presents an extensive literature review to scrutinize the sustainability of marine diving tourism and the extent to which it can create transformative experiences that may be beneficial for both the diver tourists and the host communities. It examines this based on the three pillars of sustainability including environmental, economic, and social outlook of dive tourism. Furthermore, viability of SDGs and the effectiveness of management goals are scrutinized based on a transformative paradigm.

The following three key research questions were used as starting points to initiate inquiry: What is the current sustainability outlook of marine tourism, and what are the key sustainability concerns regarding the future outlook of the industry? In what ways does marine dive tourism contribute to global marine conservation efforts? How does scuba tourism facilitate transformative experiences to ensure long-term benefit for tourists and host communities?

Methodology

A literature review was conducted to answer these key research questions and highlight some of the positive aspects of marine tourism as it relates to environmental sustainability. Additionally, this study will identify areas of sustainability concern that may be improved upon. Identifying the existing areas of weakness in regard to the sustainability of marine tourism will be useful when establishing SDGs that manage for the future of the industry. Gaps in the extant available research will be also discussed along with limitations in regard to the overall sustainability of dive tourism..

Search engines used to identify relevant literature include Google Scholar and Web of Science databases. An additional method used to gather relevant information involves an auto-ethnography approach to research. This approach includes personal observations made both as a participant of recreational diving and as a marine tourism professional employed on the Hawaiian Island of Maui. To identify and omit personal biases around the proposed Sustainable Development Goals for the future of sustainable marine tourism, a self-reflexivity method was also utilized. All images included within the chapter were personally captured throughout Maui County, HI, using various digital cameras including the Canon EOS R, Olympus TG-6, and iPhone 13.

Literature Review

The literature review section of this chapter first defines sustainable tourism and how the niche category of sustainable tourism known as ecotourism differs from other forms of less sustainable nature tourism. As recreational diving takes place entirely within potentially vulnerable underwater ecosystems, it is crucial to align with ecotourism sustainability principles. This is followed by the identification of key sustainability indicators and a suggested transformative paradigm that may be applied to marine tourism to ensure long-term success of sustainability measures.

The United Nations World Tourism Organization (UNWTO) describes sustainable tourism as a niche form of tourism that takes into account current and future economic, social, and environmental impacts while meeting the needs of visitors and host communities (UNWTO, 2023). Effective sustainable tourism presents host destinations with a valuable economic driver that has the potential to contribute significantly to the local social welfare and conservation efforts (WTO, 2005). Often mistakenly interchanged with ecotourism, nature tourism is a niche form of tourism that brings tourists into the natural environment but doesn't always exhibit the most sustainable practices.

Nature tourism does offer tourists the potential for transformative experiences through encounters with charismatic wildlife and inspiring scenery; however, the environmental impact of the nature tourism operations often does not align with sustainability standards. Alternatively, ecotourism represents a smaller niche subcategory of sustainable tourism that involves experiences within the natural environment that positively contribute to the sustainability of the host destination (Fennell, 2001). Simply put, ecotourism is sustainable nature tourism. Figure 7.2 visualizes how ecotourism lies in the intersection between sustainable tourism and nature tourism.

Fig. 7.2 The ecotourism position

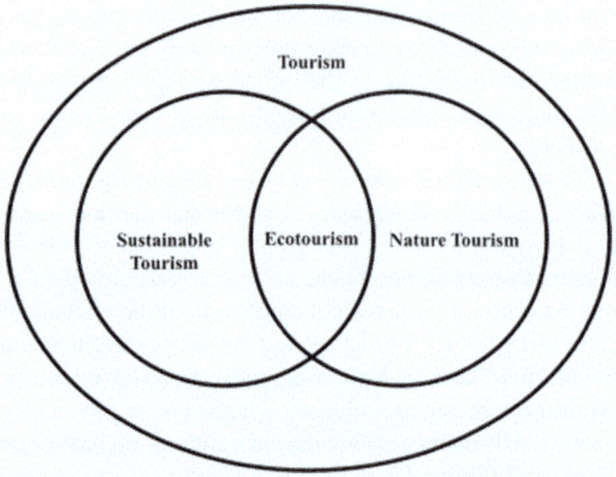

Originating in the 1980s, the concept of ecotourism was created as a way to generate revenue and awareness for conservation and sustainable development (Stronza et al., 2019). A definition for ecotourism is offered by The International Ecotourism Society (TIES) as, "responsible travel to natural areas that conserves the environment, sustains the well-being of the local people, and involves interpretation and education" (TIES, 2018). Ecotourism is intended to increase the public awareness of Sustainable Development Goals and provide the creation of a positive feedback loop between conservation and tourism (Stronza et al., 2019). A key differentiating factor that elevates ecotourism as the more ethical option over other forms of non-sustainable nature tourism is the tour operators' adherence to the principles outlined by Leave No Trace (Russell & Wallace, 2004). Based on the definition offered by TIES (2018), when held to strict sustainability standards, dive tourism, along with many other forms of marine tourism, can qualify as a form of ecotourism. All dive tourism is intrinsically a form of nature tourism as it offers tourists the opportunity to access nature and experience the natural environment in ways that differ from day-to-day life. However, only select operations that adhere to the principles of Leave No Trace while actively pursuing environmental conservation and education efforts may qualify as sustainable ecotourism operators (TIES, 2018; Russell & Wallace, 2004).

Leave No Trace (LNT) is an educational program developed with the goal of improving the public's understanding about how to recreate in nature in a responsible manner so as to minimize impact to natural areas and help ensure positive experiences for future visitors (Marion & Reid, 2001). A simplified explanation for the principles of LNT can be boiled down to the idea that when recreating in nature, responsible tourists should take nothing from the environment and leave no sign that they were ever present. It has been postulated that divers and dive tourism operators who successfully follow the principles outlined by LNT will not only have a positive impact on environmental sustainability but will also reduce amounts of visible litter at a site, which helps to sustain the continued satisfaction of future visitors

(Taff et al., 2014). This indicates that using a transformative approach to outdoor education can facilitate positive change toward more sustainable behavior. The extent to which a marine tour operator abides by the principles of LNT can therefore be used as an indicator when evaluating the effectiveness of their sustainability efforts.

Findings

The following section offers insights on the current state of sustainability efforts and how a transformative model can be integrated into the sustainable paradigm of recreational dive tourism. It first discusses different measures that have been taken to promote socio-economic and environmental sustainability. As the scuba tourism industry is predominantly aligned with the tenets of ecotourism, an emphasis is placed on the environmental aspects of sustainability. This is followed by an overview of challenges associated with marine sustainability initiatives. Next, the transformative notion of dive tourism is discussed as a tool that may be utilized to mitigate such challenges. It may accomplish this through transformative behavior of individual tourists as well as positive transformation for host communities. Key transformative models are included which may augment the success of long-term sustainability initiatives.

Socio-Economic Sustainability

In addition to fulfilling principles of responsible outdoor recreation behavior outlined by LNT, sustainable dive operations must also generate sufficient revenue that can be used to support conservation initiatives and promote equitable economic growth of local communities. Marine tourism as a whole constitutes a large sector of the global marine economic output, providing upward of 124,000 jobs and accommodating 8.9–13.6 million marine diving tourists annually (Schuhbauer et al., 2023). The large scale of the global marine dive tourism industry has generated significant revenue including roughly $18 million USD per year in Palau, which equates to roughly 8% of the country's GDP (Vianna et al., 2012). Whale watching, an adjacent marine tourism industry, has been estimated to generate around $3.4 billion USD globally on an annual basis and provides additional revenue for ocean conservation efforts of coastal communities (O'Connor et al., 2009). On a global scale, dive tourism generates upward of $0.9–3.2 billion USD per year, with a broader indirect economic impact of $8.5–20.4 billion USD when accounting for associated travel costs including food and accommodation (Schuhbauer et al., 2023). On an individual scale, dive tourists spend roughly $100–200 each day of their dive trip on average, with a large percentage of that money going directly toward dive operators and other local businesses (Saayman & Saayman, 2018).

Along with the jobs that are created by dive tourism, revenue that is generated can directly support conservation efforts through a variety of means including funding environmental education programs and supporting marine scientific research efforts.

Environmental Sustainability

Scientific Research

This section offers insights on how dive tourism can contribute to environmental sustainability through scientific research and environmental education programs such as interpretation. Extant literature posits that dive tourism is capable of contributing to the environmental sustainability of its host destination using the revenue generated to financially support marine conservation efforts. The recreational diving industry contributes to marine conservation through projects such as coral reef monitoring and ecosystem restoration programs (Hawkins & Roberts, 1992; Stolk et al., 2007). In some cases, artificial reefs have been utilized as recreational dive sites that double as underwater laboratories for scientific research and provide habitat for threatened coral species (Shani et al., 2012; Stolk et al., 2007). As a result of more divers traveling to destinations with dive sites featuring artificial reefs and coral farms, there is an opportunity to generate funding that can be used to drive further conservation efforts. The higher rate of divers attracted to artificial reef destinations may also lead to reduced overcrowding in other popular reef systems facing the threat of extinction (Jakšić et al., 2013).

Education

On top of physical conservation efforts such as the aforementioned artificial reefs, the recreational dive tourism industry contributes to marine sustainability through environmental education. Education is inherently built into the framework of modern recreational scuba diving through a series of mandatory online and in-person courses required for divers to complete before receiving a certification. A growing number of international organizations have become established sources of scuba education and skill development through structured certification processes (Johansen, 2013). Certifications from organizations like PADI (*Professional Association of Diving Instructors*) and SSI (*Scuba Schools International*), among others, provide dive tourists with information about how to behave responsibly in the marine environment. In addition to covering topics related to LNT, certification courses offer divers education about marine sciences and underwater biodiversity, along with safety and rescue training (PADI, 2022).

Environmental Interpretation

Environmental interpretation represents another aspect of the transformative educational opportunities available to dive tourism operators. Interpretation, which involves informing guests about the surrounding environment, is useful in creating transformative and sustainable tourism experiences (Moscardo, 1998; Apps et al., 2017). Interpretation offers tourists a deeper understanding of their surroundings which may also help them to feel a sense of personal connection with the environment and a stronger appreciation for the intrinsic value of the site (Townsend, 2007; Kuo, 2002). A study based on self-reported tourist data from visitors to the Great Barrier Reef suggests that reef-based interpretation programs can lead to significant changes in the visitors' understanding of relevant conservation-related topics (Madin & Fenton, 2004). Based on the findings of the 2004 Madin and Fenton study, it can be implied that interpretation programs should be considered when establishing SDGs for the management of dive sites within vulnerable ecosystems. In doing so, divers who receive interpretive information about the dive site and their own role within the ecosystem can achieve a more fulfilling and transformative experience. To accommodate this, divers should be refreshed by their guides on the standard operating procedures of a typical dive along with information about local biota and how to responsibly interact with it prior to entering the water. Post-dive debriefs are also useful to discuss wildlife that was encountered and reflect on the experience as a whole. Given the substantial level of personal engagement with tourists and thus substantial opportunities to provide transformative educational experiences, the earnings of local dive guides could be regarded as a direct factor when considering how the revenue generated from dive tourism can contribute to the educational dimension of sustainability.

Citizen Science

Alongside the financial backing that can be used to support marine conservation initiatives and environmental education, sustainable dive tourism may also contribute to environmental sustainability efforts by fostering advancements in scientific research. Scuba diving tourism significantly increases the availability for engaging citizen science (CS) opportunities that immerse volunteer tourists into new and exciting environments and provide them with a deeper understanding of marine sciences (Lucrezi et al., 2018). CS is understood to be a method of democratizing the scientific discovery process by transforming research toward a more inviting system in which the general public can partake (Heigl et al., 2019). One notable example of a positive CS project involves nonspecialist volunteer divers collecting data to study the status of the biodiversity within the Red Sea coral reef ecosystem (Branchini et al., 2015). Authors of the 2015 Branchini et al. study suggest that nonspecialized volunteer divers perform similarly to professional scientific divers in the accuracy of data collection. This implies that future CS studies could be utilized as cost-effective means of underwater scientific data collection.

Dive Against Debris® is another example of a CS project that involves volunteer divers collecting critical marine survey data to characterize the litter present on the seafloor of shallow coastal waters of the Mediterranean (Consoli et al., 2020). As the authors of the 2020 study discuss, there exists a fair amount of litter in critical marine habitats such as reef ecosystems. Much of the litter present may be directly attributed to marine tourism; however, it must also be considered that tourism-based CS projects such as Dive Against Debris® present more frequent opportunities for recreational divers to clean up the sites.

Sustainability Concerns

The following section will identify several areas of concern in regard to the sustainability of the marine dive tourism industry. This includes discussing the debris and pollution associated with tourism, potential environmental hazards associated with climate change, and issues related to overcrowding in certain destinations due to the phenomenon known as Last Chance Tourism. In addition to discussing environmental threats caused by over-tourism within host destinations, the section will also discuss how overcrowding impacts visitor's perception and overall satisfaction with their experience.

Debris and Pollution

Though sustainable dive tourism does have the potential to contribute positively to global marine conservation efforts, there are also several associated sustainability concerns. Debris and pollution in vulnerable marine ecosystems such as marine protected areas (MPAs) represent one area of concern associated with marine tourism. Increased boat traffic may be related to higher rates of pollution in vulnerable environments such as reef systems. The 2020 Dive Against Debris® CS study finds that of the marine debris that was collected, plastic was the most abundant, with single-use plastics accounting for roughly 33% of the total amount of marine litter collected by divers (Consoli et al., 2020). Similarly, da Silva et al. (2018) also suggest that plastic is the most commonly collected marine debris, with cigarette butts found to be the second most abundant source of litter. Marine debris and pollutants that result from tourists visiting the area present several threats to the ecosystems of dive destinations. Entanglement of wildlife such as whales and turtles in abandoned fishing gear or other discarded lines and nets poses risks to the biodiversity present in dive destinations (Angiolillo & Fortibuoni, 2020). Additionally, chemical pollutants from non-mineral-based sunscreens and lotions threaten the long-term sustainability of dive destinations that feature coral reef ecosystems (Schneider & Lim, 2019; Levine, 2020). Oxybenzone, a common chemical ingredient found in non-mineral-based sunscreens, has been found to be a possible contributor to coral bleaching and threatens the resiliency of coral reefs to climate change (Downs et al., 2016; Schneider & Lim, 2019; DiNardo & Downs, 2018).

Environmental Hazards

Hazards associated with global climate change, including rising marine temperatures and ocean acidification, threaten the long-term sustainability outlook of numerous tourism destinations, especially those featuring coral reefs (Anthony, 2016). When healthy, coral reefs represent one of the most biologically diverse ecosystems on the planet, yet due to strict habitat requirements, such as their need for shallow, clear, and warm water with access to sunlight, the global distribution of viable coral habitat is limited to available area that is smaller than the land area of France (Knowlton et al., 2010). The high degree of biodiversity within a healthy coral reef system is capable of providing coastal communities with a variety of valuable ecosystem services including but not limited to economic benefit, food and medicine resources, recreation opportunities, and protection against coastal erosion (Moberg & Folke, 1999; Woodhead et al., 2019). Unfortunately, mass coral bleaching events, which involve large-scale stress responses as a result of changes in light, temperature, and pH balance, have been growing in frequency over the past several decades due to rising sea levels, ocean acidification, and elevating temperatures (Hoegh-Guldberg, 2011). Healthy coral reef systems are highly productive and thus crucial to the long-term sustainability of marine tourism; however, impacts from climate change paired with continually increasing human activity along coast lines have threatened the robustness of these systems (Hoegh-Guldberg, 2011).

Climate Change and LCT

Effects of climate change on coral reef ecosystems are amplified by anthropogenic threats related to marine tourism. In recent years, popular coral reef dive destinations including the Great Barrier Reef (GBR) have seen a rise in tourism as a result of the phenomenon known as Last Chance Tourism (LCT) (Piggott-McKellar & McNamara, 2017). LCT is an emerging form of niche tourism wherein tourists opt to travel to destinations showcasing attractions in danger of disappearing, thereby allowing them to witness these natural wonders before they potentially vanish (Lemelin et al., 2010). Destinations that have seen a rise in LCT often feature endangered wildlife such as sea turtles (pictured in Image 7.1), or ecosystems on the verge of near-term disappearance such as old-growth forests, glaciers, and coral reefs. Piggott-McKellar and McNamara (2016) note that although there has been a significant increase in the number of tourists to the GBR as a result of LCT, on average the visitors were found to be more environmentally friendly and genuinely concerned about the health of the ecosystem. This trend toward LCT producing more environmentally conscious tourists could be seen as a positive in terms of tourism's ability to create beneficial transformation in the mindset of tourists. Despite the high

Image 7.1 Endangered green sea turtle (*Chelonia mydas*); Maui, HI. (Source: Cohen, 2023 (personal collection))

level of environmental awareness displayed by the visitors, the rise in mass tourism throughout vulnerable ecosystems presents challenges to sustainability associated with overcrowding and greenhouse gas emissions from transportation.Sustainability concerns associated with climate change and LCT create a positive feedback loop that poses the risk of accelerating the degradation of local habitats and biodiversity (Dawson et al., 2010).

Overcrowding

When destinations do not sufficiently manage for sustainable growth in tourism, the rise of visitors attracted to the area can lead to overcrowding or excessive activity that surpasses the area's carrying capacity (Mihalic, 2020). Overcrowding as a result of tourism has been associated with habitat degradation, higher rates of stress on local infrastructure, as well as increased litter and chemical pollutants entering the environment (Chen & Teng, 2016).

Additionally, a growing trend in irresponsible marine tourism threatens coral reefs and other ecosystems with the introduction of invasive species that are capable of outcompeting native species and altering the entire ecosystem (Keith et al., 2016; Pallewatta, 2004). As discussed, a large percentage of dive tourism exists in and around highly productive coral reef ecosystems which often provide habitat for threatened and rare species such as the endemic Hawaiian dascyllus (pictured in Image 7.2). As such, the introduction of invasives into these ecosystems could result

Image 7.2 Endemic Hawaiian dascyllus (*Dascyllus albisella*); Maui, HI. (Source: Cohen, 2023 (personal collection))

in severe biological disturbances to the communities and even loss of biodiversity. Increased marine traffic from tourist excursions including whale watching and dive boat charters raises the potential for boat collisions with wildlife and other marine resources (Denkinger et al., 2013). Vessel traffic has also been noted to influence the surface interval times of marine megafauna along with other behaviors such as foraging, resting, traveling, and mating (Machernis et al., 2018). Furthermore, higher rates of marine traffic produce loud acoustic pollution in the marine environment, which can interfere with cetacean communication (Jensen et al., 2009). As shown in Image 7.3, although the presence of anthropogenic activity within the marine environment can influence the behavior of wildlife, not all interactions are inherently negative experiences for the animals. It will be necessary, however, to further examine the sustainability of human-wildlife coexistence as the existing literature lacks comprehensive coverage on the complex relationship between overcrowding and associated anthropogenic impacts to wildlife species. Future research may include long-term case studies that examine the various impacts of overcrowding in popular marine tourism destinations and can be used to inform future Sustainable Development Goals.

Image 7.3 Humpback whale (*Megaptera novaeangliae*) spyhopping behavior near boat; Maui, HI. (Source: Cohen, 2022 (personal collection))

Visitor Perception

Not only does overcrowding threaten the environmental sustainability of marine resources; it may also affect tourists' satisfaction with their experience and perception of the destination's overall quality (Yin et al., 2020; Perera and Kotagama 2016). Papadopoulou et al. (2023) suggest that visitors who perceive high levels of overcrowding in coastal Mediterranean destinations receive low amounts of satisfaction and show reduced intention to return to the destination. Beyond the long-term economic implications associated with the decline of destination loyalty, reduced consumer satisfaction with a tourism experience can diminish the transformative potential of the experience. Dissatisfaction with the travel experience may limit the likelihood that the tourist will achieve a beneficial transformation of mindset. In addition to future research looking into the environmental impact of overcrowding in natural areas, there is room for inquiry regarding the potential of recreational diving and other niche forms of ecotourism to have a transformative role in the mindset of tourists.

Transformative Tourism

Long-term sustainability of marine tourism is affected by the tour operators' ability to provide transformative experiences. Transformative tourism may include encounters with novel biota, cultures, and/or environments that facilitate growth for individual tourists as well as host communities (Pung & Del Chiappa, 2020). According to Pung and Del Chiappa (2020), transformative learning is often utilized by sustainable tourism operations as a method of providing tourists with education regarding relevant conservation efforts and promoting cultural awareness. Effective transformative learning can ultimately lead to a shift in perspective and may positively influence the tourists' behavior upon returning home from their trip (Coghlan & Weiler, 2018). Scuba certification programs provide tourists with transformational education which help the divers acquire knowledge of marine sciences and conservation awareness. The required certification programs also facilitate the development of underwater skills including how to safely and responsibly interact with the marine environment. Additionally, pre-dive educational discussions held by dive guides allow visitors to personally engage with locals and discuss local species and standard operating procedures for a successful trip. The transformative nature of dive tourism provides authentic personal interactions between the tourists and the dive operators which can lead to lasting behavioral change upon return home.

Transformative tourism experiences hold potential to offer opportunities that foster personal growth through self-exploration and realization of the authentic self (Brown, 2013). Transformational growth in tourism is often associated with "peak experiences," which involve, "rare, emotionally charged moment[s] of extreme happiness and fulfillment that extends beyond one's personal harmony with the universe," (Maslow, 1971; Kirillova et al., 2017). Peak experiences are common during dive tourism, as divers are likely to encounter charismatic megafauna such as sharks and whales, along with other unfamiliar yet inspiring wildlife such as nudibranchs (pictured in Image 7.4). Encounters with these creatures may provide divers with profound realizations about themselves and their intrinsic relationship within the diverse web of life.

Complete immersion into underwater landscapes that vastly differ from daily routines puts divers in situations that encourage them to confront their existential authenticity and experience life in a manner that is true to one's personal values (Kirillova et al., 2017). Diving requires tourists to be self-reliant in uncomfortable situations while encouraging mindful observation of the surrounding environment. While actively engaging with their surroundings, divers are also required to continually monitor their body position and oxygen intake, essentially serving as a form

Image 7.4 Endemic gold-lace nudibranch (*Halgerda terramtuentis*); Lanai, HI. (Source: Cohen, 2023 (personal collection))

of underwater meditation and a source of mindfulness. Further lending to the transformative potential of diving tourism is the inherent digital detoxification required while underwater. As noted by Li et al. (2018), digital detoxification experiences that limit access to social media and other modern digital technologies provide tourists with the opportunity for inner personal transformation through, "...richer sensory, emotional, cognitive, and relationship experiences." Recreational scuba diving immerses tourists into uncomfortable situations that encourage exploration of the inner self in much the same way that it encourages exploration of the marine environment.

Immersive and mentally challenging travel experiences like those offered by recreational diving have the potential to lead to meaningful inner transformation and self-realization of existential authenticity (Tiberghien et al., 2020). Tourism experiences may facilitate a confrontation with the tourists' existential authenticity, through novelty of experiencing environments, cultures, and experiences that differ from day-to-day life. When a person realizes their existential authenticity, they act in ways that are true to one's own values and may find meaningful transformation (Kirillova et al., 2017) (Image 7.5).

YEP Framework of Transformative Travel

The YEP Framework of Transformative Travel as presented by Cavender et al. (2020) is a step-by-step guide of fulfilling one's existential authentic self through transformative experiences by directing energy toward oneself and actively

Image 7.5 Divemaster ascending to boat; Maui, HI. (Source: Cohen, 2023 (personal collection))

engaging with their surroundings. The YEP Framework outlines three aspects of successful transformative travel using the acronym of (Y)ourself, (E)xperiences, and (P)eople/places. Tourists initiate the first step of the YEP Framework by allowing themselves to be open-minded to new perspectives and intentionally immerse themselves in novel and challenging situations. According to Cavender et al. (2020), tourists who accept the challenges presented by the new experiences may find personal growth. Overcoming obstacles to complete a transformative journey often leads to the tourists' confrontation of their existentially authentic selves. Reflecting on the impact of a transformative experience upon returning home helps to integrate the knowledge gained into daily life and promotes lasting behavioral change (Cavender et al., 2020). After ascending from a dive, divers are encouraged to log their experiences and reflect upon what they witnessed, including the species they observed along with any other noteworthy encounters that occurred during the dive. Once filled out, the diver's logbook can be used as a tool to continue personal growth upon returning from a trip by allowing tourists to reflect upon transformative experiences that occurred during previous dives. Sustainable dive tourism may provide transformative experiences by allowing tourists to discover their authentic selves through immersive and disorienting experiences which facilitates self-fulfillment and personal growth.

In addition to providing tourists with frequent opportunities for emotionally charged peak experiences leading to personal growth, recreational dive tourism can positively contribute to the mental and physical health of participants. According to Carreno et al. (2020), physical activity within marine environments can reduce stress, promote positive social relationships, and improve mental well-being through improved feelings of connectedness with the natural environment. Digital detoxification required in the underwater environment can be attributed to further benefits to mental well-being (Li et al., 2018). In addition to the potential for positive mental transformation, dive tourism also offers divers opportunities for transformation via physical therapy. Henrykowska et al. (2021) highlight potential therapeutic benefits of recreational scuba diving for people with physical disabilities. Participants of the study indicate that recreational diving may promote an elevated sense of self-confidence, improved ability to engage in social interactions, as well as improvements to the respiratory systems, reduction of pain, and strengthened motor skills. Other studies have shown physical transformations such as spleen enlargement and increased lung volume capacities as a result of breath-hold diving (Dong et al., 2021). Currently, organizations such as *Disabled Divers International* and the *International Association for Handicapped Divers* exist to promote diving as a therapeutic tool for people with disabilities, and further developments could present novel lines of academic study. Future research may also focus on the degree to which recreational diving contributes to personal transformation, both mentally and physically (Image 7.6).

Discussion

The discussion section utilizes a case study to highlight the impact of a natural disaster on the island of Maui, which relies heavily on marine tourism. The event caused devastating impacts to the local residents and tourism operators, leading to a sharp decline in local business largely due to a halt in tourism. As a first-person witness to the impact of the events, a self-reflexivity approach was used to limit the amount of personal biases when offering suggestions for future SDGs. A transformative approach to sustainability will be crucial for the successful reintroduction of local marine tourism that takes into account the immediate and long-term environmental and social sustainability needs of interested stakeholders. The YEP Framework will be discussed as a potential transformative paradigm that

Image 7.6 Divers exploring Lanai Cathedrals; Lanai, HI. (Source: Cohen, 2023 (personal collection))

sustainable marine tourism operators may look to model their operations after as local tourism continues to be reestablished over the coming years.

Case Study

On August 8, 2023, the Hawaiian Island of Maui experienced a catastrophic wildfire that reduced the historic coastal town of Lahaina to ashes overnight. Dating back to the nineteenth century, when whaling ships frequented the harbor, Lahaina town has long served as an international hub of marine tourism on Maui (Blackford, 2001). As a result, the island's economy has been largely reliant on the revenue generated by Lahaina's tourism industry. Per the Hawaii Tourism Authority (HTA), tourism is the largest single source of private capital for Hawaii's economy, with visitor spending equating to roughly $14 million per day on the island of Maui prior to the 2020 pandemic (HTA, 2019). Key contributors to the tourism revenue included marine tourism enterprises operating out of Lahaina Harbor (pictured in Image 7.7), including *Lahaina Divers* and *PacWhale Eco-Adventures*, among many other businesses that were impacted by the fire. Following the natural disaster and subsequent loss of countless locally owned and operated businesses, the island plunged into a cycle of disturbance that mirrors the impacts witnessed in the tourism industry during the COVID-19 pandemic. In August 2023, arrivals to the island slowed down to a staggering 60% fewer visitors compared to the same time the previous year, which

Image 7.7 Sunrise over Lahaina Harbor; Maui, HI. (Source: Cohen, 2022 (personal collection))

translated to a significant decrease of approximately 49% in visitor spending (Tanji, 2023). Many local residents who were displaced by the fire also lost their jobs, as evident by the nearly 8,000 people that filed for unemployment on Maui in the 3 weeks following the disaster (McAvoy & Sinco Kelleher, 2023). On top of the numerous social aspects impacted by the disturbance, the health of the reefs and other ecosystems surrounding the burn zone may be at risk of long-term damage (Jacobo, 2023).

Roughly two months after the fire, the island officially reopened to tourism, which evoked a range of polarizing opinions among local residents and other stakeholders. With thousands of displaced residents still living in temporary housing within the island's numerous hotels, the idea of tourists reentering the area did not sit well among many locals who called for the governor to push back the reopening date (McGough, 2023). Locally owned restaurants and tourism businesses reliant on the revenue generated by tourists, however, largely argued for the reentry of tourists as many would be unable to continue operations much longer without the reintroduction of visitors. Diminished revenue generated by tourism impacted not only the bottom line of the owners of local businesses but also the numerous employees who lost their jobs as a result of the fire. With an already limited job market on the island, the destruction of Lahaina has made finding gainful employment much more challenging, resulting in many residents forced to relocate away from Maui.

With such damage caused by the wildfire, the complete rebuild of the town along with the total reintroduction of tourism to levels that existed pre-disaster will take many years. When considering how and when to rebuild the town in a sustainable manner that aligns with the interests of invested stakeholders, there are many tourism-related sustainable development goals that must be reestablished. The following suggested management goals are based on first-hand observations and interactions with stakeholders directly impacted by the natural disaster.

Management goals for the current and future outlook of the marine tourism industry on the island of Maui should consider the sustained health of the ocean and the various other natural resources around the island to be a priority. Marine resources such as charismatic megafauna including humpback whales, green sea turtles, and biologically diverse coral reef systems are the primary drivers of marine tourism around the island; and without ensuring their longevity, the tourism industry is unlikely to see sustained success. As discussed previously, repeat patronage of a tourism destination is impacted by the consumer's perception of the overall cleanliness and the amount of visible pollution within the environment. As such, it would be wise for marine tour operators around Maui to invest in policies that protect the health of their local communities and surrounding ecosystems. This may involve local dive shops hosting public education programs, community clean-up days, and citizen science projects such as data collection for coral reef monitoring.

Another method for sustainable marine ecotourism operations on the island to invest in conservation efforts is to prioritize providing transformative educational programs during their tours that discuss not only the environment but also Hawaiian culture. Increased exposure to the rich cultural history of the Hawaiian Islands improves the authenticity of the tourism experience (Agrusa et al., 2010). For dive tourism experiences around the islands, this exposure could come in the form of interpretation involving discussions around the cultural significance of various marine resources commonly encountered during dives such as sea urchins or reef sharks. Providing tourists with education about the natural and cultural resources they are interacting with may help to improve the tourists' awareness about the sensitive history and create a more authentic experience for all interested stakeholders (Agrusa et al., 2010).

Authentic experience will not only allow the tourists more opportunities to achieve positive transformative growth; they may also help to influence guest perception of the destination and improve the understanding between tourists and local residents who were impacted by the recent natural disaster (Agrusa et al., 2010; Andrade et al., 2021; Croes et al., 2013). To ensure that tourists are provided authentic experiences during sustainable marine tourism on the island, it is recommended that marine tourism operations utilize some form of transformational paradigm, such as the YEP Framework that aligns their company's SDGs with the current and future needs of the island.

These SDGs may include goals to promote sustained protection of the marine environment and strict adherence to conservation policies within local marine protected areas. SDGs to facilitate transformative change within the community should also be established such as the goal to host regularly scheduled ocean clean-up days and the goal to supply local residents with equitable employment that does not discriminate age, race, or gender. Additional transformative SDGs should be established that ensure the tour operation is providing authentic experiences that facilitate transformational changes that benefit all stakeholders including the tourists as well as local community members.

As presented previously in the chapter, the YEP Framework provides ecotourism operators an outline to follow to help ensure authenticity is achieved in a sustainable manner (Cavender et al., 2020). In the context of a dive shop on Maui, the YEP Framework may involve encouraging tourists to engage in immersive experiences that are supplemented with opportunities for citizen science projects and educational programs about the local cultural and natural resources along with their intrinsic interconnectivity within the island's ecosystem. Especially beneficial transformative education programs will produce a positive shift in perspective and conservation awareness among tourists that lasts even upon returning home from travels (Cavender et al., 2020; Pung & Del Chiappa, 2020; Coghlan & Weiler, 2018). This lasting transformative shift in perspective and conservation awareness will aid in strengthening the positive feedback loop between tourism and conservation that is crucial to the success of ecotourism.

Equally important to the role of tourism operators in promoting awareness through education is the need to generate economic revenue that can be used to help stimulate the island's economy and rebuild the community of Lahaina. Moving forward, in order for tourism operations to be sustainable on Maui, it will be necessary to provide local economic benefit by employing residents and funding conservation projects. Future SDGs for marine tourism operations on Maui may also include goals to promote awareness of Hawaiian culture in order to increase understanding between tourists and locals, adhere to LNT guidelines to provide tourists with peak experiences in nature in a responsible manner, and provide substantial financial interest to marine conservation efforts and disaster relief. To ensure that these SDGs align with the socio-economic interests of local stakeholders on the island, an emphasis should be placed on ethical marketing of locally owned tourism businesses that prioritize community involvement through events such as coastal clean-up events (Ndivo & Cantoni, 2016). Development of locally owned businesses rather than large-scale corporations will also help to reduce economic leakage and instead keep money circulating within the island's economy to aid the redevelopment of Lahaina. To reduce the impact of another halt in tourism triggered by a cycle of disturbance similar to the devastating wildfire, it is also recommended that the island diversify the economy away from such heavy reliance upon tourism. Other less volatile, yet potentially lucrative industries, such as agriculture, should be taken into consideration (Image 7.8).

Conclusion

After thorough examination of the research questions guiding this literature review, key insights emerge in regard to the sustainability of marine dive tourism and its capacity to facilitate transformational experiences. When managed for sustainability, the modern global dive tourism industry has the potential to serve as a vital

Image 7.8 Rainbow over Lahaina town at sunset; Maui, HI. (Source: Cohen, 2021 (personal collection))

economic catalyst that offers benefits to host communities in the form of employment, heightening cultural awareness, and financial support for marine conservation efforts (Schuhbauer et al., 2023; Agrusa et al., 2010). Strict adherence to the principles of ecotourism is key in elevating sustainable ecotourism over other less sustainable dive operations. This entails responsible interaction with marine resources, contribution of educational interpretation programs, and funding for other conservation efforts such as citizen science projects and habitat restoration. Although sustainability concerns associated with marine tourism such as overcrowding, climate change, and habitat destruction must be considered, the transformative nature of dive tourism provides opportunities to mitigate these challenges when ethically managed.

The transformative potential of dive tourism can have a large-scale impact and may contribute to the transformation of local communities. Through community-based tourism (CBT), dive operators can facilitate co-created spaces which may provide benefit to all interested stakeholders, including the tourists, community members, and the marine environment (Mayaka et al., 2018). Dive tourism promotes the economies of host communities by providing jobs and generating revenue for marine conservation. Additionally, scuba tourism operators can help divers meet like-minded individuals within their community by hosting community events including coastal clean-up days. August 6, 2023, was distinguished as the first-ever International Scuba Day and provided an opportunity for divers across the globe to participate in organized community underwater clean-up events (N.A., 2023). Moving forward, participating dive shops in various coastal destinations can use the day to raise public awareness about ocean conservation and help to positively transform local communities. As Greenberg (2023) writes for *Dana Point Times*, local dive shops around Dana Point Harbor in Orange County, California, are planning on continuing to use International Scuba Day as a platform to host community underwater clean-up events, which have the potential to facilitate transformation to the community as a whole.

There are a number of caveats that must be considered when evaluating marine tourism's overall impact on global sustainability. As previously discussed, it is associated with higher rates of pollution, habitat degradation, and other anthropogenic threats to the marine environment sustainability. Impacts related to the Tragedy of the Commons, such as unrestricted use of poorly managed ecosystems and irresponsible interaction with resources, pose threats to endangered marine habitats (De et al., 2020). Lack of regulation may result in overcrowding around popular dive sites, which can negatively impact the surrounding habitat.

Destinations that are economically reliant on marine tourism such as the Maldives, Honduras, and Maui are at risk from cycles of disturbances such as natural disasters or pandemics which can lead to slow tourism (Cochrane, 2010). Greenwashing, an unethical form of marketing in which companies market themselves as more eco-friendly than they actually are, poses challenges to the sustainability of the ethical consumption of ecotourism as a whole (Self et al., 2010). Certain companies may use ecotourism certifications to appeal to ethical consumers; however, a lack of universal standards and regulations surrounding sustainability certifications makes it confusing for consumers to know which choice is the most ethical (Jarvis et al., 2010).

The transformed consciousness provided by sustainable dive tourism experiences plays an important role in the mitigation of many of these limitations presented by unsustainable operations. Educational interpretation programs provided by dive tourism operators helps improve awareness of conservation issues and fosters a greater sense of responsibility in tourists when recreating in nature in the future (Townsend, 2007; Kuo, 2002). Individuals who encounter peak experiences during travel have the potential to increase their environmental awareness and become more conscious of their connectedness with nature, which may reduce future incidents related to the Tragedy of the Commons (Kirillova et al., 2017). Positive mental transformation among marine tourism consumers can also drive demand toward more clear and credible sustainability standards, making it easier for consumers to make informed choices and avoid falling victim to greenwashing. In addition to facilitating a transformation of individual self, the potential for scuba dive tourism to contribute to positive community transformation through coastal clean-up days helps to alleviate some of the pressures that are associated with overtourism in certain destinations (Pasternak et al., 2019; Greenberg, 2023). Generating higher demand for ethically run tourism operations over non-sustainable options may also lead to transformation of entire communities of invested stakeholders.

Looking Forward

Future areas of research that were revealed during this literature review include examining the extent to which increased marine tourism traffic impacts overall ecosystem health along with specific wildlife species within popular destinations. In order to gain a more comprehensive understanding of the complex relationship between over-tourism and associated anthropogenic impacts to the environment, future research may focus on human-wildlife conflict and coexistence in overcrowded destinations. Examples of anthropogenic impact studies that may be done may include how the decibel level (dB) or frequency (Hz) of the acoustics of specific tour boat motors impact wildlife acoustic communication and behavior within various marine environments. Other research may be dedicated to examining the efficacy of interpretation programs and underwater citizen science projects in destinations impacted by overcrowding or LCT. Ongoing longitudinal case studies will be useful to examine the level of anthropogenic impact caused by marine dive tourism to specific dive sites over extended periods of time. Areas of inquiry within those case studies may examine the health and biodiversity of coral reef or kelp forest ecosystems over a period of time in relation to the amount and type of human recreational use. Additionally, future studies will be useful in further examining the extent to which recreational diving may be used as a therapeutic tool for divers with disabilities, both mentally and physically. Along with studying the role diving plays on human physical health, future academia may focus on studying the extent to which sustainable recreational diving and other niche forms of ecotourism may benefit mental health and environmental awareness.

The long-term sustainability of marine dive tourism is dependent on the interconnectedness of environmental, economic, and social aspects. The future of sustainable dive tourism relies on tour operators who prioritize experiences that promote responsible interaction with marine ecosystems and operate within conservative guidelines in order to avoid succumbing to the Tragedy of the Commons and contributing to the degradation of the habitat. Beneficial regulations may include marine traffic speed limits, wildlife approach restrictions, capped tour group sizes, and the ban of harmful chemical sunscreen products within MPAs. To increase revenue created by marine tourism, policies could also be implemented that impose an environmental sustainability tax on all tourists who chose to enter marine protected areas during their travels. Sustainable marine tour operators can further directly contribute to conservation efforts by providing tourists with immersive educational experiences that facilitate lasting transformation and personal growth. SDGs should incorporate the objectives for dive tourism to foster the positive development of host communities through CBT initiatives including activities like ecosystem clean-up events. Destination marketing may also be a valuable tool in easing the effects of overcrowded destinations such as the GBR by promoting travel to less crowded regions instead. Moreover, the transformed consciousness provided by sustainable ecotourism may help to reduce the pitfalls of greenwashing and drive demand toward more ethical forms of ecotourism experiences. Continued integration of smart tourism technologies such as underwater drones and artificial intelligence technologies will also be useful in furthering developments in underwater research and marine conservation.

When managing for the improved sustainability of a dive tourism operation, it is important to consider the overall transformative potential of the operation. Management approaches should prioritize opportunities that facilitate transformation of the host communities by providing revenue to support local economic interests and conservation efforts. Furthermore, tourism operations are recommended to provide guests the potential for a transformative experience through environmental interpretation or introduction of peak experiences, which may improve environmental awareness and fulfillment of the authentic self.

Ultimately, sustainable marine ecotourism does have the potential to provide transformative experiences to both tourists and host communities. The transformed consciousness that results from sustainable experiences allows for a more mindful approach to marine ecotourism in alignment with the long-term health and well-being of the marine environment and coastal communities around the world.

References

Agrusa, W., Lema, J., Tanner, J., Host, T., & Agrusa, J. (2010). Integrating sustainability and Hawaiian culture into the tourism experience of the Hawaiian Islands. *PASOS. Revista de Turismo y Patrimonio Cultural, 8*(2), 247–264.

Andrade, G., Itoga, H., Linnes, C., Agrusa, J., & Lema, J. (2021). The economic sustainability of culture in Hawai'i: Tourists' willingness to pay for Hawaiian cultural experiences. *Journal of Risk and Financial Management, 14*(9), 420.

Angiolillo, M., & Fortibuoni, T. (2020). Impacts of marine litter on Mediterranean reef systems: From shallow to deep waters. *Frontiers in Marine Science, 7*, 581966.

Anthony, K. R. (2016). Coral reefs under climate change and ocean acidification: challenges and opportunities for management and policy. *Annual Review of Environment and Resources, 41*, 59–81.

Apps, K., Dimmock, K., Lloyd, D. J., & Huveneers, C. (2017). Is there a place for education and interpretation in shark-based tourism? *Tourism Recreation Research, 42*(3), 327–343.

Arowoshegbe, A. O., Emmanuel, U., & Gina, A. (2016). Sustainability and triple bottom line: An overview of two interrelated concepts. *Igbinedion University Journal of Accounting, 2*(16), 88–12.

Blackford, M. G. (2001). *Fragile paradise: The impact of tourism on Maui, 1959–2000*. Development of Western Resources.

Branchini, S., Pensa, F., Neri, P., Tonucci, B. M., Mattielli, L., Collavo, A., et al. (2015). Using a citizen science program to monitor coral reef biodiversity through space and time. *Biodiversity and Conservation, 24*, 319–336.

Brown, L. (2013). Tourism: A catalyst for existential authenticity. *Annals of Tourism Research, 40*, 176–190.

Brundtland, G. (1987). Our common future—Call for action. *Environmental Conservation, 14*(4), 291–294. https://doi.org/10.1017/S0376892900016805

Buhalis, D., Leung, X. Y., Fan, D., Darcy, S., Chen, G., Xu, F., et al. (2023). Tourism 2030 and the contribution to the sustainable development goals: The tourism review viewpoint. *Tourism Review, 78*(2), 293–313.

Carreño, A., Gascon, M., Vert, C., & Lloret, J. (2020). The beneficial effects of short-term exposure to scuba diving on human mental health. *International Journal of Environmental Research and Public Health, 17*(19), 7238.

Cavender, R., Swanson, J. R., & Wright, K. (2020). Transformative travel: Transformative learning through education abroad in a niche tourism destination. *Journal of Hospitality, Leisure, Sport & Tourism Education, 27*, 100245.

Chaline, E. (2017). *Strokes of genius: A history of swimming*. Reaktion Books.

Chen, C. L., & Teng, N. (2016). Management priorities and carrying capacity at a high-use beach from tourists' perspectives: A way towards sustainable beach tourism. *Marine Policy, 74*, 213–219.

Chris, C. (2006). *Watching wildlife*. U of Minnesota Press.

Cochrane, J. (2010). The sphere of tourism resilience. *Tourism Recreation Research, 35*(2), 173–185.

Coghlan, A., & Weiler, B. (2018). Examining transformative processes in volunteer tourism. *Current Issues in Tourism, 21*(5), 567–582.

Consoli, P., Scotti, G., Romeo, T., Fossi, M. C., Esposito, V., D'Alessandro, M., et al. (2020). Characterization of seafloor litter on Mediterranean shallow coastal waters: Evidence from dive against Debris®, a citizen science monitoring approach. *Marine Pollution Bulletin, 150*, 110763.

Cousteau, J. Y., & Dumas, F. (1953). *The silent world.* New York, Chicago

Cousteau, J. M., & Paisner, D. (2010). *My father, the captain: My life with Jacques Cousteau.* National Geographic Books.

Croes, R., Lee, S. H., & Olson, E. D. (2013). Authenticity in tourism in small Island destinations: A local perspective. *Journal of Tourism and Cultural Change, 11*(1–2), 1–20.

Crylen, J. (2018). Living in a world without sun: Jacques Cousteau, Homo aquaticus, and the Dream of Dwelling Undersea. *Journal of Cinema and Media Studies, 58*(1), 1–23.

da Silva, M. L., Castro, R. O., Sales, A. S., & de Araújo, F. V. (2018). Marine debris on beaches of Arraial do Cabo, RJ, Brazil: An important coastal tourist destination. *Marine Pollution Bulletin, 130*, 153–158.

Dawson, J., Stewart, E. J., Lemelin, H., & Scott, D. (2010). The carbon cost of polar bear viewing tourism in Churchill, Canada. *Journal of Sustainable Tourism, 18*(3), 319–336.

De, K., Nanajkar, M., Mote, S., & Ingole, B. (2020). Coral damage by recreational diving activities in a marine protected area of India: Unaccountability leading to 'tragedy of the not so commons'. *Marine Pollution Bulletin.*

Denkinger, J., Parra, M., Muñoz, J. P., Carrasco, C., Murillo, J. C., Espinosa, E., et al. (2013). Are boat strikes a threat to sea turtles in the Galapagos marine reserve? *Ocean & Coastal Management, 80*, 29–35.

Dimmock, K., & Cummins, T. (2013). History of scuba diving tourism. In *Scuba diving tourism* (pp. 14–28). Routledge.

DiNardo, J. C., & Downs, C. A. (2018). Dermatological and environmental toxicological impact of the sunscreen ingredient oxybenzone/benzophenone-3. *Journal of Cosmetic Dermatology, 17*(1), 15–19.

Dong, G., Brenner, R., Harris, C., Helfrich, E., & Tillmans, F. (2021). Breath-hold diving. In *DAN annual diving report 2020 edition: A report on 2018 diving fatalities, injuries, and incidents [internet]*. Divers Alert Network.

Downs, C. A., Kramarsky-Winter, E., Segal, R., Fauth, J., Knutson, S., Bronstein, O., et al. (2016). Toxicopathological effects of the sunscreen UV filter, oxybenzone (benzophenone-3), on coral planulae and cultured primary cells and its environmental contamination in Hawaii and the US Virgin Islands. *Archives of Environmental Contamination and Toxicology, 70*, 265–288.

Dwyer, L. (2022). Tourism contribution to the SDGs: Applying a well-being lens. *European Journal of Tourism Research, 32*, 3212.

Ecott, T. (2002). *Neutral buoyancy: Adventures in a liquid world.* Grove Press.

Fennell, D. A. (2001). A content analysis of ecotourism definitions. *Current Issues in Tourism, 4*(5), 403–421.

Greenberg, B. (2023, July 28). *Happy to kelp: Divers to clean local reefs on international scuba day*. Dana Point Times.

Haines-Young, R., & Potschin, M. (2010). The links between biodiversity, ecosystem services and human well-being. *Ecosystem Ecology: A New Synthesis, 1*, 110–139.

Hawaii Tourism Authority (HTA). (2019). *Fact sheet: Benefits of Hawaii's Tourism Economy*. Hawaii Tourism Authority.

Hawkins, J. P., & Roberts, C. M. (1992). Effects of recreational SCUBA diving on fore-reef slope communities of coral reefs. *Biological Conservation, 62*(3), 171–178.

Heigl, F., Kieslinger, B., Paul, K. T., Uhlik, J., & Dörler, D. (2019). Toward an international definition of citizen science. *Proceedings of the National Academy of Sciences, 116*(17), 8089–8092.

Henrykowska, G., Soin, J., & Siermontowski, P. (2021). Scuba diving as a form of rehabilitation for people with physical disabilities. *International Journal of Environmental Research and Public Health, 18*(11), 5678.

Hoegh-Guldberg, O. (2011). Coral reef ecosystems and anthropogenic climate change. *Regional Environmental Change, 11*, 215–227.

Jacobo, J. (2023). *Environmental impacts of Maui wildfires will last for years to come, experts say*. ABC News.

Jakšić, S., Stamenković, I., & Đorđević, J. (2013). Impacts of artificial reefs and diving tourism. *Turizam, 17*(4), 155–165.

Jarvis, N., Weeden, C., & Simcock, N. (2010). The benefits and challenges of sustainable tourism certification: A case study of the Green Tourism Business Scheme in the West of England. *Journal of Hospitality and Tourism Management, 17*(1), 83–93.

Jensen, F. H., Bejder, L., Wahlberg, M., Soto, N. A., Johnson, M., & Madsen, P. T. (2009). Vessel noise effects on delphinid communication. *Marine Ecology Progress Series, 395*, 161–175.

Johansen, K. (2013). Scuba diving education and training. In *Scuba diving tourism* (pp. 81–96). Routledge.

Keith, I., Dawson, T. P., Collins, K. J., & Campbell, M. L. (2016). Marine invasive species: Establishing pathways, their presence and potential threats in the Galapagos Marine Reserve. *Pacific Conservation Biology, 22*(4), 377–385.

Kirillova, K., Lehto, X., & Cai, L. (2017). Tourism and existential transformation: An empirical investigation. *Journal of Travel Research, 56*(5), 638–650.

Knowlton, N., Brainard, R. E., Fisher, R., Moews, M., Plaisance, L., & Caley, M. J. (2010). Coral reef biodiversity. In *Life in the world's oceans: diversity distribution and abundance* (pp. 65–74). Wiley.

Kuo, I. L. (2002). The effectiveness of environmental interpretation at resource-sensitive tourism destinations. *International Journal of Tourism Research, 4*(2), 87–101.

Lahtinen, K., Kurra, S., & Nissinen, A. (2015). *Freediving*. Deep Ideas Oy.

Lemelin, H., Dawson, J., Stewart, E. J., Maher, P., & Lueck, M. (2010). Last-chance tourism: The boom, doom, and gloom of visiting vanishing destinations. *Current Issues in Tourism, 13*(5), 477–493.

Levine, A. (2020). Sunscreen use and awareness of chemical toxicity among beach goers in Hawaii prior to a ban on the sale of sunscreens containing ingredients found to be toxic to coral reef ecosystems. *Marine Policy, 117*, 103875.

Li, J., Pearce, P. L., & Low, D. (2018). Media representation of digital-free tourism: A critical discourse analysis. *Tourism Management, 69*, 317–329.

Lucrezi, S., Milanese, M., Palma, M., & Cerrano, C. (2018). Stirring the strategic direction of scuba diving marine citizen science: A survey of active and potential participants. *PLoS One, 13*(8), e0202484.

Machernis, A. F., Powell, J. R., Engleby, L., & Spradlin, T. R. (2018). An updated literature review examining the impacts of tourism on marine mammals over the last fifteen years (2000–2015) to inform research and management programs.

Madin, E. M., & Fenton, D. M. (2004). Environmental interpretation in the Great Barrier Reef Marine Park: An assessment of programme effectiveness. *Journal of Sustainable Tourism, 12*(2), 121–137.
Marion, J. L., & Reid, S. E. (2001). Development of the US Leave No Trace program: An historical perspective. In *Enjoyment and understanding of the national heritage* (pp. 81–92).
Maslow, A. H. (1971). *The farther reaches of human nature* (Vol. 19711). Viking press.
Matsen, B. (2009). *Jacques Cousteau: The sea king*. Vintage.
Mayaka, M., Croy, W. G., & Cox, J. W. (2018). Participation as motif in community-based tourism: A practice perspective. *Journal of Sustainable Tourism, 26*(3), 416–432.
McAvoy, & Kelleher, S. (2023, September 8). Maui beckons tourists, and their dollars, to stave off economic disaster after wildfires. *AP News*.
McGough, W. (2023, November 14). *Hawaii travel update: Some locals believe tourism resumed too soon on Maui. Now, they're protesting on a popular tourist beach*. Forbes.
Mihalic, T. (2020). Conceptualizing overtourism: A sustainability approach. *Annals of Tourism Research, 84*, 103025.
Moberg, F., & Folke, C. (1999). Ecological goods and services of coral reef ecosystems. *Ecological Economics, 29*(2), 215–233.
Moscardo, G. (1998). Interpretation and sustainable tourism: Functions, examples and principles. *Journal of Tourism Studies, 9*(1), 2–13.
Musa, G., & Dimmock, K. (2013). *Scuba diving tourism*. Routledge.
(N.A.) August 6th – get in the water!. Scuba Day. (2023). https://www.scubaday.org/
Ndivo, R. M., & Cantoni, L. (2016). Rethinking local community involvement in tourism development. *Annals of Tourism Research, 57*, 275–278.
O'Connor, S., Campbell, R., Cortez, H. & Knowles, T. (2009). Whale watching worldwide: Tourism numbers, expenditures and expanding economic benefits, *a special report from the International Fund for Animal Welfare*. Yarmouth MA, USA, prepared by Economists at Large 228.
PADI. (2022). *2021 PADI worldwide statistics final*. 2021 PADI Worldwide Statistics. Available https://www.padi.com/sites/default/files/documents/2021-02/2021%20PADI%20Worldwide%20Statistics.pdf
Pallewatta, N. (2004). Invasive species and tourism. In Department of Zoology and Environment Science, University of Colombo, Sri Lanka. *Proceedings of a global synthesis workshop on "biodiversity loss and species extinctions: Managing risk in a changing world"sub theme: Invasive alien species—Coping with aliens*.
Papadopoulou, N. M., Ribeiro, M. A., & Prayag, G. (2023). Psychological determinants of tourist satisfaction and destination loyalty: The influence of perceived overcrowding and overtourism. *Journal of Travel Research, 62*(3), 644–662.
Pasternak, G., Ribic, C. A., Spanier, E., Ariel, A., Mayzel, B., Ohayon, S., & Zviely, D. (2019). Nearshore survey and cleanup of benthic marine debris using citizen science divers along the Mediterranean coast of Israel. *Ocean & Coastal Management, 175*, 17–32.
Perera, N., & Kotagama, S. W. (2016). *Recommendations for co-existence of coral reef conservation and tourism at*. Pigeon Island National Park.
Piggott-McKellar, A. E., & McNamara, K. E. (2017). Last chance tourism and the great barrier reef. *Journal of Sustainable Tourism, 25*(3), 397–415.
Pung, J., & Del Chiappa, G. (2020). An exploratory and qualitative study on the meaning of transformative tourism and its facilitators and inhibitors. *European Journal of Tourism Research, 24*, 2404–2404.
Robert, K. W., Parris, T. M., & Leiserowitz, A. A. (2005). What is sustainable development? Goals, indicators, values, and practice. *Environment: Science and Policy for Sustainable Development, 47*(3), 8–21.
Russell, A., & Wallace, G. (2004). Irresponsible ecotourism. *Anthropology Today, 20*(3), 1–2.
Saayman, M., & Saayman, A. (2018). Are there economic benefits from marine protected areas? An analysis of scuba diver expenditure. *European Journal of Tourism Research, 19*, 23–39.

Schneider, S. L., & Lim, H. W. (2019). Review of environmental effects of oxybenzone and other sunscreen active ingredients. *Journal of the American Academy of Dermatology, 80*(1), 266–271.

Schuhbauer, A., Favoretto, F., Wang, T., Aburto-Oropeza, O., Sala, E., Millage, K., et al. (2023). *Global economic impact of scuba dive tourism.*

Self, R. M., Self, D. R., & Bell-Haynes, J. (2010). Marketing tourism in the Galapagos Islands: Ecotourism or greenwashing? *International Business & Economics Research Journal (IBER), 9*(6).

Shani, A., Polak, O., & Shashar, N. (2012). Artificial reefs and mass marine ecotourism. *Tourism Geographies, 14*(3), 361–382.

Stolk, P., Markwell, K., & Jenkins, J. M. (2007). Artificial reefs as recreational scuba diving resources: A critical review of research. *Journal of Sustainable Tourism, 15*(4), 331–350.

Stronza, A. L., Hunt, C. A., & Fitzgerald, L. A. (2019). Ecotourism for conservation? *Annual Review of Environment and Resources, 44*, 229–253.

Taff, B. D., Newman, P., Vagias, W. M., & Lawhon, B. (2014). Comparing day-users' and overnight visitors' attitudes concerning leave no trace. *Journal of Outdoor Recreation, Education, and Leadership, 6*(2), 133–146.

Tanji, M. (2023, October 6). *Tourism return comes amid plunging arrivals, spending*. Mauinews.com

The International Ecotourism Society (TIES). (2018). *What is ecotourism?* TIES.

Tiberghien, G., Bremner, H., & Milne, S. (2020). Authenticity and disorientation in the tourism experience. *Journal of Outdoor Recreation and Tourism, 30*, 100283.

Townsend, C. (2007). Interpretation and environmental education as conservation tools. In *New frontiers in marine tourism* (pp. 189–200). Routledge.

United Nations Environment Programme, & World Trade Organization (WTO). (2005). *Making tourism more sustainable: A guide for policy makers*. https://wedocs.unep.org/20.500.11822/8741

Van Pel, H. (1961). *A guide to South Pacific fisheries*. South Pacific Commission.

Vianna, G. M. S., Meekan, M. G., Pannell, D. J., Marsh, S. P., & Meeuwig, J. J. (2012). Socio-economic value and community benefits from shark-diving tourism in Palau: A sustainable use of reef shark populations. *Biological Conservation, 145*, 267–277.

Woodhead, A. J., Hicks, C. C., Norström, A. V., Williams, G. J., & Graham, N. A. (2019). Coral reef ecosystem services in the Anthropocene. *Functional Ecology, 33*(6), 1023–1034.

World Tourism Organization. (UNWTO). (2023). https://www.unwto.org/sustainable-development

Yin, J., Cheng, Y., Bi, Y., & Ni, Y. (2020). Tourists perceived crowding and destination attractiveness: The moderating effects of perceived risk and experience quality. *Journal of Destination Marketing & Management, 18*, 100489.

Chapter 8
Impact of Hybrid Word of Mouth on Consumer Behavior: Case of Off-Highway Vehicle Recreation Markets

Eunhye Grace Kim

Abstract This study focuses on examining off-highway vehicle (OHV) recreation from a marketing perspective. It examines whether key versions of word of mouth (WOM) significantly shape consumer behavior in terms of loyalty and spending and hold potential to facilitate responsible behavior. By exploring the relationship between consumer behavior and traditional WOM and eWOM, this study stresses on the importance of integrating traditional WOM with social networking site (SNS) tools. By offering a comparative/integrated view of their impact on local spending behavior of OHV visitors, this study offers important implications from a sustainable marketing standpoint.

Keywords Off-highway vehicle (OHV) recreation · Word of mouth (WOM) · eWOM · Loyalty · Spending behavior

Introduction

Off-highway motorized vehicle (OHV) recreation activity is one of the fastest-growing segments of outdoor recreation in the United States (Hughes & Paveglio, 2019), and in fact, its popularity surged during the pandemic times. OHV refers to any vehicle that is intended to be ridden off-highway (on OHV trails), including all-terrain vehicles (ATVs), utility task vehicles (UTVs), side-by-sides, recreational off-highway vehicles (ROVs), dune buggies, sand rails, motorcycles or dirt bikes, and snowmobiles (Kil et al., 2012). OHV trails in the United States are managed by federal, state, local and/or tribal agencies, as well as private organizations (Arizona State Parks, 2018). OHV use has been a growing recreational activity in the United States for the past few decades (Cordell et al., 2005, 2008). OHV users have also

E. G. Kim (✉)
Parks, Recreation and Tourism Management, Clemson University, Clemson, SC, USA
e-mail: egkim@clemson.edu

D. Chhabra et al. (eds.), *Sustainable Development and Resilience of Tourism*,
https://doi.org/10.1007/978-3-031-63145-0_8

been identified as a recreational group with high propensity to spend on travel-related expenses (Holmes & Englin, 2010; Thomas et al., 2008). Therefore, it is important to obtain insights on sources of information used by this lucrative market in an ongoing manner, so that appropriate marketing messages (that also hold potential to promote sustainable use of resources) can be disseminated using the popular search platforms. Furthermore, sustainable behavior can be promoted using these popular communication tools.

Extent literature has recognized the power of word of mouth (WOM) communication in recreation and tourism (Díaz-Martín et al., 2000; Litvin et al., 2005; Morgan et al., 2003a, b; O'Neill et al., 2002; Shanka et al., 2002; Yu et al., 2019). As a popular source of information, it holds tremendous potential to engage interested audience in numerous ways. It is described as "informal communication directed at interested audience about the ownership, usage or characteristics of particular goods and services or their sellers" (Westbrook, 1987: p. 261). The WOM process allows customers to communicate information and perspectives that hold potential to influence their decision-making process. Particularly, the electric word of mouth (eWOM) is recognized as a persuasive marketing tool in its ability to influence consumer purchase decision process (Goldenberg et al., 2001; Ye et al., 2009). The eWOM can be defined as "any positive or negative statement made by potential, actual or former customers about a product or company that is made available to a multitude of people and institutions via the internet" (Hennig-Thurau et al., 2004, p. 39). Along similar lines, Litvin et al., (2008) described eWOM as "all informal communications directed at consumer through Internet-based technology related to the usage or characteristics of particular goods and services, or their sellers" (p. 416).

WOM can also influence responsible behavior in terms of buying local and mindful and responsible consumption. As an example, an exploratory study, by Bigne et al. (2020) examined how eWOM (as a form of online reviews) influences tourists' evaluations associated with mindfulness of economic, sociocultural, and environmental dimensions of tourism destination responsibility. In this study, it was suggested that eWOM associated with responsible environmental behavior has higher levels of perceived helpfulness, perceived influence, and destination brand equity than the other dimensions (Bigné et al., 2020). Another study investigated responsible tourist behavior using the Theory of Planned Behavior by incorporating pro-social behavior and perception of responsible tourism (Kim et al., 2018). The study results revealed that pro-social behavior had significant effects on responsible attitude of tourists, intention to travel responsibly, and their behavior in responsible tourism (Kim et al., 2018). In this study, the finding indicated that belief in the benefit of responsible tourism exerted significant impact on attitude and subjective norms of tourists. Based on the findings, authors suggested that the local governments and tourism sectors would do well to promote benefits of responsible tourism to actual and potential tourists through eWOM (Kim et al., 2018). Also, there are some other studies that have shown the significant positive effect of eWOM on the individual's behavior and the brand image of the accommodation industry (Hu & Kim, 2018). One study informs that a variety of eWOM platforms attract frequent patronization of customers (Xiang & Gretzel, 2010) and hold potential to promote sustainable behavior and well-being.

Outdoor recreation and nature-based tourism provide essential benefits to individuals, communities and society, and contribute to sustainability (Winter et al., 2019). They are widely recognized as important for improving human health and well-being, enhancing social connections, connecting people with their natural and cultural heritage, generating revenues for use in conservation, contributing to local economies, fostering local or indigenous identity, and developing a conservation consciousness (Frumkin et al., 2017; Hartig et al., 2014; Winter et al., 2019; Wolsko et al., 2019).

Tourism affects the well-being of destination communities (Uysal et al., 2016). As an example, community residents' perception of tourism impact on their quality of life (QOL) has been researched since the 1980s. Most of these studies have focused on the delineation of tourism impacts (cultural, social, economic, environmental, and ecological) and support for tourism development (Uysal et al., 2016). In general, the tourism impacts are discussed in three domains: economic, sociocultural, and environmental (Andereck et al., 2005; Sharpley, 2014; Tosun, 2002). Generally, documented literature indicates that tourism impact has both positive and negative dimensions. Community residents usually appreciate the positive economic impact of tourism because the positive dimensions contribute positively to the enhancement of residents' QOL. That said, residents differ in the manner they perceive tourism impacts. They are also concerned with negative sociocultural and environmental impacts (Perdue et al., 1990). Nevertheless, tourism holds tremendous potential to improve QOL as it contributes positively toward many domains of their life such as family life, social life, leisure life, and cultural life (Uysal et al., 2016). Such messages spread quickly through word of mouth. Responsible behavior messages (including spending at local venues) serve to promote mindfulness, particularly toward the natural environment and the host communities.

As mentioned earlier, in the United States, outdoor recreation participation has steadily increased (Cordell et al., 1999). During the pandemic, preference for outdoor recreation trips/activities increased due to safety considerations and as many people's work patterns transitioned to long online working hours at home (Chrobak, 2020; Wagner, 2022). According to a recent national panel study (titled "Who Started, Stopped and Continued Participating in Outdoor Recreation during COVID-19 Pandemic in the United States"), the pandemic has made people residing in the United States more keen to pursue outdoor recreation (Taff et al., 2021). The study shares that currently almost 50% of the adults in the United States participate in outdoor recreation, at least on a monthly basis (Taff et al., 2021). It is of no surprise then to note that popularity in OHV recreation is surging. Despite this growth, only a handful of studies have investigated the influence of information search behavior on travel and sustainable spending behavior of consumers. For the purpose of this study, sustainable spending behavior refers to spending that benefits the host/local community (Nickerson et al., 2016; Suryawardani et al., 2016) and holds potential to enhance QOL of local residents. Most studies have focused on assessing economic impact/value (Bilberman & Andereck, 2006; Rosenberger et al., 2017; Sharp et al., 2020; White et al., 2016), but marketing perspective on tools, that can be used to stimulate spending behavior, is remiss.

Undeniably WOM (particularly traditional and eWOM) has been increasingly recognized as one of the popular sources of information in the contemporary era. However, research scrutinizing the association between key categories of WOM and consumer attributes and behavior is still sparse. Such insights can have important marketing implications. To fill this gap, this study focuses on examining OHV recreation from a sustainable marketing perspective. It examines whether key versions of WOM (traditional and electronic such as social networking sites) significantly shape consumer behavior in terms of loyalty and spending in a manner that local resources are used in a sustainable manner. In the recreation and tourism field, both loyalty and mindful (in terms of buying local) spending behavior are considered important components of the consumer behavior and responsible consumption phenomenon (Alegre & Juaneda, 2006). By exploring the relationship between consumer behavior and traditional WOM and eWOM (by focusing on social networking site (SNS) platforms), this study stresses on the importance of integrating traditional WOM with SNS tools. For instance, an integrated approach can help connect with the targeted audience, communicate important sustainable messages, and promote special deals, rates, programs, well-being, and mindfulness in a timely manner. Across the globe, wellness and health have become important components of sustainable tourism (Tariyal et al., 2023). As indicated in documented literature, wellness travelers, keen to enhance their QOL, rely on electronic word of mouth (eWOM) as it offers positive/negative and neutral information from actual and potential customers (Hennig-Thurau et al., 2004; Stará & Peterson, 2017; Tariyal et al., 2023; Wang et al., 2020).

Literature Review

To understand the marketing impact of both traditional WOM and eWOM on sustainable development and economic well-being, the concepts of well-being, QOL, and the position of WOM in sustainable and responsible tourism are identified in documented literature.

In recent years the concept of well-being has received increasing attention internationally (Hanna et al., 2019.) Hanna et al. (2019) explore the form of outdoor adventure tourism as a potentially sustainable form of tourism. The study has attempted to contribute to sustainable tourism theory by highlighting the potential sustainability and well-being benefits of outdoor adventure tourism. It examined if such activities facilitate a greater understanding for and appreciation of nature (Hanna et al., 2019) and facilitate more local spending behavior. The analysis suggests that nature tourism facilitates reconnections with nature, facilitates pro-environmental attitudes and behaviors, and holds potential to improve economic well-being of the local community. The study concluded that outdoor adventure activities as a form of sustainable tourism have potential implications for our understanding of, and engagement with, sustainability and overall well-being (Hanna et al., 2019).

Such an understanding reconnects individuals with nature in a way that enables a deeper appreciation for nature, potentially providing healing for both the individuals' well-being and the natural environment (Davis, 1998), offering potential avenues for sustainable tourism practice. Further, considering the mental health crisis, Hanna et al. (2019) suggested that outdoor adventure tourism offers mental health benefits and some form of distress alleviation through active engagement with the natural environment. In addition, researchers test and empirically demonstrate that tourism can indeed improve QOL of tourism participants and non-tourism alike (Uysal et al., 2016). Moscardo (2009) points out the importance of social representations in impact studies and argues that a more explicit consideration of QOL with respect to well-being of different stakeholders can help in the development of a better framework for critically analyzing tourism impact on the well-being of individuals as tourists and providers alike. Several studies have also reported that economic benefits improve the overall well-being and QOL of local communities.

There is a study exploring the relationship between community participation in an ecotourism destination to realize QOL in the community (Ali et al., 2022). Results show that community participation depends on natural attraction and motivation of the people to engage in ecotourism activities. Also, the study indicates that the participation has been shown to have an impact on the QOL of local community (Ali et al., 2022). In nature-based tourism, visitors are motivated to experience local culture and nature (Thapa et al., 2022). Therefore, tourism contributes economically and socially to the community and brings both benefits and costs to local people. While most benefits are economic, most costs are sociocultural. The main benefits are employment, business opportunities, and income, and the main costs are acculturation and abandonment of traditional lifestyle/practices, price inflation, and conflict/crime (Thapa et al., 2022). To focus more on benefits, destinations strive for marketing strategies such as WOM communication. The Internet has provided an innovative platform for eWOM communication (Chhabra, 2015). Many people use social networking site (SNS) platforms to communicate about products, brands, and companies (Mangold & Faulds, 2009). In tourism, SNSs have become critical tools in creating and sharing travel-related information (Shirazi, 2018). They are recognized as a fundamental source of information, particularly in terms of their tenacity to share and distribute recreation experience at an OHV trail or activity and quality of ancillary services. Numerous studies in tourism state that eWOM is regarded as a trustworthy information source among friends, relatives, families, and co-workers (Shirazi, 2018). Significance of SNSs form of eWOM has been well recognized in general marketing literature (Chi & Qu, 2008; Gupta & Harris, 2010; Murphy et al., 2007; Simpson & Siguaw, 2008).

Several traditional/digital WOM models have been presented in academic literature (Harris & Prideaux, 2017; Haywood, 1989; Litvin et al., 2008). For instance, in their WOM model, Litvin et al. (2008) illustrated that several factors influence originators and receivers of messages. Originators' perspective includes positive personal relationships, a higher degree of product involvement, intensity of surprise, trustworthiness of the source, satisfaction with the product, and identity salience. The receivers' perspective highlights source of information, brand evaluation, stoichiometric integration, and memory. Customer trust and loyalty, product value, purchase decision, customer empowerment, and product acceptance are reported as meditating variables that facilitate positive WOM (Chhabra, 2015; Litvin et al., 2008).

According to Pandey and Sahu (2017), eWOM holds tremendous potential to reach a large audience, whereas the traditional WOM conversation is mostly bidirectional. Pandey and Sahu (2017) presented various factors influencing eWOM intention in tourism. In their proposed eWOM framework, several attributes such as customer experience, quality, price, value, satisfaction, technology acceptance factors, place attachment, and destination image were identified to have a significant effect in shaping positive eWOM. Their study contributed to modeling the relationship between the identified variables, to develop better marketing strategies (Pandey & Sahu, 2017).

Harris and Prideaux (2017) recognized the importance of eWOM in consumer behavior and presented a notable framework. They suggested how travelers might consider eWOM communication by outlining the integrated and supplementary roles of traditional WOM and eWOM in travel behavior. Their study indicated that eWOM offers a means for consumers to use their own filters to assess the information during their decision making process (Harris & Prideaux, 2017).

Leung et al. (2019) examined the state-of-the-art development in social media research and developed a thematic framework of social media research via content analysis and correspondence analysis based on literature review on tourism and hospitality journals. The authors argued that tourism and hospitality businesses promote marketing messages and consumers generate user-generated content using social media platforms. Social media platforms serve a useful purpose as business management tools and as conduits in shaping consumer attitudes and behavior (Leung et al., 2019).

Furthermore, several studies postulate that eWOM can promote sustainable behavior and well-being. As an instance, Kocic and Radakovic (2019) aimed to obtain relevant findings on the effect of eWOM communication on consumer behavior by using a correlation and multiple regression analysis. They examined whether the quantity, quality, and timeliness of information and the source credibility of eWOM communication affect consumer trust in wellness tourism. A regression analysis was applied to determine whether consumer trust contributed to purchase intention. The results showed that there was a statistically significant impact of eWOM communication on consumer trust and that consumer trust positively influenced purchase intention when choosing a wellness tourism package (Kocic & Radakovic, 2019).

Therefore, the role of eWOM communication in the tourism sector is very important and holds potential to influence purchase intention of consumers. By reducing risk and uncertainty, eWOM communication serves as one of the crucial sources of information (Kocic & Radakovic, 2019). It is also important to note that simultaneously, many consumers continue to be influenced by traditional WOM (Cam et al., 2019; Ishida et al., 2016; Susilowati & Sugandini, 2018). Therefore, research examining and comparing the influence of both sources of information on consumer behavior is timely. By offering a comparative/integrated view of their impact on sustainable spending behavior of OHV visitors, this study offers important implications from a marketing standpoint.

Research Methodology

The data for this study is part of a bigger project titled "Economic Impact of Off Highway Recreation in the State of Arizona." The project aimed to provide marketing information about the OHV (Off Highway Vehicles) visitors in Arizona and present a comprehensive analysis of the economic impact. A total of 3550 surveys were randomly collected (1654 were usable for data analysis) in 2017 and 2018. The survey data used in this study is from retained only ($n = 960$). Retained visitors are defined as local visitors who would have traveled outside the state if OHV trails had been absent. This particular study focuses on those who use WOM and/or SNS for their primary information sources.

Findings

The descriptive table (Table 8.1) shows that visitors who integrate both, traditional WOM and SNS (eWOM), in their information search have a higher repeat visitation rate than those who solely use traditional WOM or SNS. Their frequency of visits, to the OHV trails, is comparatively higher. Most importantly, they incur highest total spending during the OHV trip.

Table 8.2 presents OHV visitors' spending based on each expense category during the OHV trip. Interestingly, in most of the categories (10 out of 13 items—gas, food, tour, lodging camp, lodging hotel, repair, goods, entertainment, clothing, and other expenses), this cohort (who use both traditional WOM and eWOM (with SNS as its primary information source)) spends more than the other group.

Table 8.1 Spending and loyalty of WOM and SNS users

	N	Repeat %	Times visit	Day trip %	Hour	Overnight %	Nights	Average total spending (per visit)
WOM	396	80.3%	20.1	67.7%	6.7	30.6%	5.1	$1192.96
WOM ONLY	312	78.8%	19.3	67.6%	6.4	30.1%	5.4	$974.88
SNS	153	77.8%	22.6	67.3%	7.3	31.4%	3.3	$1509.22
SNS ONLY	69	68.1%	19.6	68.1%	7.1	30.4%	2.7	$908.12
WOM and SNS	84	85.7%	25.6	66.7%	8.7	32.1%	3.8	$2003.00

Table 8.2 Spending behavior—OHV items

	WOM	WOM only	SNS	SNS only	WOM and SNS
Gas	$152.64	$135.91	$174.05	$124.88	$214.40
Restaurant	$109.74	$109.04	$114.62	$117.95	$112.09
Food	$119.01	$103.04	$155.11	$128.98	$176.65
Tour	$7.60	$6.44	$9.58	$7.22	$11.23
Lodging (camp)	$131.41	$125.06	$144.91	$136.64	$150.47
Lodging (hotel)	$106.90	$97.57	$116.18	$85.43	$136.11
Repair	$338.99	$321.97	$353.74	$295.21	$396.72
Rent	$15.78	$17.10	$15.13	$20.97	$11.11
Goods	$141.79	$135.57	$115.80	$45.74	$163.44
Souvenir	$37.66	$37.33	$45.02	$53.33	$38.79
Entertain	$61.32	$59.40	$66.69	$65.29	$67.71
Clothing	$193.49	$190.91	$191.79	$177.05	$202.98
Other expenses	$479.99	$149.83	$1037.20	$135.50	$1810.09
Total spending	$1192.96	$974.88	$1509.22	$908.12	$2002.98

Table 8.3 Social demographic information—average age

	N	Average age
WOM	396	53.44
WOM only	312	54.25
SNS	153	50.52
SNS only	69	50.57
WOM and SNS	84	50.47

Table 8.4 Social demographic information—education level

Education	WOM	WOM only	SNS	SNS only	WOM and SNS
Some high school or less	1.0%	1.3%	0.7%	1.5%	0.0%
High school graduate	9.9%	10.3%	9.9%	11.8%	8.3%
Some college/technical school	41.3%	41.2%	44.7%	48.5%	41.7%
College graduate	32.7%	32.8%	31.6%	30.9%	32.1%
Postgraduate school (Masters)	12.2%	11.9%	9.9%	5.9%	13.1%
Postgraduate school (Ph.D. or equivalent)	3.0%	2.6%	3.3%	1.5%	4.8%
(% of higher than college graduate)	100.0% (47.8%)	100.0% (47.3%)	100.0% (44.7%)	100.0% (38.2%)	100.0% (50.5%)

Table 8.3 presents the average age of OHV visitors. Visitors, who integrate traditional WOM and SNS as their primary information source, are noted to be slightly younger than other groups.

Table 8.4 shows the education level of selected OHV visitors. Interestingly, the results show that this group of visitors (using both traditional WOM & SNS) are in the higher education category (holding a college graduate degree).

Table 8.5 Social demographic information—income level

Income	WOM	WOM only	SNS	SNS only	WOM and SNS
<$20,000	0.8%	1.0%	0.0%	0.0%	0.0%
$20,000–39,999	5.0%	6.0%	4.7%	5.9%	3.7%
$40,000–59,999	10.9%	11.0%	8.7%	7.4%	9.8%
$60,000–79,999	16.2%	16.6%	11.3%	10.3%	12.2%
$80,000–99,999	18.7%	18.9%	23.3%	27.9%	19.5%
$100,000–149,999	25.4%	24.6%	26.7%	23.5%	29.3%
$150,000–199,999	12.0%	11.0%	16.7%	19.1%	14.6%
$200,000–299,999	7.3%	7.3%	6.7%	4.4%	8.5%
$300,000 and above	3.6%	3.7%	2.0%	1.5%	2.4%
(% of >$100 K)	100.0% (48.3%)	100.0% (46.5%)	100.0% (52.0%)	100.0% (48.5%)	100.0% (54.9%)

With regard to annual household income, those who use WOM and SNS as their primary source of information belong to the higher income category (more than $100 K) compared to other groups. Table 8.5 shows the details.

Discussion

As mentioned earlier, identified benefits of outdoor recreation and nature-based tourism demonstrate the contributions of nature-based experiences to individual, community, and societal resilience and sustainability (Winter et al., 2019). The growing popularity of natural destinations, the tourism industry, and government organizations have taken steps to increase tourism within relevant localities and to turn natural sites into popular trip destinations (Winter et al., 2019). Nature-based tourism is one of the fastest-growing tourism sectors, and many public lands and protected areas have become important tourist destinations (Winter et al., 2019). To sustain the benefits of outdoor recreation for present and future generations, it is crucial that the recreation program must address and work toward a sustainable balance among environmental, social, and economic conditions (Arni & Khairil, 2013).

Nature-based tourism provides an important economic engine for host destinations and is a potential source of revenue for protected areas to meet conservation and biodiversity goals (Balmford et al., 2015; Snyman & Bricker, 2019). Outdoor recreation and nature-based tourism may provide an opportunity to improve ecological conditions where visitors are engaged in stewardship experiences (Schild, 2019). These contributions represent important aspects of sustainable recreation and tourism (Bricker et al., 2010). Nature-based tourism is expected to grow much faster than tourism on average (Puhakka, 2008). Tourism can play important roles in conserving nature resources through application of appropriate management strategies that promote positive synergetic relationships among natural areas, local populations, and tourism (Ross & Wall, 1999).

WOM, particularly eWOM has become a significant element of the dynamic marketing environment in the recreation and tourism industry (Chhabra, 2015). It has been indicated that the potential of eWOM, in generating revenue, is enormous (Chhabra, 2015). Many people have been losing their confidence in company-initiated communication and marketing messages, and this trend has enhanced the role of social media platforms and they have become powerful influencers (Allsop et al., 2007). Growing popularity of social media applications has boosted the patronage and significance of eWOM (Chhabra, 2015). However, eWOM is not a stand-alone phenomenon; rather, it needs to be a part of the broader marketing agenda.

Most studies on outdoor recreation lack in-depth market-related research and most are in the form of technical repots. By confirming the unique relationship between OHV activities and WOM, one of the important information sources, this study offers noteworthy insights. These results can be used as a steppingstone by future studies, to gather deeper insights and investigate the dynamic impact of hybrid (integrated) WOM communication strategies on consumer behavior with special focus on local spending patterns.

Furthermore, as tourism grows, an important goal should not only be to improve upon sustainability and development but to also try to enhance the QOL of both tourists and residents in host communities. A key challenge will be to create a comprehensive tool to measure and monitor of activities and their QOL effects (Budruk & Phillips, 2010). The long-term success and sustainability and competitiveness of tourism will certainly depend on tourism's ability to contribute to the improvement of QOL of all stakeholders. As part of the fully functioning tourism system, it is imperative to know that tourism enterprises contribute to improving the QOL not only of tourists but also residents in host communities and employees of tourism services. It is commonly accepted that sustainability, tourism activities, and QOL are all interrelated (Uysal et al., 2016). Tourism activities bringing revenue to local residents hold potential to improve their QOL, and such activities should be promoted using popular communication tools.

During the pandemic, Arizona State Parks had to limit some of the OHV services including large-scale events and tours. Some facilities also had to be closed for visitor safety. They resulted in loss of income for local businesses. Nevertheless, OHV use exploded in popularity during the pandemic. In fact, Arizona State Parks witnessed increased use of OHV trails and services across the entire State of Arizona. OHV recreation offered an opportunity to recreate while maintaining a social distance. That said, only few local businesses offered services and products due to social distancing protocol. In the post-pandemic, there are no such restrictions, and more local residents should engage in businesses to tap into the high spending propensity of OHV users.

In summary, the results of this study suggest that the OHV users, who simultaneously use both traditional WOM and eWOM (SNS) as their source of information, are more likely and more frequently to make a trip to a destination with OHV trails. The same cohort demonstrates a higher propensity to spend, locally, during the OHV trip. Overall, the study findings illustrate the significant impact of hybrid

WOM (integrating eWOM with traditional) on consumer behavior in terms of loyalty and local spending behavior. As is emphasized in this study as well as previous studies, tourism stakeholders can promote responsible tourism through both WOM and eWOM. eWOM communication promotes local purchasing behavior of current and potential tourists (Kocic & Radakovic, 2019). The destination marketing organizations should strategically utilize WOM to stimulate the local spending behavior of potential tourists to visit tourism destinations (Kim et al., 2018). This study confirms that hybrid WOM holds potential to promote sustainability and economic well-being thereby enhancing the overall quality of life of local residents.

Acknowledgments This work was funded by Arizona State Parks.

References

Alegre, J., & Juaneda, C. (2006). Destination loyalty: Consumers' economic behavior. *Annals of Tourism Research, 33*(3), 684–706.

Ali, M. B., Quaddus, M., Rabbanee, F. K., & Shanka, T. (2022). Community participation and quality of life in nature-based tourism: Exploring the antecedents and moderators. *Journal of Hospitality & Tourism Research, 46*(3), 630–661.

Allsop, D. T., Bassett, B. R., & Hoskins, J. A. (2007). Word-of-mouth research: principles and applications. *Journal of Advertising Research, 47*(4), 398–411.

Andereck, K. L., Valentine, K. M., Knopf, R. C., & Vogt, C. A. (2005). Residents' perceptions of community tourism impacts. *Annals of Tourism Research, 32*(4), 1056–1076.

Arizona State Parks: Off-Highway Vehicle Recreation Report. (2018). *Economic impact of off-highway recreation in the state of Arizona*. Retrieved from https://d2umhuunwbec1r.cloudfront.net/gallery/0004/0052/4CD7751111BA46518D5CBDE85C71493A/Final%20Report.pdf

Arni, A. G., & Khairil, W. A. (2013). Promoting collaboration between local community and park management towards sustainable outdoor recreation. *Procedia-Social and Behavioral Sciences, 91*, 57–65.

Balmford, A., Green, J. M., Anderson, M., Beresford, J., Huang, C., Naidoo, R., et al. (2015). Walk on the wild side: estimating the global magnitude of visits to protected areas. *PLoS Biology, 13*(2), e1002074.

Bigné, E., Zanfardini, M., & Andreu, L. (2020). How online reviews of destination responsibility influence tourists' evaluations: An exploratory study of mountain tourism. *Journal of Sustainable Tourism, 28*(5), 686–704.

Bilberman, J., & Andereck, K. L. (2006). The economic value of off-highway vehicle recreation. *Journal of Leisure Research, 38*(2), 208–223.

Bricker, K. S., Winter, P. L., & Schultz, J. R. (2010). Health, economy, and community: USDA Forest Service managers' perspectives on sustainable outdoor recreation. *Rural Connect, 5*(1), 38–42.

Budruk, M., & Phillips, R. (Eds.). (2010). *Quality-of-life community indicators for parks, recreation and tourism management* (Vol. 43). Springer Science & Business Media.

Cam, L. N. T., Anh, T. T., Moslehpour, M., & Thanh, X. D. T. (2019). *Exploring the impact of traditional and electronic word of mouth on travel intention* (pp. 83–87). Proceedings of the 2019 5th International Conference on E-Business and Applications.

Chhabra, D. (2015). *Strategic marketing in hospitality and tourism: Building a 'SMART' online agenda*. Nova Science Publishers, Inc.

Chi, C. G. Q., & Qu, H. (2008). Examining the structural relationships of destination image, tourist satisfaction and destination loyalty: An integrated approach. *Tourism Management, 29*(4), 624–636.

Chrobak, U. (2020, Oct 14). *National parks are logging record crowds. Here's how to visit safely*. Popular Science. Retrieved May 24, 2022, from https://www.popsci.com/story/environment/national-parks-record-crowds-covid-pandemic/

Cordell, H. K., Teasley, R. J., Bergstrom, J. C., Martin, J., Bason, J., & Leeworthy, V. R. (1999). Outdoor recreation participation trends. In H. K. Cordell, C. Betz, J. M. Bowker, et al. (Eds.), *Outdoor recreation in American life: a national assessment of demand and supply trends* (pp. 219–321). Sagamore Publishing.

Cordell, H. K., Betz, C. J., Green, G., & Owens, M. (2005). *Off-highway vehicle recreation in the United States, regions, and states: a national report from the national survey on recreation and the environment (NSRE)*. US Forest Service.

Cordell, H. K., Betz, C. J., Green, G. T., & Stephens, B. (2008). *Off-highway vehicle recreation in the United States and its regions and states: An update national report from the National survey on recreation and the environment (NSRE)*. US Forest Service.

Davis, J. (1998). The transpersonal dimensions of ecopsychology: Nature, nonduality, and spiritual practice. *The Humanistic Psychologist, 26*(1–3), 69.

Díaz-Martín, A. M., Iglesias, V., Vazquez, R., & Ruiz, A. V. (2000). The use of quality expectations to segment a service market. *Journal of Services Marketing., 14*, 132.

Frumkin, H., Bratman, G. N., Breslow, S. J., Cochran, B., Kahn, P. H., Jr., Lawler, J. J., et al. (2017). Nature contact and human health: A research agenda. *Environmental Health Perspectives, 125*(7), 075001.

Goldenberg, J., Libai, B., & Muller, E. (2001). Talk of the network: A complex systems look at the underlying process of word-of-mouth. *Marketing Letters, 12*(3), 211–223.

Gupta, P., & Harris, J. (2010). How e-WOM recommendations influence product consideration and quality of choice: A motivation to process information perspective. *Journal of Business Research, 63*(9–10), 1041–1049.

Hanna, P., Wijesinghe, S., Paliatsos, I., Walker, C., Adams, M., & Kimbu, A. (2019). Active engagement with nature: outdoor adventure tourism, sustainability, and wellbeing. *Journal of Sustainable Tourism., 27*, 1355.

Harris, A., & Prideaux, B. (2017). *The potential for eWOM to affect consumer behavior in tourism*. Handbook of Consumer Behaviour in Tourism.

Hartig, T., Mitchell, R., De Vries, S., & Frumkin, H. (2014). Nature and health. *Annual Review of Public Health, 35*, 207–228.

Haywood, K. M. (1989). Managing word of mouth communications. *Journal of Services Marketing, 3*(2), 55–67.

Hennig-Thurau, T., Gwinner, K. P., Walsh, G., & Gremler, D. D. (2004). Electronic word-of-mouth via consumer-opinion platforms: what motivates consumers to articulate themselves on the internet? *Journal of Interactive Marketing, 18*(1), 38–52.

Holmes, T. P., & Englin, J. E. (2010). Preference heterogeneity in a count data model of demand for off-highway vehicle recreation. *Agricultural and Resource Economics Review, 39*(1), 75–88.

Hu, Y., & Kim, H. J. (2018). Positive and negative eWOM motivations and hotel customers' eWOM behavior: Does personality matter? *International Journal of Hospitality Management, 75*, 27–37.

Hughes, C. A., & Paveglio, T. B. (2019). Managing the St. Anthony Sand Dunes: Rural resident support for off-road vehicle recreation development. *Journal of Outdoor Recreation and Tourism, 25*, 57–65.

Ishida, K., Slevitch, L., & Siamionava, K. (2016). The effects of traditional and electronic word-of-mouth on destination image: A case of vacation tourists visiting Branson, Missouri. *Administrative Sciences, 6*(4), 12.

Kil, N., Holland, S. M., & Stein, T. V. (2012). Identifying differences between off-highway vehicle (OHV) and non-OHV user groups for recreation resource planning. *Environmental Management, 50*(3), 365–380.

Kim, M. J., Park, J. Y., Reisinger, Y., & Lee, C. K. (2018). Predicting responsible tourist behavior: Exploring pro-social behavior and perceptions of responsible tourism. *International Journal of Tourism and Hospitality Research, 32*(4), 5–20.

Kocic, M., & Radakovic, K. (2019). The implications of the electronic word-of-mouth communication in choosing a wellness offer. *Economic Horizons, 21*(1), 43–56.

Leung, X. Y., Sun, J., & Bai, B. (2019). Thematic framework of social media research: state of the art. *Tourism Review, 74*(3), 517–531.

Litvin, S. W., Blose, J. E., & Laird, S. T. (2005). Tourists' use of restaurant webpages: Is the internet a critical marketing tool? *Journal of Vacation Marketing, 11*(2), 155–161.

Litvin, S. W., Goldsmith, R. E., & Pan, B. (2008). Electronic word-of-mouth in hospitality and tourism management. *Tourism Management, 29*(3), 458–468.

Mangold, W. G., & Faulds, D. J. (2009). Social media: the new hybrid element of the promotion mix. *Business Horizons, 52*(4), 357–365.

Morgan, N., Pritchard, A., & Piggott, R. (2003a). Destination branding and the role of the stakeholders: The case of New Zealand. *Journal of Vacation Marketing, 9*(3), 285–299.

Morgan, N., Pritchard, A., & Pride, R. (2003b). Marketing to the Welsh diaspora: The appeal to hiraeth and homecoming. *Journal of Vacation Marketing, 9*(1), 69–80.

Moscardo, G. (2009). Tourism and quality of life: Towards a more critical approach. *Tourism and Hospitality Research, 9*(2), 159–170.

Murphy, L., Mascardo, G., & Benckendorff, P. (2007). Exploring word-of-mouth influences on travel decisions: friends and relatives vs. other travellers. *International Journal of Consumer Studies, 31*(5), 517–527.

Nickerson, N. P., Jorgenson, J., & Boley, B. B. (2016). Are sustainable tourists a higher spending market?. *Tourism Management, 54*, 170–177.

O'Neill, M., Palmer, A., & Charters, S. (2002). Wine production as a service experience–the effects of service quality on wine sales. *Journal of Services Marketing, 16*, 342.

Pandey, A., & Sahu, R. (2017). Determinants of electronic word-of-mouth in tourism sector. *Proceedings of ICRBS*, 2017.

Perdue, R. R., Long, P. T., & Allen, L. (1990). Resident support for tourism development. *Annals of Tourism Research, 17*(4), 586–599.

Puhakka, R. (2008). Increasing role of tourism in Finnish national parks. *Fennia-International Journal of Geography, 186*(1), 47–58.

Rosenberger, R. S., White, E. M., Kline, J. D., & Cvitanovich, C. (2017). Recreation economic values for estimating outdoor recreation economic benefits from the National Forest System. *Gen. Tech. Rep. PNW-GTR-957. Portland, OR: US Department of Agriculture, Forest Service, Pacific Northwest Research Station, 33*, 957.

Ross, S., & Wall, G. (1999). Ecotourism: Towards congruence between theory and practice. *Tourism Management, 20*(1999), 123–132.

Schild, R. (2019). Civic recreation: Outdoor recreationists as advocates, stewards, and managers of natural resources. *Environmental Management, 63*(5), 629–646.

Shanka, T., Ali-Knight, J., & Pope, J. (2002). Intrastate travel experiences of international students and their perceptions of Western Australia as a tourist destination. *Tourism and Hospitality Research, 3*(3), 245–256.

Sharp, R., Bradley, M. J., & Maples, J. (2020). Who has the right-of-way?: Negotiating environmental and economic impacts of off-highway vehicle use in eastern Kentucky. *Journal of Appalachian Studies, 26*(2), 191–208.

Sharpley, R. (2014). Host perceptions of tourism: A review of the research. *Tourism Management, 42*, 37–49.

Shirazi, S. F. M. (2018). eWOM as an effective means to assess business administration in tourism destination. *Journal of Advanced Management Science, 6*(4), 203.

Simpson, P. M., & Siguaw, J. A. (2008). Destination word of mouth: The role of traveler type, residents, and identity salience. *Journal of Travel Research, 47*(2), 167–182.

Snyman, S., & Bricker, K. S. (2019). Living on the edge: Benefit-sharing from protected area tourism. *Journal of Sustainable Tourism, 27*(6), 705–719.

Stará, J., & Peterson, C. (2017). Understanding the concept of wellness for the future of the tourism industry: A literature review. *Journal of Tourism & Services, 8*(14), 18–29.

Suryawardani, I. G. A. O., Wiranatha, A. S., & Petr, C. (2016). Factors affecting willingness of foreign tourists to spend money in benefiting local people. *Development of Tourism and the Hospitality Industry in Southeast Asia*, 13–36.

Susilowati, C., & Sugandini, D. (2018). Perceived value, eWord-of-mouth, traditional word-of-mouth, and perceived quality to destination image of vacation tourists. *Review of Integrative Business and Economics Research, 7*, 312–321.

Taff, B. D., Rice, W. L., Lawhon, B., & Newman, P. (2021). Who started, stopped, and continued participating in outdoor recreation during the COVID-19 pandemic in the United States? Results from a National Panel Study. *Land, 10*(12), 1396.

Tariyal, A., Singh, S., Bisht, S., & Rana, V. (2023). What influences wellness customers to engage in electronic word of mouth? A total interpretive structural modelling approach. *International Journal of Spa and Wellness, 6*, 1–24.

Thapa, K., King, D., Banhalmi-Zakar, Z., & Diedrich, A. (2022). Nature-based tourism in protected areas: A systematic review of socio-economic benefits and costs to local people. *International Journal of Sustainable Development & World Ecology, 29*(7), 625–640.

Thomas, F., Bastian, C. T., Taylor, D. T., Coupal, R. H., & Olson, D. (2008). Off-road vehicle recreation in the west: Implications of a Wyoming analysis. In *Western Economics Forum* (Vol. 7, No. 1837-2016-151774, pp. 1-11).

Tosun, C. (2002). Host perceptions of impacts: A comparative tourism study. *Annals of Tourism Research, 29*(1), 231–253.

Uysal, M., Sirgy, M. J., Woo, E., & Kim, H. L. (2016). Quality of life (QOL) and well-being research in tourism. *Tourism Management, 53*, 244–261.

Wagner, A. (2022, Jan 24). *How has the COVID-19 pandemic affected outdoor recreation in America?* Penn State University. Retrieved May 24, 2022, from https://www.psu.edu/news/health-and-human-development/story/how-has-covid-19-pandemic-affected-outdoor-recreation-america/

Wang, K., Xu, H., & Huang, L. (2020). Wellness tourism and spatial stigma: A case study of bama, China. *Tourism Management, 78*, 104039.

Westbrook, R. A. (1987). Product/consumption-based affective responses and post purchase processes. *Journal of Marketing Research, 24*, 258–270.

White, E., Bowker, J. M., Askew, A. E., Langner, L. L., Arnold, J. R., & English, D. B. (2016). Federal outdoor recreation trends: effects on economic opportunities. *Gen. Tech. Rep. PNW-GTR-945. Portland, OR: US Department of Agriculture, Forest Service, Pacific Northwest Station, 46*, 945.

Winter, P. L., Selin, S., Cerveny, L., & Bricker, K. (2019). Outdoor recreation, nature-based tourism, and sustainability. *Sustainability, 12*(1), 81.

Wolsko, C., Lindberg, K., & Reese, R. (2019). Nature-based physical recreation leads to psychological well-being: Evidence from five studies. *Ecopsychology, 11*(4), 222–235.

Xiang, Z., & Gretzel, U. (2010). Role of social media in online travel information search. *Tourism Management, 31*(2), 179–188.

Ye, Q., Law, R., & Gu, B. (2009). The impact of online user reviews on hotel room sales. *International Journal of Hospitality Management, 28*(1), 180–182.

Yu, C. P., Chang, W. C., & Ramanpong, J. (2019). Assessing visitors' memorable tourism experiences (MTEs) in forest recreation destination: A case study in Xitou nature education area. *Forests, 10*(8), 636.

Chapter 9
Overtourism: A Potential Outcome in Contemporary Tourism—Concepts, Indicators, and Impacts

Kim Pham, Christine A. Vogt, and Kathleen L. Andereck

Abstract Overtourism is increasingly an issue with popular destinations. The heavy volume of visitors across multiple seasons and the resentment of residents to tourism development and visitor volumes are two key concepts of overtourism. This chapter provides a thorough review of definitions and theoretical frameworks associated with overtourism and provides a toolkit of indicators and measurement of those indicators to gauge size of tourism economy, employment, tourism visitation, development levels, media reach, and stakeholder's perspectives. Overtourism is known for when tourism exceeds social and environmental capacities of a community.

Keywords Crowding · Sustainability · Resilience · Resident quality of life · Quality-of-visitor experience

Tourism can have both positive and negative quality-of-life consequences for communities. Part of what contributes to tourism's effects on quality of life is a community's capacity to adapt to change or its resilience. This is especially true in communities where tourism is a primary factor in the local economy (Cheer & Lew, 2017). Successful tourism can lead to negative consequences that community leaders must acknowledge and manage effectively. Overdevelopment, tourist crowds, overcommercialization, and the resulting residents' dissatisfaction with the community's quality of life and eroding of community resilience are some of these consequences. Gaining a better understanding of the ways that tourism can cause these negative impacts and the ways they can be mitigated is important to build resiliency and maintain a high quality of life for residents.

Overtourism is the concept that has been attached to tourism where the volume of tourists and tourism development exceeds the social and environmental capacity

K. Pham · C. A. Vogt · K. L. Andereck (✉)
School of Community Resources and Development, Arizona State University, Phoenix, AZ, USA
e-mail: Dtpham8@asu.edu; chrisv@asu.edu; kandereck@asu.edu

D. Chhabra et al. (eds.), *Sustainable Development and Resilience of Tourism*,
https://doi.org/10.1007/978-3-031-63145-0_9

limits of a community. In the early 2000s, the term was used to address concern about over-usage of natural resources for tourism (Koens et al., 2018). Ten years later, overtourism received attention from media and social media mostly following residents' protests and resistance against tourism in many cities around the world such as Barcelona, Venice, Amsterdam, and Dubrovnik (Goodwin, 2017). "Tourists go home," "Wish you weren't here," and "We have the right to rest" were the mottos of anti-tourism protests. More recently, the World Tourism Organization considered the anti-tourist movement "a very serious situation that needs to be addressed in a serious way" (Coldwell, 2017). After decades of sustainable tourism promotion and development, the industry faced the threatening phenomenon of over- and anti-tourism (Goodwin, 2017). During the 2019 to 2021 period, tourism was in a downward spiral from the coronavirus pandemic. According to the World Tourism Organization (UNWTO), there were one billion fewer international tourist arrivals in 2020 in comparison to 2019 (UNWTO, 2021). This temporary pause cooled down discussions about overtourism. In 2022, the industry started to recover as the pandemic subsided. Discussion about overtourism re-emerged.

Recently, the Center for Responsible Tourism (CREST) identified overtourism as one of four threats to people and planet, together with climate change, biodiversity and cultural heritage loss, and inequality, the poverty and wealth widening gap (CREST, 2023). Overtourism is clearly a contemporary phenomenon in the tourism industry, communities, and destinations, and scholars must recognize its emergence as an important research topic. An understanding of overtourism as a phenomenon is critical to developing resilient and sustainable communities that provide a positive quality of life for residents.

This paper is the first in a two-paper series that considers the contemporary phenomenon of overtourism in highly visited destinations based on a literature review. Literature (e.g., articles, books, and reports) was collected via university's library, academia database systems (e.g., JSTOR, ScienceDirect, Elsevier), and search engines (e.g., Google, Google Scholar). The authors synthesized and analyzed the literature, discussed the need to update our understanding, and then differentiated, reorganized, reconceptualized, and connected different concepts together within the umbrella of overtourism.

This first paper provides a conceptual overview of overtourism within the larger tourism system, while the second paper focuses on more pragmatic concerns and proposes a comprehensive model, the *overtourism ecosystem*, within which overtourism can be conceptualized and studied. Together, these papers are a start toward moving beyond the current norm or contemporary situation to enhance knowledge. They provide "new lenses for seeing things that have not been seen before, or for seeing familiar things differently, more clearly, more simply, or more deeply" (MacInnis, 2016, p. 1). This first paper answers four questions: what is overtourism? What are the primary conceptual and theoretical frameworks within which overtourism can be viewed? How can overtourism be identified? and What are the consequences of overtourism? Definitions of overtourism provide a context for the

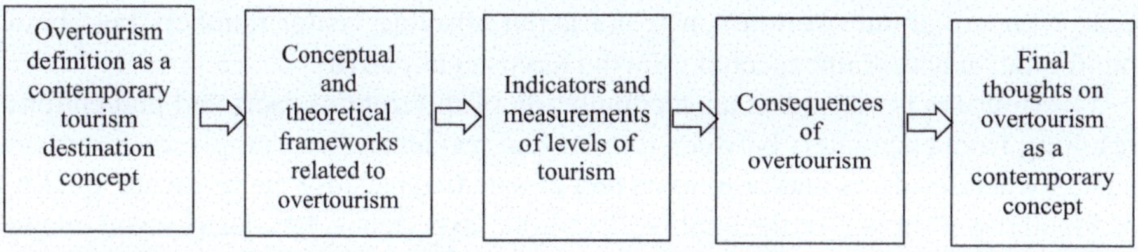

Fig. 9.1 The content flow of the book chapter

paper by clarifying the contemporary meaning of the concept. Figure 9.1 illustrates the content flow of this book chapter. Conceptual and theoretical frameworks that have been used to study the phenomenon point to the ways researchers have sought to build on understanding overtourism using explanatory models discussed, while indicators and measurements describe the ways in which variables in some models have been operationalized as well as ways of overtourism which it has been identified in communities. When overtourism has been identified, its consequences are briefly reviewed. Some final thoughts about the topics discussed in this chapter are then offered.

Defining Overtourism as a Contemporary Tourism Destination Concept

Overtourism is a complex and multifaceted concept. Several definitions have emerged by leading international tourism organizations. UNWTO (2018) defined overtourism as "the impact of tourism on a destination, or parts thereof, that excessively influences perceived quality of life of citizens and/or quality of visitors' experiences in a negative way (p. 4)." It is "the situation in which the impact of tourism, at certain times and in certain places, exceeds the physical, ecological, social, economic, psychological and/or political capacity thresholds of a destination" (Peeters et al., 2021, p. 17).

In the literature, overtourism is defined and viewed from different angles. Generally, overtourism occurs when the quantity and type of prevailing tourism exceed the destination's carrying capacity (Benner, 2019). It refers to the volume of tourists and physical condition of a tourism destination. Accordingly, overtourism happens when the tipping point or the carrying capacity of the tourism system has been reached (Goodwin, 2017), "where hosts or guests, locals or visitors feel that there are too many visitors and the quality of life in the area or the quality of the experience has deteriorated unacceptably" (Goodwin, 2017, p. 1) and when the number of tourists is higher than the number of local residents (Seraphin, 2018). Examples of overcrowded cities are Venice with 50,000 residents and 30 million tourists per year and Barcelona with 5 million residents and 11.3 million visitors per year. Concern about overtourism is not about absolute visitor numbers but about rapid relative and/or unexpected growth (Koens et al., 2018).

Overtourism is viewed as an accumulation of different impacts and perceptions resulting from encounters between tourists and residents. Accordingly, overtourism occurs when residents and/or tourists feel dissatisfied because the residents' quality of life and well-being, or the quality of visitors' experience, has deteriorated due to tourism development (Perkumienė & Pranskūnienė, 2019). It occurs when residents feel irritated with tourists' behaviors such as being drunk, vandalizing heritage sites, and disrespecting local culture (Pinke-Sziva et al., 2019). On the other hand, tourists may also feel less satisfied with the destination since they must compete with too many other tourists for places to visit or restaurants (Alexis, 2017).

Scholars refer to overtourism from destination management perspectives. The phenomenon is perceived when more and more tourists seek to consume a common resource, when destinations place tourism priorities over the interests of residents (Goodwin, 2017), or when tourism has become such a dominant sector that it may cause more problems than benefits (Oklevik et al., 2019). "Overtourism is a consequence of tourism using the destination rather than the destination using tourism" (Goodwin, 2017, p. 10).

Conceptual and Theoretical Frameworks Related to Overtourism

Definitions of overtourism are fairly new, coinciding with academic discussions about overtourism that appeared around the same time as the topic drew attention in the media in the early 2010s. The problem itself, however, is not new and has been discussed as early as in the 1960s when scholars wrote about how tourism negatively affects destinations (Perkumienė & Pranskūnienė, 2019). Noticeable related works were Pizam's description of the social costs to destination communities, Doxey's Irridex model, and Butler's tourism life cycle (Butler, 1980; Doxey, 1975; Pizam, 1978). They saw that excessive tourism development could harm the local environment and stimulate negative attitudes among residents. In 1980, the magazine GEO put forth a critical question: "How many tourists per hectare of beach?" calling for efforts to set an appropriate number of tourists at a certain destination (Erschbamer et al., 2018). Since then, different tourism researchers have established and discussed the frameworks and concepts of sustainable tourism, community resilience, residents' quality of life, residents' support for tourism, stakeholder theories, crowding, carrying capacity, limits of acceptable change, mass tourism, and touristification. Understanding these theories and concepts is important for overtourism management (Table 9.1).

Table 9.1 Overtourism-related theoretical frameworks and concepts

Theories/concepts	Implications for overtourism
Sustainable tourism	Implement sustainable/responsible tourism practices
Community resilience	Ability of a community to adapt to new conditions and challenges posed by overtourism
Residents' quality of life	Understand and consider residents' perspectives and sentiments about tourism. Involve residents in tourism planning and development decisions maximize benefits from tourism for residents
Resident's support for tourism development	
Doxey's Irridex	
Stakeholder theories	Involve stakeholders in tourism planning and development
Tourism area life cycle	Assess the status of tourism development. Diagnose and prevent overtourism before it happens
Crowding/ overcrowding	Consider the physical, social, and environmental limits in tourism development
(Tourism) carrying capacity	
Limits of acceptable change	
Mass tourism	Consider the impacts of different types of tourism on the destinations
Touristification	Tourism development to consider its impacts on the physical and social environment

Sustainable Tourism

Butler (1993) defines sustainable tourism as "tourism which is in a form which can maintain its viability in an area for an indefinite period of time" (p. 29). The UN World Tourism Organization defines it as "tourism that takes full account of its current and future economic, social, and environmental impacts, addressing the needs of visitors, the industry, the environment, and host communities" (GSTC, n.d.). Sustainable tourism has been adopted as a desirable objective for tourism development in many destinations and has become a key competitiveness variable (Jurado-Rivas & Sanchez-Rivero, 2019; Sharpley, 2000). This paradigm calls for attention to the impacts of tourism to a local destination; it aims to lessen conflicts among stakeholders and ensure long-term existence of resources. Sustainable tourism is considered a pathway to prevent and overcome overtourism issues (Cheung & Li, 2019; Maingi, 2019).

Community Resilience

Community resilience is the capacity of a community to renew or adapt to new conditions, disturbances, and challenges that render previous modes of operation unviable (Espiner et al., 2017). Community resilience is related to socio-ecological systems, the balance between social and environmental concerns, and a community's ability to adapt to shocks in the system (Cheer et al., 2019). Although communities have often been able to adapt to short-term challenges, rarely have they developed the capacity to adapt to long-term threats such as that posed by overtourism (Espiner et al., 2017). While sustainability and resilience are related concepts, the ability to adapt to a situation does not necessarily result in sustainable outcomes. Lew et al. (2016) suggest communities must be both sustainable and resilient given that both are necessary for community well-being and residents' quality of life.

Residents' Quality of Life

Spradley (1976, p. 100) defines quality of life (QOL) as "an overall state of affairs in a particular society that people evaluate positively." The overall purpose of tourism is (should be) to enhance the quality of residents' lives (McCool & Martin, 1994). Residents' QOL includes an economic domain, environmental domain, and social-cultural domain (Andereck & Nyaupane, 2011). According to the bottom-up spillover theory, residents' satisfaction with QOL is mostly influenced by evaluations of tourism impacts in those domains (Woo et al., 2015). Meanwhile, overtourism has been noted to have significant negative impacts on residents' QOL (Mihalic & Kuščer, 2022).

Residents' Support for Tourism Development

Residents' support of tourism in their community is essential for the development, competitiveness, and sustainability of destinations (Bimonte & Punzo, 2016). The goodwill and cooperation of the local community are critical for meaningful and memorable experiences for tourists. Negative attitudes or hostility towards tourists can make them feel unwelcome or intimidated and overshadow positive experiences that the tourism industry attempts to offer. Residents can affect tourism development by providing support for or oppose politicians and public policy initiatives (Woo et al., 2018). Residents' levels of support for tourism development can depend on their levels of satisfaction with the QOL in the destination under the impacts of tourism (Andereck & Nyaupane, 2011; Woo et al., 2015). In overtourism destinations, residents' QOL is often degraded due to tourism growth leading to their resentment, anger, and even backlash against tourism.

Doxey's Irridex

Doxey (1975) proposed an index based on residents' irritation levels to explain the attitudes of local people toward tourist inflow (Fig. 9.2). Accordingly, the feelings and behaviors of the host community toward tourists vary through different stages

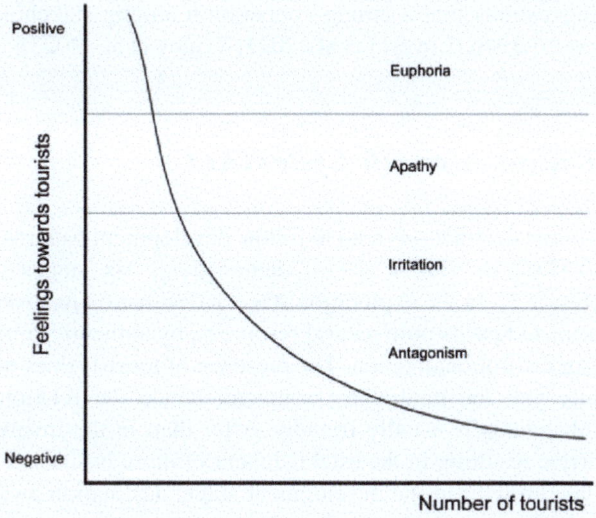

Fig. 9.2 Doxey's Irridex (Doxey, 1975)

of tourism development. When tourists first arrive in the destination (the *euphoria* stage), residents greet them with positive feelings and with no irritation toward the tourist activities. There is little planning and formalized control as there is greater potential for influence to be exerted by locals. In the second stage, this attitude of residents' changes to *apathy* when visitors are taken for granted and their relationships with residents become formal. Later, the surge in visitor numbers causes *annoyance* or *irritation* to residents due to problems such as traffic jams, visitors' disrespect to local culture, and pollution. At the final stage, *antagonism*, residents no longer perceive tourism as good for their society. Residents resent tourism development and develop different ways to cope with the situation including avoiding tourists; they may even express it in a violent way.

Stakeholder Theories

Stakeholders in tourism are "any entity that is influenced by, or that may influence, the achievement of the destination management activities" (Sheehan & Ritchie, 2005, p. 9). Main groups of stakeholders include tourists, businesses, local community, government, special interest groups, and educational institutions (Butler, 1999; Markwick, 2000). Stakeholders influence different aspects of the industry including tourism supply and demand, regulation, management, human resources, and research (Hieu & Rašovská, 2017). Sustainable tourism development cannot be separated from stakeholders' participation (Liu & Ma, 2017). Studies have attempted to identify the responsibilities of stakeholders in causing overtourism and their perspectives about it (Dodds & Butler, 2019a, b; Eckert et al., 2019). There have been suggestions that a strong cooperation among stakeholders is essential in solving overtourism (Loverio et al., 2021; Yrigoy et al., 2023).

Tourism Area Life Cycle (TALC)

Tourist destinations tend to follow development patterns. The tourism area life cycle (TALC) by Butler (1980) demonstrates six stages in a destination life cycle (Fig. 9.3). In the exploration stage, a destination receives a small number of visitors attracted by its pure natural resources, by authentic culture, and by the lack of tourism institutionalization. The presence of tourists does not disturb the local community and the natural environment; hence the relationship between tourists and inhabitants is usually friendly. After that, in the involvement stage, more tourists arrive resulting in the establishment of more businesses and facilities to accommodate growth. In the development stage, destination awareness increases, and more tourists come in large groups. New infrastructures and facilities are developed. With a well-defined tourism market, a destination attracts more outside investments. In the consolidation (or maturity) stage, tourism becomes the major economic sector of the destination. There may be more visitors than local residents. Lastly, the stagnation stage is characterized by the negative impacts of tourism on the destination's society, economy, and environment. A destination can decline with less visitors and shrinking revenues; recession of the tourism industry begins. It can also survive and rejuvenate intentionally through innovative management strategies. Overtourism is often associated with the mature stage of destinations. A downturn future caused by overtourism can be avoided by anticipating and embracing sustainability from the early stages of the life cycle.

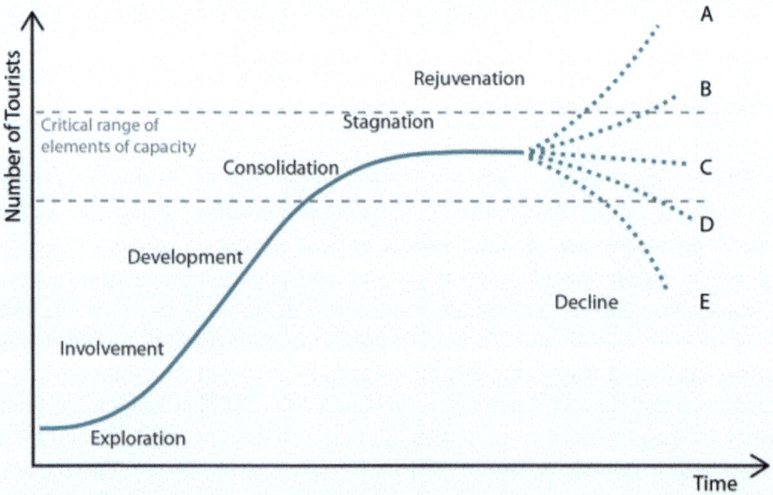

Fig. 9.3 Original tourism area life cycle model (Butler, 1980)

Crowding and Overcrowding

Crowding or overcrowding are terms often used interchangeably to indicate the most prominent symptom and main cause of overtourism. Crowding "implies exceeding the maximum number of people that can visit a destination simultaneously" and "refers to a certain level of destination saturation" (Sanz-Blas et al., 2019, p. 3). Actual or perceived levels of unacceptable use are critical to crowding evaluation (Manning, 2022). Crowding can be perceived differently depending on factors such as density of visitors, types of tourism activities, physical characteristics of the sites or facilities, and demographics of perceivers (Lee et al., 2019; Ryan & Cessford, 2003). Crowding in overtourism can refer to the spatial concentration of infrastructure, facilities, and equipment for tourism (Alonso-Almeida et al., 2019). Related to this territorial dimension, several concepts were coined such as accommodation saturation, accommodation overtourism, or accommodation overcrowding. Measurement of overcrowding can be both objective such as density and intensity (Peeters et al., 2021) and subjective such as perceptions (Jacobsen et al., 2019).

Several theories are subsumed within crowding theory that can be used to understand overtourism. Expectancy theory suggests that individuals' expectations can influence how they evaluate the crowd (Andereck & Becker, 1993). For example, tourists who plan for social activities such as participating in concerts and festivals tend to expect or prefer crowding more than those who seek for solitary activities (Wickham & Kerstetter, 2000). Social interference theory connects crowding to people's feelings of spatial restriction (Li et al., 2017). Stimulus overload theory emphasizes the psychic stress of people experiencing crowding (Lee & Graefe, 2003; Milgram, 1970). Crowding can be negative or positive depending on the perceiver (Manning, 2022).

Tourism Carrying Capacity

Tourism carrying capacity is defined as the maximum number of visitors at a destination, "the critical range of elements of capacity" that represents "the peak number of visitors" as "capacity levels for many variables have been reached or exceeded with attendant environmental, social, and economic problems" (Butler, 1980, p. 8). Carrying capacity is the level of use beyond which impacts exceed levels specified by evaluative standards (Manning, 2022).

Carrying capacity is composed of several dimensions (Zelenka & Kacetl, 2014). Physical carrying capacity (also referred to as environmental or ecological carrying capacity) reflects the impacts on ecosystems, the built cultural environment, and the infrastructure (Coccossis & Mexa, 2004). This dimension emphasizes the actual physical limitations of a space that can accommodate a certain number of persons without causing deterioration to the environment. The physical setting is comprised of fixed components belonging to a natural system that human actions cannot influence and flexible components belonging to the infrastructure systems such as the roads and utilities (Coccossis & Mexa, 2004). Social or psychological carrying capacity refers to both the quality of the experience of visitors to the area and the levels of tolerance of the host population. It is the level beyond which the negative impacts of tourism make tourists feel uncomfortable and less attracted to the destination while residents become less welcoming to tourists (Mathieson & Wall, 1982; O'Reilly, 1986). In overtourism "the upper limit is both the maximum levels of overcrowding that visitors are willing to accept and the maximum levels of tourism and its (negative) impacts that local residents are willing to accept" (Mihalic, 2020, p. 4). Social capacity thresholds are difficult to evaluate since they rely entirely on subjective judgments by visitors and residents.

Limits of Acceptable Change (LAC)

Associated with carrying capacity is the concept of limits of acceptable change (LAC). While carrying capacity focuses on the question of "how much is too much," LAC concerns "how much change is acceptable" (Weaver, 2006). The LAC was initially designed to address visitor management issues in the US's National Wilderness Preservation System. Accordingly, nature-based tourism is controlled by anticipating over-usage and proper planning (Stankey et al., 1985). LAC emphasizes relationships between existing and desired or "acceptable" conditions which are targets for the amount of change allowed in particular settings. Management is necessary to identify problems and determine suitable solutions. Defining the amount of change that is acceptable to stakeholders as well as some compromise among differing needs is at the heart of the LAC (Ahn et al., 2002). The extent of tourism impacts that may remain acceptable to local stakeholders could be influenced by different social and economic factors. When residents are under financial difficulty or crisis, for instance, they may be more tolerant of tourists and negative impacts caused by tourism development (Frauman & Banks, 2011).

The LAC process involves nine steps as illustrated in Fig. 9.4. The model helps to guide management of a resource from determining desired baseline conditions to setting up indicators and standards to decide when degradation or too much change has occurred.

Mass Tourism

Mass tourism has been a part of academic discussions for decades (Vainikka, 2013). Mass tourism is an offspring of industrialization and democracy, consumption, and globalization (Hillali, 2003) and was enabled with economic enrichment of societies, social progress, the reduction of working time, and the development of

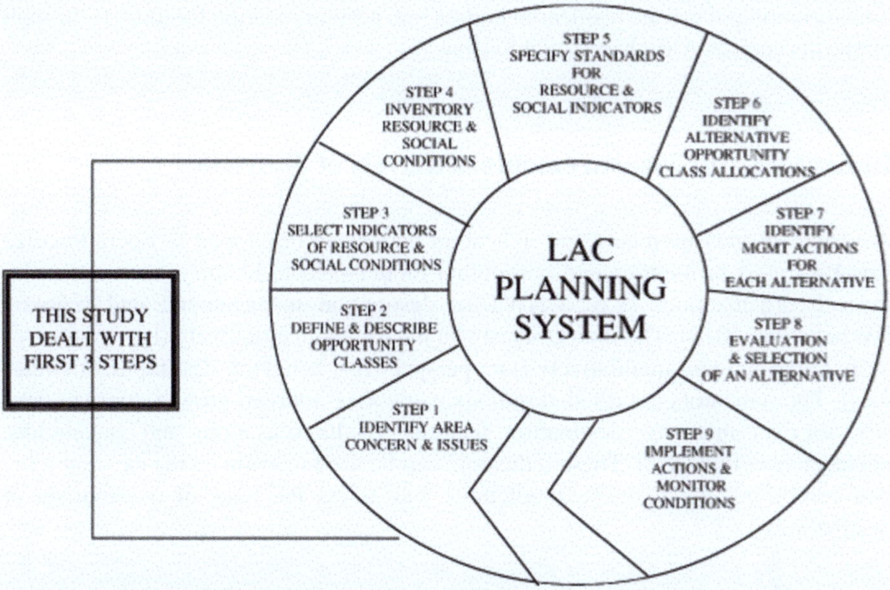

Fig. 9.4 The LAC planning system (Stankey et al., 1985)

transportation and technology (Theng et al., 2015). Mass tourism is linked to mass production, mass consumption, and mass destinations (Vainikka, 2013). It is often associated with package tours that provide multiple services at cheap price such as transportation, accommodation, food and beverage, and tour guide and accommodates a large number of travelers visiting a destination (Theng et al., 2015). Mass tourism is associated with little consideration of local norms or culture. The quality of mass tourism is poor and often placed opposite to other forms of tourism that is considered to be more positive such as ecotourism, nature-based tourism, heritage tourism, and rural tourism (Poon, 1993; Vainikka, 2013). Mass tourism brings in large numbers of tourist to communities and contributes significantly to the establishment of overtourism (Milano et al., 2019).

Touristification

Accompanying the growth of the tourism industry is touristification. Overwhelming touristification leads to overtourism. Touristification is defined as a relatively spontaneous and unplanned process to transform a space into tourism commodity; it is often related to massive tourism development (Koens et al., 2018). An example of physical touristification is the changing of retail landscape, which increasingly address tourists' demand instead of local people. Another example is the transformation of private residential houses into accommodation for tourists through platforms such as Airbnb or CouchSurfing.

Indicators and Measurements of Levels of Tourism

A variety of measurements and indicators have been developed to operationalize variables used in theories and conceptual models and to identify overtourism by tourism organizations (e.g., UNWTO), destination management, and scholars (Weber et al., 2019). The indicators can be measured quantitatively (i.e., number of tourist arrivals) and qualitatively (i.e., perspectives, behavior, development strategies). The indicators are divided into six categories: tourism growth, tourism density, tourism intensity, destination features, media sensation, and stakeholder perspectives (Table 9.2). These indicators can be used to evaluate the existing tourism conditions and to make predictions and assess the risks of overtourism at destinations.

Tourism Growth

Tourism growth indicators focus on changes in the tourism industry (tourism shares of GDP, tourism shares of employment, growth in tourist arrivals contribution to GDP, tourist's arrivals) or specific facilities of services (overnight stays, bed capacity, air transports) (McKinsey & Company and the WTTC, 2017; Peeters et al., 2021; Preveden et al., 2018). The indicators show how fast tourism is developing and its dominance in the local economy and identify the overtourism phenomenon.

Tourism Intensity

Most of the tourism density indicators calculate the number of visitors or tourism facilities and services in relation to the local population. This calculation can generate the number of each of these items per one or 1000 residents as well as the ratio over the whole population: visitors, air passengers, cruise passengers, bed-nights, beds, and overnight stays (Baños et al., 2019; McKinsey & Company and the WTTC, 2017; Martín et al., 2018; Peeters et al., 2021 ; Preveden et al., 2018). Some intensity indicators are not based on population. Examples are the number of passengers over the number of bed nights or the ratio between Airbnb booking over the general accommodation booking (Peeters et al., 2021). Tourism intensity indicators, together with tourism density ones, show the pressure of tourism development on the host community and destinations' capacity.

Table 9.2 Tourism measurements/indicators that highlight overtourism conditions

Indicators	Measurements	Sources
Tourism growth		
Tourism shares of GDP	Tourism GPD/general GDP (%)	McKinsey & Company and the WTTC (2017)
Tourism shares of employment	Tourism employment/general employment (%)	McKinsey & Company and the WTTC (2017)
Growth in tourist arrivals	%/year	McKinsey & Company and the WTTC (2017)
Growth in overnight stays	%/year	Preveden et al. (2018)
Growth in bed capacity	%	Preveden et al. (2018)
Growth of air transports	%	Peeters et al. (2021)
Tourism density		
Visitor density	Visitors/km^2	McKinsey & Company and the WTTC (2017)
	Travelers accommodated/ km^2	Cruz and Zaragoza (2019)
	Area of beach/tourist (m^2)	Cruz and Zaragoza (2019)
Accommodation density	Bed-nights/km^2	Peeters et al. (2021)
	Beds/km^2	Baños et al. (2019)
	Area of accommodation /km^2	Cruz and Zaragoza (2019)
	Accommodation / km^2	Cruz and Zaragoza (2019)
Tourism intensity		
Visitor intensity	Visitors/residents	McKinsey & Company and the WTTC (2017)
	Cruise passengers/residents	Martín et al. (2018)
	Air passengers/bed night	Peeters et al. (2021)
Accommodation intensity	Bed-nights/residents	Peeters et al. (2021)
	Bed-nights/1000 residents	Šegota et al. (2017)
	Beds/residents	Baños et al. (2019)
	Overnight stays/residents	Preveden et al. (2018)
	Airbnb prevalence/booking	Peeters et al. (2021)
Destination features		
Arrival seasonality	Difference in arriving-flight seats between high and low month (ratio)	McKinsey & Company and the WTTC (2017)
Growth of air transports	%	Peeters et al. (2021)
Airbnb average shortest distance to booking.com addresses	Km	Peeters et al. (2021)
World heritage site closeness	Number within 30 km	Peeters et al. (2021)
Cruise harbor closeness	Number within 10 km	Peeters et al. (2021)
Airport closeness	Number of arrivals within 50 km	Peeters et al. (2021)
Number of UNESCO World Heritage Sites	Number	Peeters et al. (2021)
Number of conferences	Number	Preveden et al. (2018)

(continued)

Table 9.2 (continued)

Indicators	Measurements	Sources
Media sensation		
Attraction concentration	Share of reviews limited to top 5 attractions (%)	McKinsey & Company and the WTTC (2017)
Historic site prevalence	Share of top 20 Tripadvisor attractions that are historic sites (%)	McKinsey & Company and the WTTC (2017)
Negative Tripadvisor reviews	Share of "poor" or "terrible" reviews among top attractions (%)	McKinsey & Company and the WTTC (2017)
Stakeholders' perspectives		
Visitors' perspectives	Visitors' feeling toward crowding	Jacobsen et al. (2019)
	Visitors' satisfaction with tourism experience in overtourism context	Dodds and Holmes (2019), Jacobsen et al. (2019) and Liu & Ma (2019)
	Visitor's perception about impacts of overtourism	Krajickova et al. (2022)
Residents' perspectives	Residents' perceptions about impacts of overtourism	Kim and Kang (2020), Muler Gonzalez et al. (2018) and Szromek et al. (2019)

Destination Features

Several overtourism indicators target destination features. McKinsey & Company and the WTTC (2017) measure the difference in arriving-flight seats between high and low months in destinations. Peeters et al. (2021) use several measurements on the distance of the destinations to important tourism infrastructure and attractions (airport, heritage sites, cruise harbor). The numbers of UNESCO World Heritage Sites and the number of conferences organized at destinations are considered (Peeters et al., 2021; Preveden et al., 2018). These indicators show how much destinations are prone to overtourism.

Media Sensation

There are overtourism indicators related to media sensation. McKinsey & Company and the WTTC (2017) created three indicators: share of reviews limited to top 5 attractions (%) to measure the attraction concentration at destinations, share of top 20 Tripadvisor attractions that are historic sites (%) to measure the prevalence of historic sites, and share of "poor" or "terrible" reviews among top attractions (%) to measure visitor experience. The reviews on media can help to explain the causes or impacts of overtourism.

Stakeholder Perspectives

Regarding stakeholder perspectives, indicators mostly focus on measuring overtourism through the perspectives of visitors and residents. Indicators related to visitors' perspectives include their feelings toward crowding (Jacobsen et al., 2019), their satisfaction with tourism experience in overtourism context (Dodds & Holmes, 2019; Jacobsen et al., 2019; Liu & Ma, 2019), and their perceptions about impacts of overtourism (Krajickova et al., 2022). Indicators related to residents' perspectives mostly measure their perceptions about impacts of overtourism (Kim & Kang, 2020; Muler Gonzalez et al., 2018; Szromek et al., 2019). A relative scale such as a 5-point or 9-point crowding scale is the most common type of measurement for stakeholders' perspectives indicators (Jacobsen et al., 2019; Manning, 2022).

Consequences of Overtourism

Indicators and measures have provided relatively objective ways to identify overtourism, but it has been known for a long time that tourism can cause negative impacts to a destination environmentally, economically, and culturally/socially (Cheer et al., 2019; Peeters et al., 2021). Using indicators reviewed, impacts have been documented in diverse destinations. When a destination reaches overtourism status, the impacts are much more obvious and severe than in more typical tourism settings. Residents' quality of life and well-being is reduced, destination resilience is compromised, and the destination deteriorates sometimes to the point of ruination. The negative impacts of overtourism can be felt by tourists; however, the feelings and reactions of residents and tourists toward overtourism can differ. Table 9.3 summarizes potential impacts.

Environmental Impacts

Overtourism impacts destinations environmentally, mainly as the result of increased usage of resources, infrastructure, facilities, and/or touristic sites (Peeters et al., 2021). Environmental damage includes, but is not limited to, erosion and changes in landscape, waste and litter problems, water contamination, shortage of drinking water, air pollution, and destruction of the ecosystem (Dodds & Butler, 2019a, b; González, 2018; Oklevik et al., 2019). Another negative impact of overtourism is its pressure on local infrastructure and energy consumption, causing problems such as traffic congestion and shortage of energy. In many places, overtourism adds to the problems that already exist (e.g., waste and water management) (Koens et al., 2018). Tourism can damage cultural and historic sites. Touristification, a process to change places into tourism destinations, which involves building infrastructures (e.g., hotels, resorts, roads) can change the natural landscape of the destinations.

Table 9.3 Impacts of overtourism

Domains	Impacts
Environment	Pollution (e.g., air, noise, water) Pressure on natural resources (e.g., erosion and changes in landscape, loss of habitat for local species) Pressure on public infrastructure and energy consumption (e.g., transportation, water, electricity) Damages to cultural/historic sites Touristification
Economy	Increase cost of living (e.g., goods, service, real estate) Lack of affordable housing Tourism dependency that weakens other economic sectors Tourism leakage
Culture society	Public safety issues (e.g., crime, accidents, fire danger in campsites, alcohol consumption, prostitution) Cultural clashes between tourists and residents/nuisance or illegal behaviors by visitors (e.g., trespassing, drinking) Loss of community sense/identity and cohesion Degradation of local traditions and culture/loss of authenticity Gentrification Anti-tourism or tourismphobia sentiments/lessened residents' support for tourism or protests against tourism Lessen quality of visitor experience and satisfaction

Economic Impacts

Economic impacts of overtourism are caused by increased demand for tourism services and facilities (Peeters et al., 2021). One of the issues is increased cost of living. Residents face inflated costs of services and products, especially in housing; many have to relocate to more affordable places (Capocchi et al., 2019). Shops and facilities that once met the needs of residents have been displaced by those that sell expensive goods or services to tourists. Overtourism can also lead to economic dependence on tourism, overshadowing other sectors (Goodwin, 2017). In many destinations, despite the exponential growth of tourism, revenue from the industry hardly stays in the local economy but goes to outsiders such as foreign investors, a phenomenon called leakage (Milano et al., 2022; Peterson & DiPietro, 2021).

Sociocultural Impacts

Increased demand and visitation lead to sociocultural impacts at overcrowded destinations (Peeters et al., 2021). Several public safety concerns are related to tourism such as crime, accidents, fire danger in campsites, alcohol consumption, and prostitution (Petroman et al., 2022). Cultural clashes often result from the different values and behavior of tourists relative to local norms (Goodwin, 2017). Following the spread of tourists into formerly residential neighborhoods to stay at short-term rental places, there have been concerns about nuisance behavior from tourists and the loss of community identity and cohesion (Martín Martín et al., 2018). When overtourism occurs, some community residents feel that "their identity has been subverted" either by marketers or by visitors (Goodwin, 2017, p. 2). Gentrification adds to the severity of the issue when increased cost of living and other tourism issues push residents, especially those of lower income, out of their residential areas which are then taken over by tourism facilities or others with higher status and income (Jover & Díaz-Parra, 2020). Gradually, the destinations lose their traditions and authenticity which once charmed tourists.

In many destinations, overtourism results in discontent and decreased support of residents toward tourism. Protests against tourism have happened in Venice and other European cities (Coldwell, 2017). The phenomenon/movement is known as tourismphobia or anti-tourism (Milano et al., 2019). It is the outburst of resentment among residents toward the uncontrolled growth of tourism, signaling a change of paradigm when locals place more importance on their quality of life than tourism income (Koens et al., 2018).

Those experiencing overtourism are not only residents but also tourists. High levels of occupancy at hotels, restaurants, and other public places together with the degraded environmental quality at many destinations make tourists feel uncomfortable (Cruz & Zaragoza, 2019). Simmonds et al. (2018) note dissatisfaction among visitors to several American national parks that have become overcrowded in recent years (e.g., Yellowstone, Zion, Grand Tetons). Visitors are tired of waiting in long lines to enter national parks or have difficulty in finding a parking spot.

In comparison to destinations with modest numbers of visitors, the effects of overtourism are felt stronger among residents than tourists. Generally, tourists are more tolerant of the crowd when visiting a culture other than their own since they only deal with the situation for a short time (Li et al., 2017). Tourismphobia or anti-tourism forms differently in residents than in visitors. Goodwin (2017) suggests residents tolerate but grumble about the crowd of tourists for a long time. When visitation levels become intolerable with significant negative impacts, anger erupts, and residents launch protests such as what occurred in Barcelona, Berlin, and Venice. Goodwin (2017) suggests that tourists' dissatisfaction with the destination tends to be delayed; "each generation of tourists discovers the destination afresh—they have no experience or memory of how it was before to fuel their dissatisfaction" (p. 8).

Overall Impacts of Overtourism

Overall, impacts from overtourism degrade residents' quality of life, visitors' quality of experience, community sustainability and resilience, and places destinations on the verge of being ruined physically and reputationally. Panayiotopoulos and Pisano (2019) call this "overtourism dystopia" or "the paradox of tourism risks destroying the very thing that tourists come to see (p. 7)." Maya Bay in Thailand is one example; the damage of tourism on the island's environment was so severe that it was closed to the public indefinitely for its coral reef to recover (CNN Travel, 2018). Concern about how destinations could be lost through overtourism, González (2018, p. 37) writes: "Cities can die by three different ways: when they're destructed by an enemy, when a new civilization set up by force, expelling the natives and their gods and thirdly, when the inhabitants themselves lose their memory and become foreigners in their own city." Social media recently has made several lists of places that travelers may want to avoid due to overtourism (Phi, 2019). The negative impacts of overtourism can extend beyond one destination's limit, affecting the broad travel supply chain across destinations (Cheer et al., 2019).

Final Thoughts on Overtourism as a Contemporary Concept

This paper lays out overtourism as a concept that has been attached to communities with a high volume of tourism that is coupled with tourism development which exceeds social and environmental capacity limits of a community. A plethora of classic and new conceptual models, theories, and measures or indicators have merged in recent years to guide research and assist communities to identify the evolution of overtourism and associated impacts. The next paper goes further with this base knowledge to provide a suite of tools and solutions for communities and scholars to moderate crowding and other negative impacts from overtourism. Many of these solutions are grounded in sustainability and resilient community approaches.

References

Ahn, B., Lee, B., & Shafer, C. S. (2002). Operationalizing sustainability in regional tourism planning: An application of the limits of acceptable change framework. *Tourism Management, 23*(1), 1–15.

Alexis, P. (2017). Overtourism and anti-tourist sentiment: An exploratory analysis and discussion. *Ovidius University Annals, Economic Sciences Series, 17*(2), 288–293.

Alonso-Almeida, M. D. M., Borrajo-Millán, F., & Yi, L. (2019). Are social media data pushing overtourism? The case of Barcelona and Chinese tourists. *Sustainability, 11*(12), 3356.

Andereck, K. L., & Becker, R. H. (1993). Perceptions of carry-over crowding in recreation environments. *Leisure Sciences, 15*(1), 25–35.

Andereck, K. L., & Nyaupane, G. P. (2011). Exploring the nature of tourism and quality of life perceptions among residents. *Journal of Travel Research, 50*(3), 248–260.

Baños, C. J., Hernández, M., Rico, A. M., & Olcina, J. (2019). The hydrosocial cycle in coastal tourist destinations in Alicante, Spain: Increasing resilience to drought. *Sustainability, 11*(16), 4494.

Benner, M. (2019). From overtourism to sustainability: A research agenda for qualitative tourism development in the Adriatic. *German Journal of Economic Geography, 2*(64), 74–87.

Bimonte, S., & Punzo, L. F. (2016). Tourist development and host–guest interaction: An economic exchange theory. *Annals of Tourism Research, 58*, 128–139.

Butler, R. (1980). The concept of a tourist area cycle of evolution: Implications for management of resources. *Canadian Geographer, 24*, 5–12.

Butler, R. W. (1993). Tourism–An evolutionary perspective. In J. G. Nelson, R. W. Butler, & G. Wall (Eds.), *Tourism and sustainable development: Monitoring, planning, managing* (pp. 25–39). Univ. of Waterloo Press.

Butler, R. W. (1999). Sustainable tourism: A state-of-the-art review. *Tourism Geographies, 1*(1), 7–25.

Capocchi, A., Vallone, C., Amaduzzi, A., & Pierotti, M. (2019). Is 'overtourism' a new issue in tourism development or just a new term for an already known phenomenon? *Current Issues in Tourism, 23*, 1–5.

Cheer, J. M., & Lew, A. A. (2017). Understanding tourism resilience: Adapting to social, political, and economic change. In *Tourism, resilience and sustainability* (pp. 3–17). Routledge.

Cheer, J. M., Milano, C., & Novelli, M. (2019). Tourism and community resilience in the Anthropocene: Accentuating temporal overtourism. *Journal of Sustainable Tourism, 27*(4), 554–572.

Cheung, K. S., & Li, L. H. (2019). Understanding visitor–resident relations in overtourism: Developing resilience for sustainable tourism. *Journal of Sustainable Tourism, 27*, 1–20.

CNN Travel. (2018, Oct 2). *Thailand bay made popular by 'The Beach' closes indefinitely.* https://www.cnn.com/travel/article/maya-bay-closure-thailand/index.html

Coccossis, H., & Mexa, A. (2004). *Tourism and carrying capacity*. Continuum.

Coldwell, W. (2017, August 10). First Venice and Barcelona: Now anti-tourism marches spread across Europe. *The Guardian*. Retrieved from https://www.theguardian.com/travel/2017/aug/10/anti-tourism-marches-spread-across-europe-venice-barcelona

CREST (Center for Responsible Tourism). (2023, February 15). *Power point slides in the academic affiliate annual meeting*.

Cruz, M. S., & Zaragoza, M. P. P. (2019). Analysis of the accommodation density in coastal tourism areas of insular destinations from the perspective of overtourism. *Sustainability, 11*(11), 1–19.

Dodds, R., & Butler, R. (2019a). The phenomena of overtourism: A review. *International Journal of Tourism Cities, 5*(4), 519–528.

Dodds, R., & Butler, R. (Eds.). (2019b). *Overtourism: Issues, realities and solutions* (Vol. 1). Walter de Gruyter GmbH & Co KG.

Dodds, R., & Holmes, M. R. (2019). Beach tourists; What factors satisfy them and drive them to return. *Ocean & Coastal Management, 168*, 158–166.

Doxey, G. (1975). A causation theory of visitor-resident irritants: Methodology and research inferences. In *Travel and Tourism research association sixth annual conference proceedings* (pp. 195–198).

Eckert, C., Zacher, D., Pechlaner, H., Namberger, P., & Schmude, J. (2019). Strategies and measures directed towards overtourism: A perspective of European DMOs. *International Journal of Tourism Cities, 5*(4), 639–655.

Erschbamer, G., Innerhofer, E., & Pechlaner, H. (2018). *Overtourism: How much tourism is too much*. Eurac Research.

Espiner, S., Orchiston, C., & Higham, J. (2017). Resilience and sustainability: A complementary relationship? Towards a practical conceptual model for the sustainability–resilience nexus in tourism. *Journal of Sustainable Tourism, 25*(10), 1385–1400.

Frauman, E., & Banks, S. (2011). Gateway community resident perceptions of tourism development: Incorporating importance-performance analysis into a limits of acceptable change framework. *Tourism Management, 32*(1), 128–140.

Goodwin, H. (2017). The challenge of overtourism. *Responsible tourism partnership, 4*, 1–19.

González, A. (2018). Venice: The problem of overtourism and the impact of cruises. *Investigaciones Regionales=Journal of Regional Research, 42*, 35–51.

GSTC (Global Sustainable Tourism Council). (n.d.). *What is sustainable tourism?*. https://www.gstcouncil.org/what-is-sustainable-tourism/

Hieu, V. M., & Rašovská, I. (2017). Developing cultural tourism upon stakeholders' perceptions toward sustainable tourism development in Phu Quoc Island, Vietnam. *Research & Science Today, 2*(14).

Hillali, M. (2003). *Le tourisme international vu du sud: Essai sur la problématique du tourisme dans les pays en développement, Québec*. Presse de l'Université du Québec.

Jacobsen, J. K. S., Iversen, N. M., & Hem, L. E. (2019). Hotspot crowding and overtourism: Antecedents of destination attractiveness. *Annals of Tourism Research, 76*, 53–66.

Jover, J., & Díaz-Parra, I. (2020). Gentrification, transnational gentrification and touristification in Seville. *Urban Studies, 57*(15), 3044–3059.

Jurado-Rivas, C., & Sanchez-Rivero, M. (2019). Willingness to pay for more sustainable tourism destinations in world heritage cities: The case of Caceres, Spain. *Sustainability, 11*(21), 5880.

Kim, S., & Kang, Y. (2020). Why do residents in an overtourism destination develop anti-tourist attitudes? An exploration of residents' experience through the lens of the community-based tourism. *Asia Pacific Journal of Tourism Research, 25*(8), 858–876.

Koens, K., Postma, A., & Papp, B. (2018). Is overtourism overused? Understanding the impact of tourism in a city context. *Sustainability, 10*(12), 4384.

Krajickova, A., Hampl, F., & Lancosova, E. (2022). Visitors' perception of overtourism impacts in a small destination. *Anatolia, 33*(2), 236–246.

Lee, H., & Graefe, A. R. (2003). Crowding at an arts festival: Extending crowding models to the front country. *Tourism Management, 24*(1), 1–11.

Lee, K. J., Morgan, M., & Shim, H. (2019). Crowding, race, and ethnicity: A case study at Onondaga Cave State Park. *Journal of Interpretation Research, 24*(1), 49–61.

Lew, A. A., Ng, P. T., Ni, C.-C., & Wu, T.-C. (2016). Community sustainability and resilience: Similarities, differences and indicators. *Tourism Geographies, 18*(1), 18–27.

Li, L., Zhang, J., Nian, S., & Zhang, H. (2017). Tourists' perceptions of crowding, attractiveness, and satisfaction: A second-order structural model. *Asia Pacific Journal of Tourism Research, 22*, 1250–1260.

Liu, J., & Ma, Y. (2017). The perceptual differences among stakeholders in the tourism supply of Xi'an City, China. *Sustainability, 9*(2), 214.

Liu, A., & Ma, E. (2019). Travel during holidays in China: Crowding's impacts on tourists' positive and negative affect and satisfactions. *Journal of Hospitality and Tourism Management, 41*, 60–68.

Loverio, J. P., Chen, L. H., & Shen, C. C. (2021). Stakeholder collaboration, a solution to overtourism? A case study on Sagada, The Philippines. *Tourism Geographies*, 1–22.

MacInnis, D. J. (2016). Developing conceptual articles for JCR: Research curation. *Journal of Consumer Research*.

Maingi, S. W. (2019). Sustainable tourism certification, local governance and management in dealing with overtourism in East Africa. *Worldwide Hospitality and Tourism Themes, 11*(5), 532–551.

Manning, R. E. (2022). *Studies in outdoor recreation*. Oregon State University Press.

Markwick, M. C. (2000). Golf tourism development, stakeholders, differing discourses and alternative agendas: The case of Malta. *Tourism Management, 21*(5), 515–524.

Martín Martín, J., Guaita Martínez, J., & Salinas Fernández, J. (2018). An analysis of the factors behind the citizen's attitude of rejection towards tourism in a context of overtourism and economic dependence on this activity. *Sustainability, 10*(8), 2851.

Mathieson, A., & Wall, G. (1982). *Tourism, economic*. Physical and Social Impacts.

McCool, S. F., & Martin, S. R. (1994). Community attachment and attitudes toward tourism development. *Journal of Travel Research, 32*(3), 29–34.

McKinsey & Company and the WTTC (World Travel & Tourism Council). (2017). *Coping with success—Managing overcrowding in tourism destinations*. https://www.mckinsey.com/industries/travel-logistics-and-infrastructure/our-insights/coping-with-success-managing-overcrowding-in-tourism-destinations

Mihalic, T. (2020). Conceptualizing overtourism: A sustainability approach. *Annals of Tourism Research, 84*, 103025.

Mihalic, T., & Kuščer, K. (2022). Can overtourism be managed? Destination management factors affecting residents' irritation and quality of life. *Tourism Review, 77*(1), 16–34.

Milano, C., Novelli, M., & Cheer, J. M. (2019). Overtourism and tourismphobia: A journey through four decades of tourism development, planning and local concerns. *Tourism Planning & Development, 16*(4), 353–357.

Milano, C., Novelli, M., & Cheer, J. M. (2022). Overtourism. In *Encyclopedia of Tourism Management and Marketing* (pp. 413–416). Edward Elgar Publishing.

Milgram, S. (1970). The experience of living in cities: A psychological analysis. In F. F. Korten, S. W. Cook, & J. I. Lacey (Eds.), *Psychology and the problems of society* (pp. 152–173). American Psychological Association.

Muler Gonzalez, V., Coromina, L., & Gali, N. (2018). Overtourism: Residents' perceptions of tourism impact as an indicator of resident social carrying capacity-case study of a Spanish heritage town. *Tourism Review, 73*(3), 277–296.

Oklevik, O., Gössling, S., Hall, C. M., Steen Jacobsen, J. K., Grøtte, I. P., & McCabe, S. (2019). Overtourism, optimization, and destination performance indicators: A case study of activities in Fjord Norway. *Journal of Sustainable Tourism*, 1–21.

O'Reilly, A. M. (1986). Tourism carrying capacity: Concept and issues. *Tourism Management, 7*(4), 254–258.

Panayiotopoulos, A., & Pisano, C. (2019). Overtourism dystopias and socialist utopias: Towards an urban armature for Dubrovnik. *Tourism Planning & Development, 16*, 1–18.

Peeters, P. M., Gössling, S., Klijs, J., Milano, C., Novelli, M., Dijkmans, C. H. S., et al. (2021). *Research for TRAN committee-overtourism: Impact and possible policy responses*. European Parliament, Directorate General for Internal Policies, Policy Department B: Structural and Cohesion Policies, Transport and Tourism.

Perkumienė, D., & Pranskūnienė, R. (2019). Overtourism: Between the right to travel and residents' rights. *Sustainability, 11*(7), 21–38.

Peterson, R., & DiPietro, R. B. (2021). Is Caribbean tourism in overdrive? Investigating the antecedents and effects of overtourism in sovereign and nonsovereign small Island tourism economies (SITEs). *International Hospitality Review, 1*(35), 19–40.

Petroman, I., Văduva, L., Marin, D., Sava, C., & Petroman, C. (2022). Overtourism: Positive and negative impacts. *Quaestus, 20*, 171–182.

Phi, G. T. (2019). Framing overtourism: A critical news media analysis. *Current Issues in Tourism*, 1–5.

Pinke-Sziva, I., Smith, M., Olt, G., & Berezvai, Z. (2019). Overtourism and the night-time economy: A case study of Budapest. *International Journal of Tourism Cities, 5*(1), 1–16.

Pizam, A. (1978). Tourism's impacts: The social costs to the destination community as perceived by its residents. *Journal of Travel Research, 16*(4), 8–12.

Poon, A. (1993). *Tourism, technology and competitive strategies*. CAB International.

Preveden, V., Mirkovic, G., Gratzer, M., & Schenk, O. (2018). *Protecting your city from overtourism: European city tourism study 2018*.

Ryan, C., & Cessford, G. (2003). Developing a visitor satisfaction monitoring methodology: Quality gaps, crowding and some results. *Current Issues in Tourism, 6*(6), 457–507.

Sanz-Blas, S., Buzova, D., & Schlesinger, W. (2019). The sustainability of cruise tourism onshore: The impact of crowding on visitors' satisfaction. *Sustainability, 11*(6), 1510.

Šegota, T., Mihalič, T., & Kuščer, K. (2017). The impact of residents' informedness and involvement on their perceptions of tourism impacts: The case of Bled. *Journal of Destination Marketing & Management, 6*(3), 196–206.

Seraphin, H. (2018). Destination branding and overtourism. *Journal of Hospitality and Tourism Management, 38*(December 2018), 1–4.

Sharpley, R. (2000). Tourism and sustainable development: Exploring the theoretical divide. *Journal of Sustainable Tourism, 8*(1), 1–19.

Sheehan, L. R., & Ritchie, J. B. (2005). Destination stakeholders exploring identity and salience. *Annals of Tourism Research, 32*(3), 711–734.

Simmonds, C., McGivney, A., Reilly, P., Wilkinson, T., Canon, G., Wright, M., and Whaley, G. (2018, November 20). *Crisis in our national parks: How tourists are loving nature to death.* https://www.theguardian.com/environment/2018/nov/20/national-parks-america-overcrowding-crisis-tourism-visitation-solutions.

Spradley, J. P. (1976). The revitalization of American culture: An anthropological perspective. In E. J. Logue (Ed.), *Qualities of life: Critical choices for Americans* (Vol. 7, pp. 99–121). Lexington Books.

Stankey, G. H., Cole, D. N., Lucas, R. C., & Petersen, M. E. (1985). *Wilderness Planning.*

Szromek, A. R., Kruczek, Z., & Walas, B. (2019). The attitude of tourist destination residents towards the effects of overtourism—Kraków case study. *Sustainability, 12*(1), 228.

Theng, S., Qiong, X., & Tatar, C. (2015). *Mass tourism vs alternative tourism? Challenges and new positionings* (pp. 31–32). Études Caribéennes.

UNWTO (World Tourism Organization). (2018). *'Overtourism'?—Understanding and managing urban tourism.* https://www.e-unwto.org/doi/pdf/10.18111/9789284420070

UNWTO (World Tourism Organization). (2021, Jan 28). *2020: Worst year in tourism history with 1 billion fewer international arrivals.* https://www.unwto.org/news/2020-worst-year-in-tourism-history-with-1-billion-fewer-international-arrivals

Vainikka, V. (2013). Rethinking mass tourism. *Tourist Studies, 13*(3), 268–286.

Weaver, D. B. (2006). *Sustainable tourism: theory and practice*. Routledge.

Weber, F., Eggli, F., Meier-Crameri, U., & Stettler, J. (2019). Measuring overtourism: Indicators for overtourism: Challenges and opportunities. In *Report for the World Tourism Forum*. Luce.

Wickham, T. D., & Kerstetter, D. L. (2000). The relationship between place attachment and crowding in an event setting. *Event Management, 6*(3), 167–174.

Woo, E., Kim, H., & Uysal, M. (2015). Life satisfaction and support for tourism development. *Annals of Tourism Research, 50*, 84–97.

Woo, E., Uysal, M., & Sirgy, M. J. (2018). Tourism impact and stakeholders' quality of life. *Journal of Hospitality & Tourism Research, 42*(2), 260–286.

Yrigoy, I., Horrach, P., Escudero, L., & Mulet, C. (2023). Co-opting overtourism: Tourism stakeholders' use of the perceptions of overtourism in their power struggles. *Journal of Sustainable Tourism, 1*(5), 1–17.

Zelenka, J., & Kacetl, J. (2014). The concept of carrying capacity in tourism. *Amfiteatru Economic Journal, 16*(36), 641–654.

Chapter 10
Overtourism: A Potential Outcome in Contemporary Tourism—Causes, Solutions, and Management Challenges

Kim Pham, Kathleen L. Andereck, and Christine A. Vogt

Abstract This chapter reviews causes of and solutions to overtourism. Solutions include a suite of tools and management and development approaches for communities and scholars to moderate crowding and other negative impacts from overtourism. Many of these solutions are grounded in sustainability and resilient community approaches, as well as destination capacity management. An overtourism ecosystem model integrates from concepts to solutions with a keen focus on stakeholders. Developing a tourism plan is recommended as a roadmap for communities to identify, manage, and mitigate overtourism.

Keywords Paradigm shifts in tourism industry · Technology · Tourism growth management · Overtourism ecosystem · Tourism plan

Sustainable tourism as a concept and practice came to the fore in the 1970s as negative impacts from too many visitors became more evident. More recently, resiliency has emerged as a parallel concept with which to view the tourism system as it manifests itself in communities (Espiner et al., 2017). Resilience, as it relates to tourism specifically, refers to the ability of social, economic, and environmental systems to recover from the stresses tourism causes. A complex network of interrelated factors can influence resilience thresholds beyond which systems change irreversibly. One of the most influential factors influencing resilience thresholds is the numbers of people in an area (Tyrrell & Johnston, 2008). The number of people traversing the world is becoming increasingly problematic. International tourist arrivals increased from 25 million in 1950 to over 1.5 billion in 2019 (Behsudi, 2020). The tourism sector is predicted to continue its rapid growth (WTTC, 2023).

At a destination scale, overtourism occurs when communities' carrying capacity has been exceeded. Researchers (e.g., Dodds & Butler, 2019; Goodwin, 2017) and

K. Pham · K. L. Andereck (✉) · C. A. Vogt
School of Community Resources and Development, Arizona State University, Phoenix, AZ, USA
e-mail: Dtpham8@asu.edu; kandereck@asu.edu; chrisv@asu.edu

D. Chhabra et al. (eds.), *Sustainable Development and Resilience of Tourism*,
https://doi.org/10.1007/978-3-031-63145-0_10

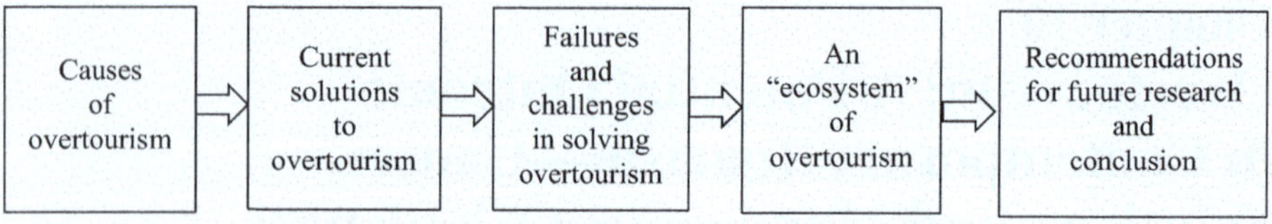

Fig. 10.1 The content flow of the book chapter

nonprofit organizations such as the Center for Responsible Tourism (CREST), Sustainable Travel International (STI), and others are amassing evidence that too much tourism can generate negative outcomes for communities such as pressure on infrastructure, energy overconsumption, environmental problems, exploited natural resources, increased prices of goods and services, inflated real estate market, crime, degraded local cultures, and generally decreased quality of life of residents.

This conceptual paper is the second in a series of two papers that consider overtourism as a contemporary global phenomenon. In the first chapter, current definitions of overtourism are discussed as a context for the remainder of the two chapters. The ways in which overtourism has been conceptually and theoretically framed is presented, as are measures and indicators used to operationalize variables in explanatory models to objectively identify and explain overtourism. The chapter closes with some thoughts on the consequences of overtourism as a contemporary concept.

As with the first chapter, the content of this chapter was identified through a comprehensive literature review. Literature (e.g., articles, books, and reports) was collected via university's library, academia database systems (e.g., JSTOR, ScienceDirect, Elsevier), and search engines (e.g., Google, Google Scholar). The literature review focused on pragmatic concerns including overtourism causes, solutions, and management challenges, followed by a new *overtourism ecosystem* that integrates the two papers into a conceptual model for future research and destination management. This paper answers the questions: what are the causes of overtourism? What solutions have been proposed for overtourism? and What are the challenges inhibiting solutions to overtourism? Figure 10.1 illustrates the content flow of this book chapter.

Causes of Overtourism

Overtourism is caused by various factors that can be categorized by global trends or development, destinations, stakeholders, and media. They can also be cross-classified into those within the tourism industry and out of the tourism industry, or within destination and out-of-destination boundaries (Table 10.1). The categories are not mutually exclusive.

It is important for destination managers and related agencies to identify the causes of overtourism for possible solutions. The classification helps to narrow down the areas of focus and decide which areas the destination and local tourism

Table 10.1 Causes of overtourism

Causes of overtourism	Within tourism industry	Beyond tourism industry	Within destination boundary	Outside destination boundary
Global trend/development factors				
Absence of large-scale wars and military conflicts (e.g., World War) or change in political environment/visa policy (e.g., Hong Kong returned to China) leading to more opened destinations		x		x
(Global) population growth leading to increase of potential tourists		x		x
(Global) economy growth or increase of middle class and above who can afford to travel		x		x
Change in people's lifestyle and working style enabling them to travel (e.g., value experience more than materials, working remotely, earn living by traveling)		x		x
Technology development making traveling easier, more affordable, and arranged within a short time, free from mid-agent/ tour operators, travel to more destinations in short time (e.g., Google Maps, peer-to-peer platforms such as Uber and Airbnb, cheap airlines)	x	x	x	x
Destination factors				
Destination governance (e.g., tourism dependency, growth vs. development, improper infrastructure, lack of strategies, insufficient coordination of stakeholders, ignorance of resident's sentiments, poor management of short-term rentals)	x	x	x	
Successful marketing and branding by DMOS, travel companies, etc.	x		x	
Destination features (e.g., iconic destinations, proximity to big ports or other popular destinations, last-chance tourism)	x	x	x	x
Mega/cultural/ political events (e.g., Olympics, movies, immigration policy)	x	x	x	
Seasonality/peak times of tourism	x		x	
Stakeholders' factors				
Tourists' changing travel patterns (e.g., seeking authentic experiences, off-beaten-track) leading to their appearance everywhere	x		x	
Tourists' inappropriate behavior causing annoyance to local community	x		x	

(continued)

Table 10.1 (continued)

Causes of overtourism	Within tourism industry	Beyond tourism industry	Within destination boundary	Outside destination boundary
Residents' lack of understanding about tourism (e.g., benefits, tourism management) or vocal about tourism issues	x		x	
Media factors				
Media and social media creating growth in visitation	x	x	x	x
Media and social media influencing public view of overtourism	x	x	x	x

agencies can tackle on their own and which areas require cooperation with stakeholders outside of the tourism industry or outside of the destination. By taking action with consideration to factors and actors that contribute to the problem, the process can be efficient, thus saving time and resources. Knowledge about causes can facilitate communication and understanding among stakeholders. For example, residents may be angry with and confront the local tourism office about their marketing and management activities that attract more visitors; however, the destination's DMO might not be the only organization engaged in marketing. There are many factors contributing to overtourism issues that the local DMO cannot control. Educating residents on different causes can enhance the relationship between the host community and local tourism offices and strengthen their cooperation.

Global Trend/Development Factors

Overtourism can be caused by global situations. The absence of large scale or world wars since the end of World War II and Vietnam War has created a relatively peaceful environment for people to travel around the world and the development of new attractive destinations which used to be in war zones (Becker, 2016). The world's ever-increasing population is part of the overtourism problem resulting in more tourists (Dodds & Butler, 2019). The world's population was one billion in 1800 and reached eight billion in 2023 (Worldometer, 2023). The United Nations predicts that the number will increase to 8.5 billion in 2030 and 9.7 billion in 2050 (UN News, 2022).

Changes in economies are accompanied with changes in people's lifestyle. People are enjoying more holidays and use them for traveling more than before (Goodwin, 2017). Technological advances and changes in business environments have created flexible working arrangements that enable more people to travel while working (Goodwin, 2017). There has been a change in people's attitudes toward traveling, especially among young generations. Today, people put less importance on staying at one place or owning property; instead, they value experience, including travel, more than previous generations (Cavagnaro et al., 2018).

Global growth of tourism is also enabled by advanced technology that makes traveling easier and more affordable than before. The industry now has large travel corporations and cruises to provide service for large groups of tourists (Alexis, 2017). Low-cost carriers allow many people to travel by air for the first time and others to travel more regularly (Dodds & Butler, 2019). The Internet enables instant access to travel information. Free and convenient services to facilitate travel, such as Google Translate and Google Maps, help people travel easily (Pencarelli, 2020). Peer-to-peer platforms such as Airbnb and Uber extend tourism services to every corner of the world. These advances make it easier for people to arrange trips and travel independently without intermediaries.

Destination Factors

Destination governance, development, and management strategies influence overtourism. Dependency of many destinations on tourism as their major economic sector coupled with a "boosterism" approach prioritizing tourism growth and increases in tourist arrivals make those destinations likely to experience overtourism (Dwyer et al., 2010; Peterson & DiPietro, 2021). In these destinations, tourism services expand on a large scale, as do international chain resorts and other related tourism infrastructure. Overtourism has become a critical issue resulting from improper development of tourism while there is a lack of facilities and infrastructure. Without sufficient preparation, growth causes problems (Weber, 2017). Even a small increase of visitors in newly developing areas can have great negative impacts (Koens et al., 2018). Other ill governance issues come into play as overtourism enablers (e.g., lack of strategic approach, insufficient coordination of stakeholders).

Successful marketing by travel companies and destinations is another cause of overtourism. With the Internet and digital technology, marketing reaches more people; the popularization of Facebook, Instagram, and other forms of social means that word of mouth or individual experience can be shared and spread wider and quicker (Milano, 2018). Destinations' branding strategy contributes to overtourism (Seraphin, 2018). Some branding strategies bring a flux of visitors to a destination, especially when they capture emotions and connections visitors develop with the destination (Chacko & Marcell, 2008).

Some destinations are more prone to overtourism than others due to distinguishing features. Some are iconic destinations that many tourists want to visit thus resulting in heavily visited destinations such as Venice or Paris. Within each destination, there are must-see sites that most first-time visitors want to experience creating a concentration of people in certain areas (Jacobsen et al., 2019). There are proximity factors that enable overtourism such as being located near big airports, big cities, and other famous destinations.

Several destinations are subject to last-chance tourism, another cause of overtourism. Those destinations are at risk of vanishing or being closed to tourists due to reasons such as climate change, a new tourism policy, or damaged ecosystem. They attract many tourists before the destination is gone or inexorably damaged and changed (Lemelin et al., 2010).

Overtourism can result from a single event outside of tourism such as a country's visa policies, a famous movie related to a destination, or mega sport events (Benner, 2019; Cheung & Li, 2019). Seasonality and peak times are another influential factor to overtourism. The temporary surge puts pressure on the local infrastructure and services and social and ecological systems (Cheer et al., 2019).

Stakeholders' Factors

Travel patterns of tourists influence tourism growth, leading to overtourism. In recent decades, the interaction between local people and tourists increased due to development of authentic traveling ideology. Tourists' desire to seek to see "real" and "authentic" everyday life or "off-the-beaten" track experiences has caused tourism activities to become further intertwined with local life, mostly outside of the main tourist areas in cities (Goodwin, 2017). Short-term rental such as Airbnb enable tourists to spread out and go deeper into local communities (Plichta, 2018). These types of interactions make residents feel that there are more and more tourists in their area and feel suffocated by their presence. The "living like a local" traveling trend becomes questionable in terms of sustainability (Goodwin, 2017; Perkumienė & Pranskūnienė, 2019). Tourists' inappropriate behavior adds to residents' feelings that there is too much tourism (Koens et al., 2018).

Overtourism depends on residents' awareness and perspectives about tourism (Gössling et al., 2020) and attracts attention from management and media when local community residents complain or protest (Dhiraj & Kumar, 2021). Residents may not be aware of tourism, especially, those who are not involved with the industry for a living. They are more likely to feel the negative impacts, become upset with tourism development, and publicly vocal about the issues (Park & Kovacs, 2020).

Media and Social Media Factors

Scholars have explored and discussed the relation of media and social media with overtourism. Media and social media encourage visitation to destinations as effects of people sharing their travel experience on Instagram, travel blogs, and travel articles (Alexis, 2017) and influence how the public views overtourism. Media coverage brings overtourism as a new phenomenon to public attention (Clark & Nyaupane, 2020). Social media contributes to the exaggeration of overtourism phenomenon, thus, triggering local discontent and anti-tourist sentiments (Koens et al., 2018).

Current Solutions to Overtourism

Many strategies have been adopted around the globe to solve overtourism: manage tourism growth and development; manage and control the crowd of tourists; manage stakeholders' behaviors, perspectives, and involvement; manage destination capacity and tourism practices; and develop and implement strategic tourism development plans (Table 10.2). The solutions can be applied to specific sites or large scale. Solutions can be reactive to an overtourism situation to reduce its negative impacts or be proactive/preventative before it happens. They can be executed and expected to be effective in short, medium, and long terms. The categories are not mutually exclusive.

Awareness of the variety of solutions offers destination management the flexibility to take actions. Classifying solutions into categories can be helpful to allocate resources and form different taskforces that involve relevant and specialized partners. Since overtourism is a complicated issue, solving it requires systematic efforts from different directions and levels. A list and categories of solutions can be used as a checklist for destination managers to develop a comprehensive action plan.

Manage Tourism Growth and Development

To address overtourism, there have been suggestions whether destinations should follow a "sustainable degrowth" pathway (Büscher & Fletcher, 2017). The process of sustainable degrowth reorients unsustainable and inappropriate tourism practices as well as promotes less production and consumption in a smaller economy (Andriotis, 2014). Hall (2009) suggests application of the four Rs principles in tourism consumption: reduce, reuse, recycle, and regulate. Degrowth is revolutionary and anti-capitalist, aiming for "sustainable change and results from an interest in locality and place, small and medium-sized enterprises, job creation and reduction of working hours, ecology and quality of life, reduction of tourism activities, carbon reduction in the transport sector, a change in production and consumption patterns, and the high priority of the travel experience (Andriotis, 2014, p. 11)."

Efforts to address overtourism have been made in destination marketing. Some destinations apply a demarketing (or demotion) approach which is defined as "that aspect of marketing that deals with discouraging customers in general or a certain class of customers in particular on either a temporary or permanent basis (Kotler, 1971, p. 76)." The purposes are to increase financial returns from tourism while maintaining or even decreasing tourist arrival numbers. Tools for demarketing destinations can apply are categorized into four groups (Gülşen et al., 2021): price increases (increasing fees, increasing entry prices), place (reducing the fields of consumption and distribution, putting restraints on sales), products (restricting the utility of the product, increasing the availability of alternatives, highlighting the damage factor of the product, reducing the attractiveness), and promotion (reducing

Table 10.2 Solutions to overtourism

Solutions	Explanation/tools	Types of solutions or how solutions are applied						
		Reactive	Proactive/ preventative	Site level	Destination level	National level	Short/ medium term	Long term
Manage tourism growth and development								
Sustainable degrowth	Reorient to a smaller economy with less production and consumption	x	x	x	x	x	x	x
Demarketing/demotion	Reduce destination marketing activities	x	x		x	x		x
Tourist market segmentation/optimization/ repositioning approach/ promote niche tourism	Target tourists with limited impacts on destinations, high spending capacity, repeat visitation (e.g., overnight vs. day visitors, luxury market vs. backpackers)	x	x	x	x	x		x
Restrict/control tourism facilities development	Limit the establishment of new facilities (e.g., hotels, short-term rentals)	x	x		x	x		x
Restrict tourism activities	Restrict harmful/unsuitable activities (e.g., beer bikes, red-light districts activities)	x	x	x	x	x	x	x
Promote multi-pillar economy	Tourism is not the main or only economic sector		x		x	x		x
Manage and control the crowd of tourists								
Control/capping number of tourist arrivals	Apply reservation and ticketing, timed/limited entry; cap number of tourist visas, cruises, or flights; apply pricing differentiation, minimum spending, tax on services that bring large volume of tourists (e.g., cruises, short-term rentals)	x	x	x	x	x	x	x
Manage tourist flow	Navigate flow via turnstiles, timing visitation, virtual queueing, pedestrian zones	x	x	x	x	x	x	x
Zoning	Separating touristic areas from residential areas	x	x	x	x	x	x	x
Tourist distribution across seasons or destination	Promote low season or less well-known areas of the destinations, develop new routes and attractions	x	x	x	x	x		x

Monitoring crowd control	Increase presence of tourism patrols/security officers/traffic controllers/surveillance	x	x	x	x	x	x	x
Manage stakeholders' behaviors, perspectives, and involvement								
Influence and control tourists' behaviors	Soft measures: Education via code for tourists, pledge, information on visitors' behavior; hard measures: Regulation, fines	x	x	x	x	x	x	x
Enhance public awareness about tourism	Communicate with the public about tourism issues and benefits, increase employment opportunities for locals		x	x	x	x		x
Ensure tourism benefit for residents	Tax incentives, employment and recreation opportunities, reduced entrance fees, community events	x	x		x	x		x
Engage residents in tourism development	Involve the public in decision-making/tourism projects (e.g., community meetings and forums, volunteers)	x	x		x	x	x	x
Exchange/share information among destinations and agencies	Webinars, seminars, etc.	x	x	x	x	x	x	x
Manage destination capacity and tourism practices								
Infrastructure and facilities improvement	More parking areas or parking outside of popular areas, restrooms, road expansion, shuttle bus, public transportation, biking/walking routes	x	x	x	x	x	x	x
Promote sustainable tourism practices	No plastic policy, efficient and green energy, green/electric transportation	x	x	x	x	x	x	x
Site/destination closure	Close specific parts of or whole destination for a period of time	x		x	x		x	x
Develop and implement tourism plan								
(sustainable) tourism plan	Research on overtourism issues, incorporate different solutions, involve a variety of stakeholders, apply technology to forecast and control overtourism, sustainability assessment and certification		x		x	x		x

advertisement field, adding mandatory warning labels) (Gerritsma, 2019; Kern & Armstrong, 2007).

Another approach to curb tourism is optimization of markets. This strategy is opposite to the maximization strategy to attract higher numbers of tourists. Destinations identify markets that can bring in more profits including stable economic revenue, longer length of stay, a lower carbon cost, or visit during another season (Schiff & Becken, 2011; Weaver & Oppermann, 2000). Day-trippers are often less favored by destinations (Briguglio & Avellino, 2019) as are markets such as backpackers (Dodds & Butler, 2019).

Repositioning approach can help to curb overtourism. Applying this approach, destinations focus on certain highlighted attributes to rejuvenate themselves or fix the destination's image, influence visitors' preferences, and increase competition capacity (Chacko & Marcell, 2008; Foxell & De Trafford, 2010). Destinations subject to overtourism can brand each region differently by focusing on particular types of tourism products or experiences (Seraphin, 2018). Repositioning will still encourage visitation to the destination, just reduce the number of visitors (Seraphin, 2018). Destinations can choose to restrict certain types of tourism or activities that are harmful (Goodwin, 2021).

Another solution for overtourism is for destinations to develop a multi-pillar economy and lessen their dependence on tourism (Goodwin, 2021). Destinations consider developing alternative sectors such as astronomy, green energy, and agriculture (Oda, 2020).

Manage and Control Tourist Crowds

Many destinations manage the number of tourist arrivals as a way to reduce overtourism (Nyaupane & Timothy, 2010). Alternative options may include application of air passenger duties and departure taxes which generate more revenue and limit tourism growth (Hall, 2013). Several destinations have capped the numbers of ships arriving or docking at their ports over a period of time and the numbers of cruise passengers or sizes of ships (Dodds & Butler, 2019).

Some places attempt to manage the flow of tourists. Measures include turnstiles to control the pedestrian and aquatic traffic (González, 2018), diversion of large cruise ships (González, 2018), virtual queuing with pre-booked participation time (Visit Disney, n.d.), and timed admissions and limits on length of stay or visitation (Butler & Dodds, 2022). Technologies (e.g., mobile apps, camera) are helpful for destinations to control tourist flows.

Zoning or territorial strategy is another recognized tactic to solve overtourism. Zoning can be applied to regulate traffic (Koens et al., 2018). Some destinations use this tactic to contain accommodation growth (Calle-Vaquero et al., 2020). Several destinations focus on distributing or spreading tourist volume across the entire destination or seasons to reduce visitors at popular sites and time (Alexis, 2017; Capocchi et al., 2019).

Monitoring the crowd and flow of tourists is another important strategy. Increasing the presence of tourism patrols, security officers, traffic controllers, and surveillance devices are tools (Goodwin, 2017; Wall, 2020). The measures can be useful for intervention to stop or reroute traffic or walking tourists, ensure the tourists follow rules and regulations at sites, and create a safe environment for both tourists and the host community.

Manage Stakeholders' Behaviors, Perspectives, and Involvement

Besides strategies to manage the crowd of tourists, there are ways to improve tourist behavior (Muler Gonzalez et al., 2018). When policies aimed at controlling the volume of visitors are not adequate to address overtourism, stimulating behavioral change by visitors is necessary (Benner, 2019). This is termed a "nudging approach" which aims to "to steer citizens towards making positive decisions as individuals and for society while preserving individual choice" (Hall, 2013, p. 1098). Examples of nudging include installing signs or codes of conduct to guide tourists' behavior, creating a feedback mechanism about visitors' behavior, and creating pledges for visitors to sign up for (Hall, 2013; Siegel et al., 2023). In dealing with tourist behavior there are "soft" measures such as promoting awareness among tourists about appropriate behaviors and creating codes of conduct to guide tourists, and "hard" policy interventions in forms of rules and regulations (Benner, 2019).

Overtourism risk can be managed by managing residents' perspectives. Understanding the factors that negatively influence residents' perceptions is necessary for proper management tactics (Kuščer & Mihalič, 2019). Destination managers are encouraged to listen to residents' concerns about impacts of tourism, enhance awareness of residents about social and economic benefits of tourism, increase employment opportunities in tourism for residents, promote conservation and preservation initiatives, and involve residents in tourism planning and development.

Involvement of a large network of stakeholders is important in dealing with overtourism (Loverio et al., 2021). To facilitate their involvement, sharing data about overtourism with relevant stakeholders, among destinations, and among different levels of tourism management is recommended (Adams & Sanchez, 2020; Goodwin, 2019; Jamieson & Jamieson, 2019). Public meetings, webinars, seminars, conferences, publication, articles, and blogs are tools used by destinations and organizations to disseminate knowledge about overtourism.

Manage Destination Capacity and Tourism Practices

Many destinations attempt to improve infrastructure and facilities to accommodate increasing volumes of tourists. Promoting sustainable tourism practices is a route for destinations to minimize the impacts of overtourism including measures like responsible lighting to reduce light pollution, promoting leave-no-trace principles to protect the environment, increasing usage of green energy, and supporting local businesses to prevent tourism leakage (SCC&TB, 2019). A small number of destinations that experience severe consequences from overtourism have resorted to more extreme measures of closing some specific sites or an entire destination to tourists due to pollution and damaged ecosystems (Clark & Nyaupane, 2020; Johnston, 2019).

Develop and Implement Strategic Tourism Plan

Overall, overtourism solutions should be applied strategically using sustainable planning (Milano et al., 2019). The process involves research to assess and measure overtourism, cooperation among different stakeholders in and out of the destination or the tourism industry, short- and long-term goals and objectives to manage tourism growth, and tactics to solve overtourism.

Failures/Challenges in Solving Overtourism

Strategies to control overtourism are not always successful and can lead to more problems. For example, destinations promote traveling in lower seasons to reduce overcrowding or to bring in more tourism revenue has prolonged the tourist season; the appearance of tourists in many destinations now is observed year-round (Jacobsen et al., 2019). In other cases, the effect of the strategy is not good for residents such as increased parking fees also paid by residents (Koens et al., 2018). The luxury market includes products that generate a unique experience away from mass consumption and seem to be more economically and environmentally beneficial to destinations but can generate few local benefits but still produce negative impacts (González, 2018). The approach of segmentation to a select type of tourist market has been met with criticism of inhibiting tourist diversity and inclusivity. Measures to limit the number of tourists may make tourists feel unwelcome and affect destination growth in the long term. The segmentation approach may not be feasible if there is insufficient volume of the targeted tourist market for the destination to thrive. Pricing policies that offer incentives for residents may face negative reactions from tourists (Dodds & Butler, 2019).

There are several reasons why overtourism is difficult to solve (Table 10.3). Destinations are different in *context*, and their overtourism issues have varied causes and impacts. The approach to limit tourists' arrivals is more feasible in small and isolated destinations such as islands. Urban destinations might find it impossible to apply the same approach since there are many routes and transportation modes coming in and out (Dodds & Butler, 2019). There are no one-size-fit-all measurements and solutions for overtourism, and destination customization is needed.

Table 10.3 Challenges in solving overtourism

Challenges	Explanation
Context	Overtourism is contextual. There cannot be a one-size-fits-all strategy for destinations
Multidisciplines and multi-stakeholders	Tourism is multidisciplinary and involves a large network of stakeholders who are not always on the same page, sometimes even contradicting one another. A unified collaboration is difficult to achieve
Global issues	Overtourism is a global issue where there is no central authority
Causal chains	Many issues in tourism happen in causal chains. It is difficult to get to the root of the issue and identify responsibilities of involved parties
Boundary	Many causes of overtourism are outside the boundary of the tourism industry or destinations
Data and measurement	Lack of data about overtourism and complicated measurements/ indicators
Application of solutions	Solutions are often applied reactively rather than preventatively, site-specific rather than at destination or national/global level. Lacking strategic tourism plans
Awareness and practice	Lack of awareness and willingness to adapt sustainable practices by practitioners and governments
Iconic destinations/ sites	Tourists always want to see iconic sites making it difficult to manage tourist flow
Facilities and services	Infrastructure and services for tourists are also shared by community, businesses making it difficult for zoning activities

Tourism is a complex *multi-sector* industry with involvement of *multiple stakeholders* making an integrated action agenda difficult (Butler & Dodds, 2022). Stakeholders are varied in their awareness of overtourism issues or sustainability, priority missions, and capacity. They may not reach mutual agreement on solutions leading to lack of cooperation. Tourism is an economic activity where most participants expect and depend on economic gains for survival (Walmsley, 2017). It is "generally unrealistic and ineffective to expect those agencies to have much regard for anything beyond their own economic benefit" (Butler & Dodds, 2022, p. 38). Many DMOs still focus on destination marketing rather than management or sustainability (Butler & Dodds, 2022). A lack of cooperation between national governments that continue to aim for tourism growth and local governments that suffer from and try to mitigate overtourism exists (Nepal & Nepal, 2021). Overtourism is a collective *global issue* that lacks a central leadership (Adams & Sanchez, 2020). Relation theorists describe this status as an international anarchy (Waltz, 2001). The anarchy makes it immensely challenging to solve overtourism with no agreed upon rules and enforcement mechanisms (Adams & Sanchez, 2020).

Issues in tourism often occur in *causal chains* making it difficult to identify the root causes as well as identify and coordinate responsibilities of involved parties to solve the issues effectively (Butler & Dodds, 2022; Koens et al., 2018; Peeters et al., 2018). An example is technology advancement that led to the development of airlines which in turn led to the availability of cheap flights and an increase in travelers. Another example is the combination of technology, a lack of housing market regulations, and residents/entrepreneurs seeking income which led to the exponential growth of peer-to-peer platforms such as Uber and Airbnb which in turn increased the arrival of tourists in communities.

Many causes of overtourism are out of tourism or destination *boundaries* where local tourism agencies and governments have little or no control (Butler & Dodds, 2022). Examples increase of the world's population and growing economy leading to an increase in potential tourists. Other examples include air travel and cruise travel that are often controlled by air/cruise companies and national governments rather than local ownership. Airports and cruise ship ports can be located outside of the destination (Goodwin, 2019).

Solving overtourism is challenging with lack of *data* and holistic systems *indicators and/or measurements* leading to dubious assumptions and management. Collecting data can be costly, and many DMOs do not have sufficient research budgets (Butler & Dodds, 2022). Many measurements/indicators are difficult to calculate. The numbers of accommodations or bed nights can be hard to count/estimate since many are not registered with local government or tourism agencies, especially short-term rentals. Measurements that are related to carrying capacity are even more complicated due to circumstances constantly changing making calculations out of date in a short time; difficulty in estimating human attitudes and preferences, especially when there are many stakeholders involved; and difficulty in predicting the number of tourist arrivals (Shelby et al., 1989; Wall, 2020).

The ineffectiveness of prospective overtourism solutions is partly because those *solutions* are often reactive rather than proactive or applied at small scale rather than large (Ivars-Baidal et al., 2019). Measures to tackle overtourism are often applied when destinations are already overcrowded. Many measures are applied individually at specific attraction sites rather than whole destinations or as a part of a strategic plan (Butler & Dodds, 2022). Few destinations have a long-term sustainable tourism plan to prevent and solve overtourism.

Though sustainable development has been widely supported by international organizations and politicians across the globe, *awareness* and its *practice* are still questionable (Dodds & Butler, 2019). While more and more private businesses have adopted sustainability initiatives, the management authorities of many destinations, especially in the public sectors, do not recognize the existence of overtourism and are unwilling to address the issue. For many destinations and leading tourism agencies at international and national levels, growth in tourism is still the desired goal (Wall, 2020), and they focus more on managing tourist volume than reducing it. Destinations worry more about their reputation being damaged and decrease in tourist arrivals than suffering from too much success, while businesses are unlikely to cease advertising the destinations (Dodds & Butler, 2019).

An approach to promote less popular sites and attractions in the destination is difficult because tourists, especially first-time visitors, often want to see *iconic places* (Butler & Dodds, 2022). Finally, a zoning approach is not always feasible because it is difficult or sometimes impossible to separate often-shared *facilities and services* for tourists from those for residents.

An "Ecosystem" of Overtourism

Overtourism is a complicated phenomenon with many interrelated aspects. It is important for tourism researchers and practitioners to be aware of this complexity for effective study and management strategies and to facilitate sustainable and resilient communities to maintain positive quality of life and well-being in communities. Scholarly work suggests a development plan should start with a mutual agreement among stakeholders on the definitions or concept of the targeted topic. Additionally, a thorough understanding of the causes and impacts of overtourism leads to better solutions that address the right issues and involve the right decision-makers and implementers. A good strategy can benefit from awareness of the potential failures and challenges that may come along. Related theories and concepts are helpful for scholars to identify research problems and for practitioners to understand the underlying causes of overtourism issues. With conceptual frameworks, academics and practitioners work more effectively with each other on development projects since they can share and use the same language.

Based on the reviews in papers 1 and 2, an "ecosystem" model of overtourism is proposed that can be used as a general guideline for researchers and practitioners to consider while studying and solving overtourism (Fig. 10.2). This "ecosystem" shows overtourism as a broad subject composed of several specific areas. Solving overtourism could be more effective if each area is well understood and addressed. By segmenting the big problem of overtourism into smaller parts following the

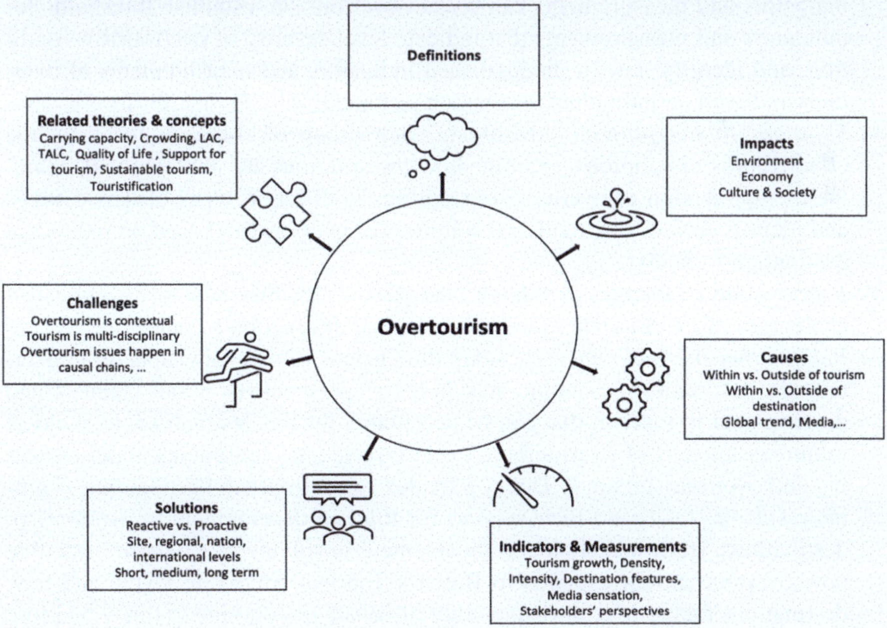

Fig. 10.2 An "Ecosystem" of overtourism

suggested model, destination managers might find the problem less overwhelming or vague. The model can be helpful to manage overtourism systematically and strategically. For researchers, the model is useful to position and estimate the significance of their research in the overall overtourism scenario.

Recommendations for Future Research

Several directions for future research can provide more robust and updated empirical evidence to expand or modify the overtourism "ecosystem" model. Prospective directions for research corresponding to the ecosystem of overtourism developed in this study include the following:

(1) *Causes of overtourism:* study each group of causes, identify and study related actors or stakeholders that are responsible for the causes, compare individual or groups of causes, and identify new or distinguished causes of overtourism based on destination's context.
(2) *Impacts of overtourism:* investigate each group of impacts, identify and study related actors or stakeholders that are impacted, compare the severity of individual or groups of impacts, and identify new or distinguished causes of overtourism based on individual destination's contexts.
(3) *Indicators and measurements of overtourism:* examine the application of the indicators and measurements in different types of destinations, examine how indicators and measurements can predict overtourism, establish thresholds for indicators and measurements that indicate level/severity of overtourism condition, and identify new or distinguished indicators and measurements of overtourism based on individual destination's context.
(4) *Solutions to overtourism:* investigate each group of solutions, evaluate the effectiveness of solutions, examine how the solutions are implemented, examine the application of overtourism solutions in different types of destinations, and identify new or distinguished solutions of overtourism based on individual destination's context.
(5) *Failures and challenges in solving overtourism:* identify new or distinguished challenges of overtourism based on individual destination's context.
(6) *Related theories and concepts:* study dimensions of carrying capacity in overtourism destinations, examine stakeholders' perspectives about changes and development in tourism that can be acceptable for the destinations to avoid or minimize impacts of overtourism, explore residents' sentiments about crowding and overtourism; apply Doxey's Irridex to overtourism destinations, examine residents' QOL and their support for tourism development in overtourism destinations, study stakeholders' involvement in solving overtourism, examine how overtourism corresponds to Butler's Tourism Area Life Cycle and how destinations facing overtourism could rejuvenate or continue to thrive, explore how a sustainable tourism approach could help to prevent and mitigate the issues of overtourism, and study how mass tourism and touristification contribute to causing overtourism.

Conclusion

Overtourism is an ongoing global issue. Crowded destinations face damaging impacts on the environment, economy, culture, and society. Overtourism is a threat to community sustainability and resilience, quality of life, and well-being. A community and its tourism management organization and tourism business members need to be acutely aware of concerns that relate to quality of life and well-being.

The COVID pandemic particularly highlighted that travel can rapidly spread any disease and tourism economics and all stakeholders' well-being and quality of life are at risk. Many destinations prioritized the health and well-being of residents over tourists visiting their community during the pandemic (Pocinho et al., 2022). Pocinho et al. (2022) reported that no studies were yet published on residents' well-being in tourism destinations related to COVID. Tourism businesses were found to be vulnerable to severe economic challenges with global tourism volumes plummeting during the pandemic (Alonso et al., 2020). The well-being of employees was also a concern during the pandemic in terms of spreading COVID and the ability of a business to pay and retain workers. In the end, tourists avoided overtourism destinations and instead sought out destinations with low visitation levels, outdoor activities, and destinations with COVID protection policies and practices (Santos et al., 2020).

Due to its complexity, solving overtourism is challenging and requires understanding the phenomenon from many different perspectives. These two linked papers provide an overall review of the phenomenon with the most important points that destination management should consider while addressing overtourism. Based on the review, the papers propose an "ecosystem" of overtourism model. The ongoing pursuit for scientific evidence and knowledge is significant to develop a more thorough strategic plan to prevent and solve overtourism. Researchers can benefit from the papers as they offer a variety of recommendations for future research.

References

Adams, K. M., & Sanchez, P. M. (2020). Tourism as industry and field of study: Using research and education to address overtourism. In *Overtourism and tourism education* (pp. 43–59). Routledge.

Alexis, P. (2017). Overtourism and anti-tourist sentiment: An exploratory analysis and discussion. *Ovidius University Annals, Economic Sciences Series, 17*(2), 288–293.

Alonso, A., Kok, S., Bressan, A., O'Shea, M., Sakellarios, N., Koresis, A., et al. (2020). COVID-19, aftermath, impacts, and hospitality firms: An international perspective. *International Journal of Hospitality Management., 91*, 102654.

Andriotis, K. (2014). *Tourism development and the degrowth paradigm* (pp. 37–45). In Proceedings of the 2nd Belgrade International Tourism Conference (BITCO 2014) Thematic Tourism in a Global Environment: Advantages, Challenges and Future Developments.

Becker, E. (2016). *Overbooked: The exploding business of travel and tourism*. Simon and Schuster.

Behsudi, A. (2020, Dec). *Tourism-dependent economies are among those harmed the most by the pandemic*. International Monetary Fund. https://www.imf.org/en/Publications/fandd/issues/2020/12/impact-of-the-pandemic-on-tourism-behsudi.

Benner, M. (2019). From overtourism to sustainability: A research agenda for qualitative tourism development in the Adriatic. *German Journal of Economic Geography, 2*(64), 74–87.

Briguglio, L., & Avellino, M. (2019). *Has overtourism reached the Maltese Islands?* (Vol. 1). Occasional Papers on Islands and Small States.

Büscher, B., & Fletcher, R. (2017). Destructive creation: Capital accumulation and the structural violence of tourism. *Journal of Sustainable Tourism, 25*(5), 651–667.

Butler, R. W., & Dodds, R. (2022). Overcoming overtourism: A review of failure. *Tourism Review, 77*(1), 35–53.

Calle-Vaquero, M. D. L., Garcia-Hernandez, M., & Mendoza de Miguel, S. (2020). Urban planning regulations for tourism in the context of overtourism. Applications in historic centres. *Sustainability, 13*(1), 70.

Capocchi, A., Vallone, C., Amaduzzi, A., & Pierotti, M. (2019). Is 'overtourism' a new issue in tourism development or just a new term for an already known phenomenon? *Current Issues in Tourism*, 1–5.

Cavagnaro, E., Staffieri, S., & Postma, A. (2018). Understanding millennials' tourism experience: Values and meaning to travel as a key for identifying target clusters for youth (sustainable) tourism. *Journal of Tourism Futures, 4*(1), 31–42.

Chacko, H. E., & Marcell, M. H. (2008). Repositioning a tourism destination: The case of New Orleans after hurricane Katrina. *Journal of Travel & Tourism Marketing, 23*(2–4), 223–235.

Cheer, J. M., Milano, C., & Novelli, M. (2019). Tourism and community resilience in the Anthropocene: Accentuating temporal overtourism. *Journal of Sustainable Tourism, 27*(4), 554–572.

Cheung, K. S., & Li, L. H. (2019). Understanding visitor–resident relations in overtourism: Developing resilience for sustainable tourism. *Journal of Sustainable Tourism*, 1–20.

Clark, C., & Nyaupane, G. P. (2020). Overtourism: An analysis of its coverage in the media by using framing theory. *Tourism Review International, 24*(2–3), 75–90.

Dhiraj, A., & Kumar, S. (2021). Overtourism: Causes, impacts and solution. In *Overtourism as destination risk* (pp. 49–56). Emerald Publishing Limited.

Dodds, R., & Butler, R. (2019). The phenomena of overtourism: A review. *International Journal of Tourism Cities, 5*(4), 519–528.

Dwyer, L., Forsyth, P., & Dwyer, W. (2010). *Tourism economics and policy* (Vol. 3). Channel View Publications.

Espiner, S., Orchiston, C., & Higham, J. (2017). Resilience and sustainability: A complementary relationship? Towards a practical conceptual model for the sustainability–resilience nexus in tourism. *Journal of Sustainable Tourism, 25*(10), 1385–1400.

Foxell, E., & De Trafford, A. (2010). Repositioning Malta as a cultural heritage destination. *International Journal of Culture, Tourism and Hospitality Research, 4*(2), 156–168.

Gerritsma, R. (2019). Overcrowded Amsterdam: Striving for a balance between trade, tolerance and tourism. In *Overtourism: Excesses, discontents and measures in travel and tourism* (pp. 125–147). CAB International.

González, A. T. (2018). Venice: The problem of overtourism and the impact of cruises. *Journal of Regional Research, 42*, 35–51.

Goodwin, H. (2017). The challenge of overtourism. *Responsible Tourism Partnership, 4*, 1–19.

Goodwin, H. (2019). Barcelona–crowding out the locals: A model for tourism management. *Overtourism: Issues Realities and Solutions, 1*, 125–138.

Goodwin, H. (2021). City destinations, overtourism and governance. *International Journal of Tourism Cities, 7*(4), 916–921.

Gössling, S., McCabe, S., & Chen, N. C. (2020). A socio-psychological conceptualization of overtourism. *Annals of Tourism Research, 84*, 102976.

Gülşen, U., Yolcu, H., Ataker, P., Erçakar, İ., & Acar, S. (2021). Counteracting overtourism using demarketing tools: A logit analysis based on existing literature. *Sustainability, 13*(19), 10592.

Hall, C. M. (2009). Degrowing tourism: Décroissance, sustainable consumption and steady-state tourism. *Anatolia, 20*, 46–61.

Hall, C. M. (2013). Framing behavioral approaches to understanding and governing sustainable tourism consumption: Beyond neoliberalism, "nudging" and "green growth"? *Journal of Sustainable Tourism, 21*(7), 1091–1109.

Ivars-Baidal, J., García Hernández, M., & Mendoza de Miguel, S. (2019). Integrating overtourism in the smart tourism cities agenda. *E-Review of Tourism Research, 17*(2), 122–139.

Jacobsen, J. K. S., Iversen, N. M., & Hem, L. E. (2019). Hotspot crowding and overtourism: Antecedents of destination attractiveness. *Annals of Tourism Research, 76*, 53–66.

Jamieson, W., & Jamieson, M. (2019). Overtourism management competencies in Asian urban heritage areas. *International Journal of Tourism Cities, 4*(5), 581–597.

Johnston, Z. (2019, April 12). *Updates on iconic destinations closed due to overtourism (and where to go instead).* UPROXX. https://uproxx.com/life/destinations-closed-from-overtourism-alternatives/

Kern, C. L., & Armstrong, E. (2007). Demarketing as a tool for managing visitor demand in national parks-an Australian case study. *CAUTHE 2007:* Tourism-Past Achievements, Future Challenges, 415.

Koens, K., Postma, A., & Papp, B. (2018). Is overtourism overused? Understanding the impact of tourism in a city context. *Sustainability, 10*(12), 4384.

Kotler, P. L. (1971). Demarketing. *Harvard Business Review*, 74–80.

Kuščer, K., & Mihalič, T. (2019). Residents' attitudes towards overtourism from the perspective of tourism impacts and cooperation—The case of Ljubljana. *Sustainability, 11*(6), 1823.

Lemelin, H., Dawson, J., Stewart, E. J., Maher, P., & Lueck, M. (2010). Last-chance tourism: The boom, doom, and gloom of visiting vanishing destinations. *Current Issues in Tourism, 13*(5), 477–493.

Loverio, J. P., Chen, L. H., & Shen, C. C. (2021). Stakeholder collaboration, a solution to overtourism? A case study on Sagada, The Philippines. *Tourism Geographies*, 1–22.

Milano, C. (2018). Overtourism, social unrest and tourismphobia. A controversial debate. *PASOS: Revista de Turismo y Patrimonio Cultural, 16*(3), 551–564.

Milano, C., Novelli, M., & Cheer, J. M. (2019). Overtourism and tourismphobia: A journey through four decades of tourism development, planning and local concerns. *Tourism Planning & Development, 16*(4), 353–357.

Muler Gonzalez, V., Coromina, L., & Gali, N. (2018). Overtourism: Residents' perceptions of tourism impact as an indicator of resident social carrying capacity-case study of a Spanish heritage town. *Tourism Review, 73*(3), 277–296.

Nepal, R., & Nepal, S. K. (2021). Managing overtourism through economic taxation: Policy lessons from five countries. *Tourism Geographies, 23*(5–6), 1094–1115.

Nyaupane, G. P., & Timothy, D. J. (2010). Power, regionalism and tourism policy in Bhutan. *Annals of Tourism Research, 37*(4), 969–988.

Oda, D. (2020, May 11). *Lack of clear plan to replace tourism leaves Hawaii's economic future in doubt*. Star Advertiser. https://www.staradvertiser.com/2020/05/11/hawaii-news/lack-of-clear-plan-to-replace-tourism-leaves-hawaiis-economic-future-in-doubt/

Park, H., & Kovacs, J. F. (2020). Arts-led revitalization, overtourism and community responses: Ihwa Mural Village. *Seoul. Tourism Management Perspectives, 36*, 100729.

Peeters, P. M., Gössling, S., Klijs, J., Milano, C., Novelli, M., Dijkmans, C. H. S., et al. (2018). *Research for TRAN Committee-Overtourism: Impact and possible policy responses*. European Parliament, Directorate General for Internal Policies, Policy Department B: Structural and Cohesion Policies, Transport and Tourism.

Pencarelli, T. (2020). The digital revolution in the travel and tourism industry. *Information Technology & Tourism, 22*(3), 455–476.

Perkumienė, D., & Pranskūnienė, R. (2019). Overtourism: Between the right to travel and residents' rights. *Sustainability, 11*(7), 21–38.

Peterson, R., & DiPietro, R. B. (2021). Is Caribbean tourism in overdrive? Investigating the antecedents and effects of overtourism in sovereign and nonsovereign small Island tourism economies (SITEs). *International Hospitality Review, 1*(35), 19–40.

Plichta, G. (2018). The problem of short-term rental as a manifestation of the overtourism effect on the housing market in the historical city of Krakow1. *Świat Nieruchomości, 4*(106), 57–64.
Pocinho, M. S., Garces, S., & Neves de Jesus, S. (2022). Wellbeing and resilience in tourism: A systematic literature review during COVID-19. *Frontiers in Psychology, 12*, 748947.
Santos, A., Gonzalez, C., Haegeman, K., & Rainoldi, A. (2020). *Behavioral changes in tourism in times of COVID-19*. Publications Office of the European Union.
SCC&TB (Sedona Chamber of Commerce & Tourism Bureau). (2019). *Sustainable tourism plan.* https://sedonachamber.com/wp-content/uploads/2018/06/SED-STP_4-17-19-HR-UPDATE.pdf
Schiff, A., & Becken, S. (2011). Demand elasticity estimates for New Zealand tourism. *Tourism Management, 32*(3), 564–575.
Seraphin, H. (2018). Destination branding and overtourism. *Journal of Hospitality and Tourism Management, 38*(December 2018), 1–4.
Shelby, B., Vaske, J. J., & Heberlein, T. A. (1989). Comparative analysis of crowding in multiple locations: Results from fifteen years of research. *Leisure Sciences, 11*(4), 269–291.
Siegel, L. A., Tussyadiah, I., & Scarles, C. (2023). Exploring behaviors of social media-induced tourists and the use of behavioral interventions as salient destination response strategy. *Journal of Destination Marketing & Management, 27*, 100765.
Tyrrell, T. J., & Johnston, R. J. (2008). Tourism sustainability, resiliency and dynamics: Towards a more comprehensive perspective. *Tourism and Hospitality Research, 8*(1), 14–24.
UN News. (2022, July 11). *World population to reach 8 billion this year, as growth rate slows*. https://news.un.org/en/story/2022/07/1122272
Visit Disney. (n.d.). *Virtual queues*. https://disneyworld.disney.go.com/guest-services/virtual-queue/
Wall, G. (2020). From carrying capacity to overtourism: A perspective article. *Tourism Review, 1*(75), 212–215.
Walmsley, A. (2017). Overtourism and underemployment: A modern labour market dilemma. *Proceedings of the Responsible Tourism in Destinations*, 13.
Waltz, K. N. (2001). *Man, the state, and war: A theoretical analysis*. Columbia University Press.
Weaver, D., & Oppermann, M. (2000). *Tourism management*. John Wiley and Sons.
Weber, F. (2017). *Tourism destinations under pressure*. Working Paper. Institute of Tourism ITW, Lucerne University of Applied Sciences and Arts.
Worldometer. (2023). *Current world population*. https://www.worldometers.info/world-population/#:~:text=World%20Population%20Clock%3A%208%20Billion,(LIVE%2C%202023)%20%2D%20Worldometer
WTTC (World Travel & Tourism Council). (2023, April 23). *Global travel & tourism catapults into 2023 says WTTC*. https://wttc.org/news-article/global-travel-and-tourism-catapults-into-2023-says-wttc

Chapter 11
Transformative Learning and Experiences and Their Role in Promoting Well-Being and Quality of Life

Deepak Chhabra

Abstract This chapter begins with offering some highlights from the previous chapters. It discusses the manner in which the work of the contributors is aligned with the relevant SDGs. Next, the notions of transformation, transformative experiences, and transformative learning are discussed and contextualized using a wellness/well-being standpoint. The chapter closes with insights on the potential of viable interdependencies between inner transformation and sustainability.

Keywords Transformative learning and experiences · Inner transformation · Sustainability · Sustainable development goals · Well-being

The contributing authors present a broad spectrum of scenarios and issues associated with sustainable tourism, health/well-being, and quality of life. One of the prevailing issues in tourism across the globe, particularly in the pre-COVID times, was overtourism. The pandemic brought overtourism to a halt. The tourism industry has started booming again in places which were weary of tourists in the pre-pandemic times although proactive initiatives, to counter negative impacts of tourism, are being noted. For instance, Venice is imposing taxes on day trippers and planning to limit group sizes (Pitrelli, 2023). The two sequential chapters on overtourism by Pham et al. (2024a, b) present a toolkit to assess economic-related factors (such as magnitude of economic impact, jobs generated, and visitor counts), levels of development, media coverage, and views of different stakeholders. The authors suggest solutions to check overtourism by integrating sustainability and resilient perspectives and capacity management strategies for destinations. They suggest a community-centered integrated overtourism model geared towards strengthening collaborative efforts between important stakeholders. Pham et al.'s work is aligned with several sustainable development goals (SDG) such as SDG 9

D. Chhabra (✉)
School of Community Resources and Development, Watts College of Public Service and Community Solutions, Arizona State University, Phoenix, AZ, USA
e-mail: Deepak.chhabra@asu.edu

D. Chhabra et al. (eds.), *Sustainable Development and Resilience of Tourism*,
https://doi.org/10.1007/978-3-031-63145-0_11

(industry, innovation, and infrastructure), SDG 11 (sustainable cities and communities), and SDG 12 (responsible consumption and production). Parallel attention is drawn by Ayazlar et al. towards the concept of mindfulness and well-being. The authors focus on the employees of the tourism industry and call for strategies to improve employee motivation and well-being to optimize performance. Ayazlar and Ayazlar's (2024) study stresses on the importance of mindfulness and offers pathways to decrease pessimism and generate enthusiasm/upbeat emotions in tourism employees. Their work lends credence to SDG 3 (good health and well-being) and SDG 8 (decent work and economic growth).

Cohen (2024), Joson et al. (2024) and Pham et al. (2024b) use popular forms of nature-based tourism to stress on the need to develop site-specific sustainability measures. For instance, Cohen examines sustainability of marine diving tourism and argues that a transformative state of mind is required to promote sustainable behavior that can benefit the recreational scuba diving industry, tourist divers, and the local residents. The author also discusses the viability of dive tourism in the context of environmental, economic, and social components of sustainability. He uses a transformational paradigm to discuss viability of SDGs in the context of dive tourism. Using a catastrophic event (the destruction caused by the wildfire in Maui) as an example, he discusses the transformative potential of the marine tourism industry. He maintains that a transformative paradigm is crucial to promote long-term sustainable behavior. Cohen's chapter lends credence to SDG 8 (decent work and economic growth), SDG 11 (sustainable cities and communities), SDG 12 (responsible consumption and production), SDG 17 (partnerships), and SDG 14 (life below water).

Joson et al. (2024) discuss the significance of wellness tourism to stimulate the physical and emotional well-being of tourists. The authors use one of the oldest and patronized forms of wellness tourism (spa tourism) as a case in point. Hot springs in Australia are the focus of their study. Gaelle et al. contend that water treatments are popularly sought by wellness tourists around the globe. A sustainable roadmap is suggested to promote and gauge the importance of Australian hot springs by developing a set of indicators. Joson et al.'s chapter promotes SDG 3 (good health and well-being) and SDG 12 (responsible consumption and production). Kim's work promotes SDG 8 (decent work and economic growth), SDG 11 (sustainable cities and communities), and SDG 17 (partnerships). Kim (2024) examines the significance of off-highway vehicle recreation in Arizona from an economic sustainability viewpoint. The author argues that WOM and eWOM can facilitate responsible behavior of OHV recreationists. Sustainable marketing strategies are suggested and Kim's work supports SDG 3, SDG 12, and SDG 17.

Gogoi and Maheshwari (2024) scrutinize motorcycle tourism, from a sustainability standpoint, in the rural and remote areas of the Himalayas in the northernmost state of India (Jammu and Kashmir). They suggest strategies in which motorcycle tourism can benefit the rural and remote regions in the Himalayas. The authors also discuss a viable roadmap for developing innovative initiatives to foster inclusivity and support pro-poor tourism. A sustainable livelihood model is proposed. Gogoi and Maheshwari's (2024) work promotes several SDGs such as SDG

8 (decent work and economic growth), SDG 10 (reduced inequality), SDG 11 (sustainable cities and communities), and SDG 12 (responsible consumption and production). Atal and Atal bring to light the healing potential of herbs and the importance of promoting herbal tourism to travelers who are seeking a cure for their existing ailment. Moreover, authentic herbal tourism can also attract serious leisure travelers who are keen to enhance their knowledge and obtain deeper insights on non-pharmaceutical remedies and products to help maintain health robusticity (2024). One of the co-authors (late Dr. CK. Atal) was recognized by the United Nations and medical practitioners across the globe for discovering Risorine (herbal cure for tuberculosis). Scientific endorsement by worldwide organizations can serve to promote remedies that hold potential to produce minimum side effects for treatment of severe diseases. Innovative success stories and narratives on cutting-edge impactful research in non-pharmaceutical fields, such as these, can stimulate admiration of travelers and motivate them to learn. Learning is an important step to inspire a transformative state of mind (Ponder & Holladay, 2013). As discussed earlier in this book, endorsement by recognized institutions can draw the attention of tourists who are not only seeking treatment but also desire to transform towards a sustained state of well-being (Mezirow, 1991, 2000). Atal and Atal's (2024) work lends credence to SDG 3 (good health and well-being) and SDG 4 (quality education).

All the contributors of this book promote, directly or indirectly, a broad spectrum of SDGs. Aligning tourism with the SDGs illustrates that long-term community-based well-being and sustainability solutions are a key step towards building resilient tourism and resilient communities. It has become critical to transform the consciousness of all key tourism stakeholders to address the challenges encountered in developing and promoting sustainable tourism (Sheldon, 2020). Sheldon writes that the "living life from a transformed or awakened state has profound significance for the individual, for society, and for tourism" (2020, p. 2). As pointed by Sheldon (2020), to ensure long-term success of sustainability and resilience measures, a transformation approach is crucial. Therefore, it is important for the stakeholders and policymakers to develop transformative learning tools to reach an elevated sense of consciousness and mindfulness. Sullivan describes the notion of transformative learning as "a shift of consciousness that dramatically and irreversibly alters our way of being in the world" (2002, p. 11). According to Yang et al., a transformative state of mind holds tremendous potential to initiate "positive changes leading to a natural status free of physical and mental disease" (2015, p. 1884). The next section delves deeper into the transformation phenomenon. Various dimensions of transformative experiences and learning and their outcomes are discussed.

Transformation

Extant literature reiterates that transformation is an evolving and dynamic process (Chhabra, 2022; Coghlan & Weiler, 2018; Pung & Chiappa, 2020; Sheldon, 2020) that strives to expand "consciousness whereby individuals become critically aware of old and new self-views and choose to integrate these views into a new self-definition" (Holland-Wade, 1998, p. 713). Walter conceptualizes the transformation process as a learning cycle comprising of "concrete experience," "reflective observation," "abstract conceptualization," and "active experimentation" (Walter, 2016, p. 163). Inner transformation relates to a higher sense of consciousness. Some transformations are transitory and last for a few moments, whereas other transformations can occur and grow over a period of time and eventually become long-lasting or sustained. According to Sheldon, "for potentially transformative interchanges, spaces must be conducive to powerful and revealing conversations, and ideally create a sense of mindfulness and reverence" (2020, p. 6). For instance, natural landscapes (especially in the wilderness) are nourishing places where mindfulness and deep-rooted transformations can happen. Furthermore, the spaces in the post-transformation phase are also important as they can be designed to inspire reflection and facilitate a sustained sense of higher consciousness (Chhabra, 2022; Ponder & Holladay, 2013; Soulard, 2020).

Transformative experiences (TEs) can trigger attention towards self-existence and prompt emotional connections, facilitate exploration of the self, and stimulate functioning at an optimal physiological and mental levels (Cavender et al., 2020; Coghlan & Weiler, 2018; Dillette et al., 2019). People are also able to enhance their sense of spirituality and well-being through TEs. Some scholars posit that altruistic engagements during travel hold potential to facilitate sustained inner transformation leading to a higher sense of consciousness and well-being (Huang et al., 2023; Nandasena et al., 2022; Pope, 2018; Pung et al., 2020). As an instance, adventure tourism holds potential to test the physical and psychological threshold of visitors who are in quest for self-improvement (Smith & Diekmann, 2017), and this type of experience can result in enhancing overall well-being. According to Alahakoon et al. (2021), well-being facilitates transformative experiences. As an example, an elated state of psychological well-being can motivate tourists to consistently patronize transformative forms of tourism. Several studies report that people who are psychologically strong are more receptive towards seeking transformative experiences (Kim et al., 2015).

Transformative Learning

Coghlan and Gooch write that "engagement in transformative learning appears to be driven by an albeit subconscious disillusion with existing sets of reference and the desire to seek something novel" (2011, p. 719). People, particularly in the west experience a sense of alienation as they feel restricted in their daily mundane settings and work time (Lange, 2004). As pointed out by Taylor (2007), a context to facilitate reflection is important and critical reflection can be facilitated by distancing oneself from daily routine and engaging in experiential learning opportunities. According to O'Sullivan (1999), in the postmodern times, it is important to inculcate a sense of community because of which people feel alienated and displaced.

This sense of displacement can be overcome by fostering a sense of place through study of the territory along with the history of various communities who have inhabited it. Most of the documented literature lends credence to Mezirow's theory of TE.

Mezirow's theory of TE suggests a ten-step transformation process. It posits that transformative learning helps a person become aware of his or her position within the broader context of environment forces, many of which are not under his/her control. These include political, economic, sociocultural, and spiritual forces. Mezirow's (1978) ten steps to achieve a transformed state of mind include the following:

1. Experience a disorienting dilemma.
2. Undergo self-examination.
3. Critical appraisal of epistemic, sociocultural, and psychic assumptions.
4. Recognize and analyze personal discontent and share similar experiences with others.
5. Explore options for new roles, relationships, and action.
6. Plan a course of action.
7. Acquire knowledge and skills for actions.
8. Try new roles and assess feedback.
9. Build competence and self-confidence in new roles and relationships.
10. Reintegrate into one's life and society with a new personal perspective (Kitchenham, 2008).

As a case in point, Coghlan and Gooch (2011) employ the core tenets of Mezirow's theory of TE to volunteer tourism. The authors identify five steps based on a review of documented literature: (1) the pre-requisites and context for transformative learning to occur (i.e., seeking deeper involvement with the social and natural world and a change in context that facilitates transformative learning opportunities), (2) disorienting dilemmas and the emergence of confusing emotions (relating to the first three of Mezirow's steps), (3) the importance of dialogue and reflection (Steps 4 and 5), (4) self-actualization as an outcome of volunteer tourism (Steps 6 and 8), and finally (5) reintegration into society (Step 10) (year, p.). Two more themes are added to this list: disillusion and the search for new reference points which can facilitate interest in participating in a volunteer tourism project. The second theme stresses on the importance of creating a context and space to aid in progressing through the different phases (Coghlan & Gooch, 2011). Clearly, alternative forms of tourism such as volunteer tourism and spiritual tourism provide idyllic "othered" spaces and deep experiences to facilitate transformation.

Several authors reiterate that self-actualization is a key consequence of TE (relating to Steps 6 and 8 of Mezirow's theory). In fact, this element is an integral part of the transformational learning process. As participants experience progress through different learning steps, they reassess their personal values, cultivate new proficiencies, and poise with additional skills and knowledge (Coghlan & Gooch, 2011). Cranton and King (2003) also stress on the significance of personalization, authenticity, and a sense of autonomy in facilitating transformative learning. Distinctiveness and authenticity signify the manner in which a person can "gradually become conscious of our own unique psychological makeup over our lifetime" (Cranton & King, 2003, p. 91) and the manner in which people convey their novel and authentic self to the community. These traits are recognized as anticipated outcomes that drive the quest to pursue volunteer tourism (Coghlan & Gooch, 2011; Devereux, 2008; Wearing, 2002).

Transformative learning can happen when tourists immerse themselves in alternative forms of tourism or slow tourism such as study tours, backpacker tourism, volunteer tourism, and study tours (Kirillova et al., 2017; Pung et al., 2020). According to Huang et al., these "facilitate positive transformation in tourists, by stimulating self-reflection, knowledge integration, self-development, cross-cultural views……. and subjective well-being. Specifically, backpacker tourism involves extended trips ………………… focused on social interaction, flexible schedules, and participatory activities"; and study tours offer a learning opportunity, where traveling plays a key role in expanding the knowledge of participants, introducing them to new cultures, and offering relief from daily chores (2023, p. 2).

Clearly, tourists who are traveling to seek deep transformative experiences cherish occasions for self-inquiry and self-reflection to understand their inner selves, find a sense of purpose, and realize their inner truth while being distanced from their routine life (Chhabra, 2022; Dillette et al., 2019; Voigt et al., 2011). Sheldon (2020) presents three ideal scenarios to inspire transformation: deep human connections in cultural settings, deep connection with the environment in natural spaces, and engagements that stimulate self-inquiry, are self-reflective, and generate the desire to enhance knowledge of the self, seek learning and creative experiences, and offer a dedicated service to the visited destination. Inspiring stories can emphasize on richness of cultural, historical and individual diversity or generate awareness for shared benevolence and harmony between communities in visited destinations (Chhabra, 2022). The aforementioned TE tools offer an ideal platform to bring several SDGs (3,11,12, and 17) to fruition. As mentioned earlier, transformative potential of a destination can be heightened by designing spaces and scenarios to facilitate deep interaction between residents and tourists (e.g., sharing deep personal tales and narratives). Creative and naturally pristine landscapes are ideal settings to nurture deep human bonds. In fact, "immersion in nature is a common stimulus for peak experiences" (Sheldon, 2020, p. 6).

Optimal flow type of experiences can offer glimpses of transformed consciousness particularly during mindful and slow travel to remote or authentic places. As stated by Soulard, "a sense of awe or wonder is a common peak experience that brings heightened perception, greater appreciation of beauty, acceptance, connection and harmony with the surrounding world. The power of unknown places and people can create strong emotional responses which loosen identification with the small self. Mindfulness is a fertile ground for transformation, and is most likely to be found in slower paced trips—slow travel" (Soulard, 2020, p. 8,9). Gill et al., 2019) write that flow experiences hold potential to transform the visitors.

Once the tourists are on a transformative path and are inspired to sustain it, they patronize or build their transformative enterprises or groups upon return home which enable them to live in harmony with their inner self and the external world while at the same time promoting transformation to others (Chhabra, 2022; Ponder & Holladay, 2013). The state of inner transformation enhances the desire to serve other living beings and mindfulness towards the environment. In fact, conservation of nature practices is a common path of transformation in many indigenous communities which are more spiritually inclined (Sharmer, 2016). This transformative shift from ego-consciousness (obsession of personal welfare) to eco-consciousness (mindfulness towards the well-being of other living beings) is an important step in the transformation journey (Sharmer, 2016). To facilitate transformative moments, several opportunities of dedicated service exist in destinations across the globe such as "volunteer tourism, traveler philanthropy, citizen science, Peace Corps Program, Organic farms etc." (McGehee & Santos, 2005). Huang et al. (2023) opine that activity-specific information is important to analyze the connection between transformative tourism and well-being, for instance, the manner in which some activities can collectively or separately result in multiple dimensions of well-being. As an example, retreats aid in self-inquiry to expand consciousness of the inner self and are situated across the world but more exclusively in remote and or naturally pristine locations. Other stimulating factors relate to ability to deeply interact with others and immerse in nature in addition to the layout and structure of the retreat centers and the staff and teachers (Buzinde, 2020; Gill et al., 2019; Page et al., 2023).

According to Teoh et al. (2021), transformative tourism experiences hold potential to change lifestyles, of both the guests and the hosts, in a positive manner. Extant literature posits that tourism enhances well-being and offers opportunities to pursue healthy (psychological and physical) engagements. Such experiences make tourists pause in a space away from their daily life and reflect on themselves and their way of life, in general (Chhabra, 2022; Lean, 2009; Lean et al., 2014; Pine & Gilmore, 2013). TE is subject to numerous activators, progressions, and outcomes. Several scholars have viewed TE from an existential perspective. For instance, Kirillova et al. (2017) opine that epiphanic moments play a key role in transforming people. Walter, on the other hand, uses a learning perspective to show that visitors proactively self-examine and develop sustainable behaviors.

Teoh et al. call for a co-created approach that relates to both consumers and the well-being of the visited destination. According to the authors, landscape, its social attributes and tangible aspects hold potential to transform tourists. Experience during the encounter can stimulate cognitive and emotional responses which can initiate changes at the psychological, physical, and social levels and openness to accumulate new knowledge. Some consumers might acknowledge the value of their experience immediately; others might note impact on their lives during postexperience reflections. In deep experiences, triggered by the interaction with experience facilitators, both consumer and facilitator can become transformed (Rydzik et al., 2013). The authors contend that "while consumer transformation results in changes to their circumstances in most cases, transformative outcomes in facilitators may impact on how the overall experience is managed or provided" (Rydzik et al., 2013, p. 185).

Facilitators such as hosts, tour guides, and instructors (on an educational trip) play a key role in molding experience of tourists (Lloyd et al., 2015). At the same time, cognitive and emotional interactions can also generate transformative experiences in the facilitators (Walker & Weiler, 2017). In the case of tourists attending retreats, “class flexibility, the packages offered, health regime designs, and the compassionate, friendly, and personable service delivered by experience-facilitators can contribute to the relaxing atmosphere that holds potential to stimulate transformation” (Fu et al., 2015, p. 85). Another example is the site managers at a dark tourism site who can use strategies to “convey reconstructed socially symbolic meanings to consumers, from a place of death and suffering to an educational place for the future” in a manner that consumers can develop favorable dispositions towards others (Magee & Gilmore, 2015). TEs can also be triggered by authentic ecotourism experiences where local experts teach local culture, environment, and livelihood aspects to inform and educate consumers (Walter & Reimer, 2012).

TEs contribute significantly towards psychological well-being such as enhanced self-confidence and sense of tranquility, autonomy, and empowerment (Brown, 2009; McWha et al., 2018). Profound TEs, particularly in culturally different settings, can result in important lifestyle changes and transformed sense of identity leading to restructuring of priorities in life and changes in social behavior (Lukoff, 2018; Ourahmoune, 2016; Willson et al., 2013). Teoh et al. write that “social changes refer to consumers’ mindset towards others and potentially impacts consumers’ social surroundings when gaining broad-mindedness and tolerance for others. An essential outcome of TE is not only self-actualisation but developing civic attitudes, openness, harmony, altruism,……, trust, and cross-cultural understanding” (Teoh et al., 2021, p. 183).

Transformative Experiences and Well-Being

A plethora of literature has gauged well-being based on two well-being theories (flow experience theory and SDT (self-determination theory)) and the PERMA model of well-being. Several studies have used SDT to examine the impact of spiritual and wellness tourism experiences on fundamental emotional needs. According to SDT, three core needs need to be satisfied to promote overall well-being: autonomy (freedom and desire to feel free), competence (i.e., aspiring to maneuver consequences and become a maestro), and relatedness (desire to closely bond with others) (Chhabra, 2022; Buzinde, 2020; Wray-Lake et al., 2019). Flow experience theory postulates that the balance between challenges associated with an activity or experience and proficiencies, at an individual level, holds potential to nurture well-being (Csikszentmihalyi, 1990). Transformative experiences which energize focus and self-inquiry on the meaning of life can result from a moderately challenging tourism activity that holds potential to gratify desire for daring experiences and proficiencies (Huang et al., 2023; Wu & Liang, 2011). Flow experiences enhance eudaimonic well-being (Vada et al., 2020) particularly in the context of serious leisure pursuits and adventure/risk tourism (such as sky diving, surfing, and mountain climbing) (Holm et al., 2017).

All theories and paradigms of well-being postulate that well-being has multiple connotations, and therefore there are multiple pathways that help in achieving well-being goals (Farkić et al., 2020). SDT stresses on the importance of personal will-power and motivation to attain desired transformation, whereas flow experience theory is based on the significance of alignment between the nature of experiential sites and competencies of the participants (Holm et al., 2017). Huang et al. note that "fulfilling tourism-specific needs bridge transformative tourism activities with well-being. For instance, escaping one's routine inspires new lifestyle perspectives (Mannell & Iso-Ahola, 1987); the opportunity to pursue education during travel spur self-introspection and personal growth (Hwang & Lee, 2019)" (Huang et al., 2023, p. 4).

Transformative Tourism and Well-Being

In the post-pandemic (COVID-19) times, demand has surged for transformative tourism. As discussed earlier, wellness tourism, a subset of transformative tourism, can be designed "by tourist destinations by integrating the concept of tourism services with the concept of 'rest' and health services" (Pratiwi et al., 2021, p. 140). Wellness-related travel seeks a combination of healthcare, exercise and relaxation, spiritual connectedness with nature, consumption of healthy food, and gathering knowledge and competencies and engaging at a personal level with "other" cultures, devotions, and communities (Chhabra, 2021; Dillette et al., 2019). Harmonious therapeutic settings are sought for optimal mental purification and spiritual healing/enlightenment (Chhabra, 2022).

Travel to enhance well-being generates benefits such as transcendence, physical health and appearance, escapism and relaxation, importance of community, re-establishing self-esteem, and indulgence (Voigt et al., 2011). Wellness experiences can be measured on a scale stretching from ordinary to extraordinary. With regard to extraordinary experiences, if they are immersive, deep, and optimal, they can trigger enduring transformation (Chhabra, 2022; Sheldon, 2020) such as in the case of serious leisure and its slow pace. Transformation tourism is often pursued with a desire to bond with the self. As postulated by Ponder and Holladay (2013), transformational experiences can help achieve Maslow's higher-level needs of self-satisfaction and self-actualization at an optimum level. These transformations can be enduring or ephemeral. Chhabra writes that "the ephemeral changes refer to experiences stimulated by, for instance, a retreat environment. Enduring changes, on the other hand, are the ones that go beyond ………… and continue to shape a person's lifestyle upon return home" (Chhabra, 2022, p. 130).

Sustained Transformation and the Cyclical Process

Most transformational frameworks of well-being are drawn from Seligman's PERMA (positive emotion, engagement, relationships, meaning, and accomplishment) theory of well-being. Dillette et al. (2019) employed the PERMA model to develop an integrative transformational framework. The aim was to examine the linkages between psychological well-being and transformative experiences. The authors confirmed that the PERMA dimensions trigger transformative benefits albeit some are transitory.

Sustained transformative potential of wellness tourism (e.g., yoga tourism) was confirmed by Ponder and Holladay (2013). The authors presented a cyclical model and highlighted the psychological, emotional, and spiritual capitals gathered by yoga travelers (yogis). These travelers acquired a higher sense of consciousness and knowledge at the retreat site. When they returned home, they shared their newly acquired knowledge and views with their community and inspired other people to embrace and seek advanced knowledge by journeying to an authentic locale. The yogis continued to advance their knowledge by enthusiastically accruing psychological capital (related to self-assurance, positive disposition, and buoyancy), emotional capital (developing capabilities for personal, social, and managerial skills development), and spiritual capital (based on moral virtues and faith/conviction). Chhabra also scrutinized the transformational potential of wellness tourism by looking at serious yoga travelers at a retreat hub in Rishikesh, India. A three-step approach was used: (1) pre-trip anticipation of benefits, (2) onsite transformation effects, and (3) enduring transformation after returning home. The author developed a cyclical transformation model of well-being to show that "after one a lag time of one year, some well-being dimensions from the PERMA model continue to remain robust. Enduring transformational benefits are cemented by a sense of accomplishment and engagement. Altruist behaviour is also noted as almost half of the cohort informs that they are volunteering their time to hold online/in person yoga/relaxation sessions for extended groups, based on donations or complementary" (Chhabra, 2022, p. 142). Furthermore, spirituality is found to be a sustained benefit and hold potential to stimulate ongoing desire to seek more opportunities to for a higher state of consciousness by pursuing different programs and readings.

Notably, the integration of transformative experiences upon return home is important for sustained transformation albeit difficult to accomplish due to the stress of work and other life commitments (Gill et al., 2019). If a tourist is part of an organized retreat or group experience, the post-trip context is easier and can encourage the continuation of practices learned during travel (Chhabra, 2022). Electronic communications and social media, journaling, and reunions can also be effective ways to integrate the awakenings to more permanent transformation (Soulard, 2020; Soulard et al., 2023).

Sustained Transformation and Sustainability

Dwyer (2023a) suggests a "Beyond GDP" approach to integrate well-being dimensions in policies associated with sustainable development of tourism. He refers to the "Better Life" paradigm emerging from the "Beyond GDP" agenda. It rests on three pillars for knowing and gauging current and future well-being of residents: material living conditions (comprising of income, wealth, jobs and earnings, and housing), quality of life (based on multiple dimensions that shape people's lives such as health, work-life balance, education/training, social connections, civic engagement and governance, environmental quality, personal security, and subjective well-being), and sustainability (measurement metrics are economic, human, social, and natural capitals) (2023a, p. 2401). In support of the "Better Life" framework, Dwyer writes that its "distinguishing features are its recognition of the importance of both subjective and objective dimensions of well-being and its distinction of the importance of both subjective and objective dimensions of well-being and its distinction between the drivers of current and future well-being, this embedding sustainability considerations into the framework" (2023a, p. 2402).

A sustainable approach towards well-being can lend support to the materialistic environment and quality of life dimensions by ensuring that future generations enjoy the same level of well-being human beings enjoy today (Dwyer, 2023b). The capitals (economic, human, social, and natural) approach looks at how the various capital stocks evolve overtime to support overall well-being. Sheldon (2020) points out that the integration of transformative experience design into a destination's sustainable tourism portfolio is not an easy task. What is required is a well-rounded "fundamental value shift by all decision makers away from short-term profitability towards concern for the long-term well-being of the community. A critical part of this paradigm shift is related to the re-structuring the role of tourism employees. Transformative tourism experiences cannot be delivered by the current system which tends to promulgate low income, low skilled employment" (2020, p.8). Conducive working conditions and appropriate education and training should include thoughtfulness and empathy towards of tourist well-being and familiarity of different ways to promote transformative consciousness. As pointed out by Ayazlar and Ayazlar (2024), friendly working environment can help nurture the well-being of employees and inculcate mindfulness and elevated sense of consciousness. In other words, the transformation should begin at the grass-roots level so that sustained transformation can become a reality and facilitate mindful and responsible production and consumption of tourism resources.

Aligned with the aforementioned, Lean writes that "to fully realize ideals like sustainability, the industry must work towards inspiring enduring changes of behavior that ensure the health and well-being of the individual and their economic, sociocultural and ecological environments. These changes of action will help deliver individual and global wellness" (Lean, 2009, p. 1). Wellness and well-being terms are occasionally associated with a person although they are the outcome of their environment. At a macro level, wellness is reliant on the healthiness of the planet

Earth and its ecological, sociocultural, and economic ecosystems (Lean, 2009). These ideas are captured in the sustainable development and health/wellness notions established with the intent of nurturing the utilitarian and intrinsic values of life (2009, p. 2). Lean proposes the notion of sustainability ambassadors. He argues that "for tourism to make a greater contribution to sustainable development, it must use its unique position as an interpreter within meaningful settings to aid the transformation of one's thinking and behavior upon returning home. Tourism should have as a core objective the development of 'sustainability ambassadors.' They should fulfill a number of objectives: return to their home environments, exemplify the principles of sustainability, and encourage others to do likewise; reduce their impact on destinations both during their travels and, most importantly, upon their home environments" (2009, p. 4). As evidenced in the aforementioned paragraphs, the equation between well-being, QOL, sustainability, and resilience is dynamic and shaped by specifics of different tourism settings and locations. As discussed in the first chapter, sustainable initiatives can be gauged with the help of SDGs. The latter can serve as effective platforms to promote well-being/QOL and resilience. However, a transformative state of mind of all stakeholders in the tourism system is critical to sustain well-being, QOL, and sustainability.

Inner transformation can be referred to beliefs and practices of people acquired in a place, ingrained in the cultural cognizance of daily life (Meyer, 2003). According to Meyer, three elements of existence mold the existence of human beings: gross, subtle, and causal which can be labeled "place, space and time- To see; to know; to be given insight" (2003, p. 57). Drawing from Meyer's work and the core message of Vedanta (Sivananda, 1960), Chhabra (2021) developed a conceptual framework titled "Towards an Inner Path of Being" proposing a route to transform to a higher state of consciousness. The paradigm encapsulates an internal path of "Being" that harmonizes gross (the physical body) and subtle (mental) bodies and transcends to the causal (ego), an optimal state of being. Brahmaprana (2001) writes that a person represents the gross, subtle, and causal bodies, and these require nurturing to optimize their intrinsic potential. For example, "food and exercise are needed to keep our physical body healthy. The subtle body requires nourishment from healthy/positive emotions and refinement of our mental faculties……. positive emotions and flourishing state of mind can be stimulated through artistic and intellectual pursuits" (Chhabra, 2021, p. 765). The causal state can be reached with the help of alternative tourism daises guided by theories such as transformative learning and existentialist (Meyer, 2003) to co-create flow of optimal moments or B-cognition (triggered by disorientation and/or positive emotions), energized by tourists and different stakeholders from the visited destination such as suppliers, hosts, and facilitators (Chhabra (2021). The reoccurring of peak moments and experiences can cement introspective and deep learning needs thereby infusing an enduring need to move forward in the transformative journey. An ideal experiential stage includes the physical features of a place or landscape, social engagement, and bonding thereby stimulating immersive state of mind, leading to the state of self-inquiry and a higher state of consciousness. The next (concluding) chapter suggests a roadmap to develop/promote enduring transformation, enhanced quality of life, and sustainability of visited and home destinations of tourists.

References

Alahakoon, T., Pike, S., & Beatson, A. (2021). Transformative destination attractiveness: An exploration of salient attributes, consequences, and personal values. *Journal of Travel & Tourism Marketing, 38*(8), 845–866.

Atal, N., & Atal, C. (2024). Healing potential of herbs and herbal tourism. In D. Chhabra, N. Atal, & A. Maheshwari (Eds.), *Sustainable development and resilience of tourism: Wellbeing and quality of life perspectives*. Springer. In Press.

Ayazlar, R., & Ayazlar, G. (2024). The importance of mindfulness for tourism employees. In D. Chhabra, N. Atal, & A. Maheshwari (Eds.), *Sustainable development and resilience of tourism: Wellbeing and quality of life perspectives*. Springer. In Press.

Brahmaprana, P. (2001). Vedanta: Death and the art of dying. *CrossCurrents*, 337–346.

Brown, L. (2009). The transformative power of the international sojourn: An ethnographic study of the international student experience. *Annals of Tourism Research, 36*(3), 502–521.

Buzinde, C. N. (2020). Theoretical linkages between well-being and tourism: The case of self-determination theory and spiritual tourism. *Annals of Tourism Research, 83*, 102920.

Cavender, R., Swanson, J. R., & Wright, K. (2020). Transformative travel: Transformative learning through education abroad in a niche tourism destination. *Journal of Hospitality, Leisure, Sport & Tourism Education, 27*, 100245.

Chhabra, D. (2021). Transformative perspectives of tourism: Dialogical perceptiveness. *Journal of Travel & Tourism Marketing, 38*(8), 759–768.

Chhabra, D. (2022). Transformational model of well-being for serious travellers. *International Journal of Spa and Wellness, 5*(2), 129–146.

Coghlan, A., & Gooch, M. (2011). Applying a transformative learning framework to volunteer tourism. *Journal of Sustainable Tourism, 19*(6), 713–728.

Coghlan, A., & Weiler, B. (2018). Examining transformative processes in volunteer tourism. *Current Issues in Tourism, 21*(5), 567–582.

Cohen, D. (2024). Transformative potential of sustainable Scuba Dive tourism. In D. Chhabra, N. Atal, & A. Maheshwari (Eds.), *Sustainable development and resilience of tourism: Wellbeing and quality of life perspectives*. Springer. In Press.

Cranton, P., & King, K. (2003). Transformative learning as a professional development goal. *New Directions in Adult and Continuing Education, 98*, 31–38.

Csikszentmihalyi, M. (1990). Literacy and intrinsic motivation. *Daedalus*, 115–140.

Devereux, P. (2008). International volunteering for development and sustainability: Outdated pater-nalism or a radical response to globalization? *Development in Practice, 18*(3), 357–370.

Dillette, A. K., Douglas, A. C., & Andrzejewski, C. (2019). Yoga tourism–A catalyst for transformation? *Annals of Leisure Research, 22*(1), 22–41.

Dwyer, L. (2023a). Tourism development and sustainable well-being: A beyond GDP perspective. *Journal of Sustainable Tourism, 31*(10), 2399–2416.

Dwyer, L. (2023b). Resident well-being and sustainable tourism development: The 'capitals approach'. *Journal of Sustainable Tourism, 31*(9), 2119–2135.

Farkić, J., Filep, S., & Taylor, S. (2020). Shaping tourists' wellbeing through guided slow adventures. *Journal of Sustainable Tourism, 28*(12), 2064–2080.

Fu, X., Tanyatanaboon, M., & Lehto, X. Y. (2015). Conceptualizing transformative guest experience at retreat centers. *International Journal of Hospitality Management, 49*, 83–92.

Gill, C., Packer, J., & Ballantyne, R. (2019). Spiritual retreats as a restorative destination: Design factors facilitating restorative outcomes. *Annals of Tourism Research, 79*, 102761.

Gogoi, P., & Maheshwari, A. (2024). Mapping the impact of motorcycle tourism in trans-Himalayas: A study on alternative livelihood creation in rural Ladakh, India. In D. Chhabra, N. Atal, & A. Maheshwari (Eds.), *Sustainable development and resilience of tourism: wellbeing and quality of life perspectives*. Springer. In Press.

Holland-Wade, G. (1998). A concept analysis of personal transformation. *Journal of Advanced Nursing, 28*(4), 713–719.

Holm, M. R., Lugosi, P., Croes, R. R., & Torres, E. N. (2017). Risk-tourism, risk-taking and subjective well-being: A review and synthesis. *Tourism Management, 63*, 115–122.
Huang, X., Wang, P., & Wu, L. (2023). Well-being through transformation: An integrative framework of transformative tourism experiences and hedonic versus eudaimonic well-being. *Journal of Travel Research*, 00472875231171670.
Hwang, J., & Lee, J. H. (2019). A strategy for enhancing senior tourists' well-being perception: Focusing on the experience economy. *Journal of Travel & Tourism Marketing, 36*(3), 314–329.
Joson, G., Pyke, J., Delacy, T., & Davidson, C. (2024). Wellness in the spa industry: an investigation of hot springs tourism in Australia. In D. Chhabra, N. Atal, & A. Maheshwari (Eds.), *Sustainable development and resilience of tourism: Wellbeing and quality of life perspectives*. Springer. In Press.
Kim, E. (2024). Impact of hybrid word of mouth on consumer behavior: Case of off-highway vehicle recreation markets. In D. Chhabra, N. Atal, & A. Maheshwari (Eds.), *Sustainable development and resilience of tourism: Wellbeing and quality of life perspectives*. Springer. In Press.
Kim, H., Woo, E., & Uysal, M. (2015). Tourism experience and quality of life among elderly tourists. *Tourism Management, 46*, 465–476.
Kirillova, K., Lehto, X., & Cai, L. (2017). What triggers transformative tourism experiences? *Tourism Recreation Research, 42*(4), 498–511.
Kitchenham, A. (2008). The evolution of John Mezirow's transformative learning theory. *Journal of Transformative Education, 6*(2), 104–123.
Lange, E. (2004). Transformative and restorative learning: A vita dialectic for sustainable societies. *Adult Education Quarterly, 54*, 121–139.
Lean, G. L. (2009). *Transformative travel: Inspiring sustainability* (pp. 191–205). Mind, body, spirit, place.
Lean, G. L., Staiff, R., & Waterton, E. (Eds.). (2014). *Travel and transformation*. Ashgate Publishing Company.
Lloyd, K., Suchet-Pearson, S., Wright, S., Tofa, M., Rowland, C., Burarrwanga, L., et al. (2015). Transforming tourists and "culturalising commerce": Indigenous tourism at Bawaka in Northern Australia. *The International Indigenous Policy Journal, 6*(4).
Lukoff, M. M. (2018). Immersion in indigenous agriculture and transformational learning. *Anthropology and Humanism, 43*(1), 90–106.
Magee, R., & Gilmore, A. (2015). Heritage site management: From dark tourism to transformative service experience? *The Service Industries Journal, 35*(15–16), 898–917.
Mannell, R. C., & Iso-Ahola, S. E. (1987). Psychological nature of leisure and tourism experience. *Annals of Tourism Research, 14*(3), 314–331.
McGehee, N. G., & Santos, C. A. (2005). Social change, discourse and volunteer tourism. *Annals of Tourism Research, 32*(3), 760–779.
McWha, M., Frost, W., & Laing, J. (2018). Travel writers and the nature of self: Essentialism, transformation and (online) construction. *Annals of Tourism Research, 70*, 14–24.
Meyer, M. (2003). Hawaiian hermeneutics and the triangulation of meaning: Gross, subtle, causal. *Social Justice, 30*(4 (94)), 54–63.
Mezirow, J. (1978). Perspective transformation. *Adult Education, 28*, 100–110.
Mezirow, J. (1991). *Transformative dimensions of adult learning*. Jossey-Bass.
Mezirow, J. (2000). Learning to think like an adult. In *Learning as transformation: Critical perspectives on a theory in progress* (pp. 3–33).
Nandasena, R., Morrison, A. M., & Coca-Stefaniak, J. A. (2022). Transformational tourism–A systematic literature review and research agenda. *Journal of Tourism Futures, 8*(3), 282–297.
O'Sullivan, E. (1999). *Transformative learning: Education vision for the 21st century*. Zed Books.
Ourahmoune, N. (2016). Narrativity, temporality, and consumer-identity transformation through tourism. *Journal of Business Research, 69*(1), 255–263.
Page, S. J., Connell, J., Price, S., Owen, S., Ledingham, K., & Clare, L. (2023). Operationalizing transformative tourism: Creating dementia-friendly outdoor and nature-based visitor experiences. *Journal of Travel Research*, 00472875231217735.

Pham, K., Vogt, C., & Andereck, K. (2024a). Overtourism; A Potential outcome in contemporary tourism, Part I: Concepts, indicators and impacts. In D. Chhabra, N. Atal, & A. Maheshwari (Eds.), *Sustainable development and resilience of tourism: Wellbeing and quality of life perspectives*. Springer. In Press.

Pham, K., Andereck, K., & Vogt, C. (2024b). Overtourism: A potential outcome in contemporary tourism. Part 2: Causes, solutions and management challenges. In D. Chhabra, N. Atal, & A. Maheshwari (Eds.), *Sustainable development and resilience of tourism: Wellbeing and quality of life perspectives*. Springer. In Press.

Pine, B. J., & Gilmore, J. H. (2013). The experience economy: Past, present and future. *Handbook on the experience economy, 1*, 21–44.

Pitrelli, M. (2023). Years in the making: Venice approves tax on daytrippers. https://www.cnbc.com/2023/09/13/new-tax-to-visit-venice-travelers-must-pay-to-visit-during-te-day.html. CNBC Travel 2023.

Ponder, L. M., & Holladay, P. J. (2013). The transformative power of yoga tourism. In *Transformational tourism: Tourist perspectives* (pp. 98–107). CABI.

Pope, E. (2018). Tourism and wellbeing: Transforming people and places. *International Journal of Spa and Wellness, 1*(1), 69–81.

Pratiwi, R., Rama, R., & Sulistiyanti, N. (2021). Building the trust for the tourism destination resiliency in new normal society (the role of wellness tourism system). *IKRA-ITH HUMANIORA: Jurnal Sosial Dan Humaniora, 5*(1), 1–9.

Pung, J., & Chiappa, G. (2020). An exploratory and qualitative study on the meaning of transformative tourism and its facilitators and inhibitors. *European Journal of Tourism Research, 24*, 2404–2404.

Pung, J. M., Gnoth, J., & Del Chiappa, G. (2020). Tourist transformation: Towards a conceptual model. *Annals of Tourism Research, 81*, 102885.

Rydzik, A., Pritchard, A., Morgan, N., & Sedgley, D. (2013). The potential of arts-based transformative research. *Annals of Tourism Research, 40*, 283–305.

Sharmer, O. (2016). *Theory U: Leading from the future as it emerges*. Berrett-Koehler Publishers.

Sheldon, P. J. (2020). Designing tourism experiences for inner transformation. *Annals of Tourism Research, 83*, 102935.

Sivananda, S. (1960). Vedanta for beginners.

Smith, M. K., & Diekmann, A. (2017). Tourism and wellbeing. *Annals of Tourism Research, 66*, 1–13.

Soulard, J. M. (2020). *An investigation of transformative tourism outcomes and strategies* (Doctoral dissertation, Virginia Tech).

Soulard, J., Stewart, W., Larson, M., & Samson, E. (2023). Dark tourism and social mobilization: Transforming travelers after visiting a Holocaust Museum. *Journal of Travel Research, 62*(4), 820–840.

Sullivan, S. (2002). *How sustainable is the communalizing discourse of 'new' conservation* (pp. 158–197). Berghahn Press.

Taylor, E. (2007). An update of transformative learning theory: A critical review of the empirical research (1999–2005). *International Journal of Lifelong Education, 26*(2), 173–191.

Teoh, M. W., Wang, Y., & Kwek, A. (2021). Conceptualising co-created transformative tourism experiences: A systematic narrative review. *Journal of Hospitality and Tourism Management, 47*, 176–189.

Vada, S., Prentice, C., Scott, N., & Hsiao, A. (2020). Positive psychology and tourist well-being: A systematic literature review. *Tourism Management Perspectives, 33*, 100631.

Voigt, C., Brown, G., & Howat, G. (2011). Wellness tourists: In search of transformation. *Tourism Review, 66*(1/2), 16–30.

Walker, K., & Weiler, B. (2017). A new model for guide training and transformative outcomes: A case study in sustainable marine-wildlife eco-tourism. *Journal of Ecotourism, 16*(3), 269–290.

Walter, P. G. (2016). Catalysts for transformative learning in community-based ecotourism. *Current Issues in Tourism, 19*(13), 1356–1371.

Walter, P. G., & Reimer, J. K. (2012). The "ecotourism curriculum" and visitor learning in community-based ecotourism: Case studies from Thailand and Cambodia. *Asia Pacific Journal of Tourism Research, 17*(5), 551–561.

Wearing, S. (2002). Re-centering the self in volunteer tourism. In G. Dann (Ed.), *The tourist as a metaphor of the social world* (pp. 237–262). CAB International.

Willson, G. B., McIntosh, A. J., & Zahra, A. L. (2013). Tourism and spirituality: A phenomenological analysis. *Annals of Tourism Research, 42*, 150–168.

Wray-Lake, L., DeHaan, C. R., Shubert, J., & Ryan, R. M. (2019). Examining links from civic engagement to daily well-being from a self-determination theory perspective. *The Journal of Positive Psychology, 14*(2), 166–177.

Wu, C. H. J., & Liang, R. D. (2011). The relationship between white-water rafting experience formation and customer reaction: A flow theory perspective. *Tourism Management, 32*(2), 317–325.

Yang, J. Y., Paek, S., Kim, T., & Lee, T. H. (2015). Health tourism: Needs for healing experience and intentions for transformation in wellness resorts in Korea. *International Journal of Contemporary Hospitality Management, 27*(8), 1881–1904.

Chapter 12
Conclusion: Building Sustainable Transformative Quality of Life Frameworks

Deepak Chhabra

Abstract This chapter revisits key messages conveyed by the contributors of this book and traces the scholarly progress in the quality of life (QOL) paradigms. Emerging key areas of inquiry, related to QOL, are emphasized such as developing appropriate mechanisms to promote SDGs (Sustainable Development Goals), the remedial potential of non-pharmaceutical interventions, and surge in wellness tourism. It is further pointed out that sustained transformations and enhanced quality of life of both visited and visiting communities are key to strengthening resilience capacity of tourism systems. A roadmap, to boost QOL and sustainability, is recommended. The chapter closes by proposing a forward-looking cyclical transformative quality-of-life model which underlines the significance of transformation and heightened state of consciousness to build and strengthen resilient tourism systems by promoting holistic well-being and healthfulness.

Keywords Quality of life · Sustainability · Sustainable Development Goals (SDGs) · Holistic well-being · Transformation

The present and future success of sustainable tourism calls for a deep-rooted paradigm shift by embracing four pillars of sustainability. With so many environmental, political, economic, and social changes observed in the contemporary times, relevancy and integration of the health sector (as one core dimension of human sustainability) have become paramount. According to Marques, Serraqueiro, and Nogueira, "the popular expression of 'I'm in good health, the rest is secondary', holds zero meaning today" (Marques et al., 2021, p.2). Dwyer reminisces that most studies on resident well-being have been more inclined to focus "on current well-being outcomes rather than a detailed study of the links between well-being and sustainability" (2023, p. 2399). Sirgy and Uysal note that only a handful of studies "explore

D. Chhabra (✉)
School of Community Resources and Development, Watts College of Public Service and Community Solutions, Arizona State University, Phoenix, AZ, USA
e-mail: Deepak.chhabra@asu.edu

D. Chhabra et al. (eds.), *Sustainable Development and Resilience of Tourism*,
https://doi.org/10.1007/978-3-031-63145-0_12

how co-creation and design for QOL impact the well-being of tourists," and there is also meager research on "how the design of the experience setting (the workplace, the hotel, restaurant, the entertainment place, etc.) may affect the well-being of not only customers but also providers of services such as hotel and restaurant employees" (Sirgy & Uysal, 2023, p. 596). Creating conducive dais for tourism settings should not only aid in promoting the well-being of tourists/customers but also enhance the destinations from a livable standpoint (Uysal et al. (2020).

This book aims to fill a gap in existing literature by scrutinizing intersections between quality of life, well-being, sustainability, and resilience notions. It conceptualizes a sustained transformative approach, to reinforce the linkages, between these concepts. In-depth insights are offered on the supporting role of non-pharmaceutical interventions, such as herbal healing, to promote the well-being of visited and visiting communities and the subsequent efforts to link them with sustainability. Furthermore, the shifting consumer preference and the era of new travelers offer multiple opportunities to explore new paths and promote/accomplish various SDGs (Sustainable Development Goals), particularly those related to responsible production and consumption, health, and well-being (Lew, 2020; Santos et al., 2020). It is argued that key stakeholders in the tourism system need to be proactively engaged in co-creating sustainable solutions, promoting Sustainable Development Goals (SDGs), and building and fortifying resilience capacity (Pocinho et al., 2022).

Recent research suggests the emergence of conscious consumers, who are keen to "balance between experiencing the authentic offerings from a destination and reciprocating back equitably" (Mondal & Samaddar, 2021, p. 6). Mindful visitors are more aware and concerned of others in their home and travel environments and practice ethical consumption (Barber & Deale, 2014; Chhabra, 2010; Sheth et al., 2011). In fact, the pandemic has inspired quest for sustained transformations toward an enhanced sense of "being" (Mondal & Samaddar, 2021; Moscardo et al., 2017; Sheldon, 2020).

Unpredictable catastrophes such as the pandemic and wars will continue to challenge/barricade initiatives associated with the development of sustainable tourism. Ongoing need will remain for innovative and resilient strategies, and products as tourism systems remain in a state of flux (Lew, 2020). The contemporary environment calls for adaptivity and resilience and inner transformation to stay anchored/grounded as our human systems adapt to unpredictable phases of "floating fixity." In other words, due to unexpected or expected interventions, the systems might remain in a state of equilibrium for a brief period of time but are likely to shift again to accommodate interventions and disturbances. From the demand side, significance of sustainable and healthful consumption is becoming increasingly important (Dwyer, 2022; Saarinen, 2021; Sirgy & Uysal, 2023; Um & Yoon, 2021). This implies that it has become crucial to design and develop tourism strategically and sustainably by seeking trade-offs between environmental, economic, social, and health/wellness goals of different societies across the globe (Streimikiene et al., 2021).

In fact, the quality of life (QOL) of host communities is receiving unprecedented attention due to the pandemic, as local communities are displaying more keenness to become involved in sustainable tourism planning and development decisions associated with the shaping of tourism in a manner that benefits them without compromising their health and safety prioritizations. As mentioned earlier in the book, QOL is a multidimensional concept and refers to personal well-being, happiness, and life satisfaction (Hu et al., 2024). A rich body work has delved into this concept and dissected it mostly from three standpoints: visitors/tourists, employees of the tourism industry, and host community/residents (Sirgy & Uysal, 2023; Uysal et al., 2016). The historical underpinnings of the QOL/well-being notion are rooted in hedonic and eudaimonic dichotomies (Uysal et al., 2016; Sirgy & Uysal, 2023).

Sirgy and Uysal (2023) propose four QOL theoretical paradigms based on perusal of existing research. These frameworks focus on tourists, hosts, tourism industry employees, and the significant role of ICT (information and technology). The QOL model for tourists scrutinizes the manner in which experiences enrich the QOL of tourists. It endeavors to scrutinize the influence of attributes associated with tourism on their well-being (delineated into objective and subjective well-being). Objective well-being focuses on various dimensions of life such as leisure, sociocultural, job, domestic, travel, food, and spiritual. Subjective indicators are related to satisfaction with life, perceived QOL, and satisfaction in different aspects of life such as leisure, social, familial, and job-related. The model is divided into three parts: independent variables (such as tourist and trip attributes and interaction between both), moderators and mediators, and dependent variables (positive impacts and externalities, satisfaction in various domains, satisfaction with life and eudaimonia related to factors such as happiness and meaningfulness in life).

Although most scholars refer to life satisfaction as the absolute dependent attribute, Sirgy and Uysal (2023). argue that it is equally important to consider other factors such as positive and negative impacts, eudaimonia, and absence of deficiency in health, happiness, and/or prosperity. In the tourist QOL model, the authors describe the independent variables as tourist and trip attributes and relationship between the two characteristics. Tourist attributes relate to factors such as sociodemographic, psychographic, and sociocultural traits. Their proposition is based on Genç's (2023) work who views the probable influence of age, gender, country of residence, and social status and cultural background on QOL of tourists. Genç (2023) also notes the influence of psychological traits such as personality, self-perception, self-consistency, and social recognition. Motivation can shape QOL (Dann, 2012). Along similar lines, other studies have examined the tourism behavior and QOL of senior tourists (Zielińska-Szczepkowska, 2023) and perceptions based on gender on QOL in medical tourism (Sevuk, 2023). In summary, existing body of work supports Uysal and Sirgy's (2023) argument that individual differences exist in the manner tourism influences QOL of tourists at the individual level.

With regard to trip attributes, type and nature of the trip can impact QOL in a direct or indirect manner. As an instance, Sirgy and Uysal (2023) describe how the type of treatment and the manner in which the medical facility abroad offers services influences the QOL of patients. Genç (2012) reported that medical tourism offers economic, health-related, and psychological benefits to patients thereby influencing their QOL in a positive manner. Different types of tourism hold potential to contribute to well-being such as aesthetic tourism (Koc & Ayyildiz, 2023) and gastronomy tourism (Kesgin, 2023). Numerous elements encapsulate the interaction between tourists and trip attributes such as place attachment and the manner in which it can shape QOL (Williams & McIntyre, 2011). Interaction factor between destination image and the tourists' perception of self can significantly impact the sense of well-being (Karagöz & Kama, 2023). Intense human connectivity and meaningful bonding between guests and hosts can stimulate positive emotions and contribute toward holistic well-being (Mitta et al., 2023).

Sirgy and Uysal (2023) also present a QOL and host community framework, capturing the to-date research, to scrutinize and understand the influence of tourism on the well-being of the host community. Well-being is delineated into objective and subjective well-being. It is posited that "community resources and conditions play an important role in influencing overall community well-being. The relationship between community resources/conditions and community well-being is moderated by a host of other constructs such as involvement of community stakeholders, competitive strategy and destination management, sustainable development, and type of tourism. Mediators include perceived living conditions, perceived tourism impact, residents' support of tourism, and resident satisfaction with life domains. Dependent characteristics are community well-being (e.g., satisfaction with one's community at large; objective indicators of community quality of life; place attachment)" (2023, p. 579–580).

Kim pioneered a study in 2002 and noted the association between effects of tourism activities and satisfaction with specific domains of life. This association is noted to vary at different phases of the tourism life cycle (Kim, 2002, Kim et al., 2003; Moscardo, 2009). With regard to community resources and infrastructure, Uysal and Sirgy (2023) point out that there are several aspects that need to be taken into consideration such as developing and using relevant resources and then managing them. Mckercher and Ho (2012) reported a link between resources and their state and QOL while examining the association between cultural tourism and QOL of local residents. It is noted that "regardless of the type of resource and tourism product, the resource base needs to be augmented by a range of different forms of capital, including social capital of the destination community" (2023, p. 581).

Furthermore, numerous QOL studies contend that subjective well-being (overall satisfaction with life) of an individual is closely associated with social exchange theory (SET) (Ferriss, 2006; Hu et al., 2024). SET has been extensively used to obtain dynamic insights on the manner in which stakeholder support/interact with each other alongside other theories such as the TALC theory, the tourism carrying capacity (TCC) theory, the emotional solidarity notion (Suess et al., 2021), and the liveability theory (Lindberg et al., 2021). Butler's (1980) TALC theory has often been employed "to discern tourism-related effects on residents' happiness between countries with different TD (tourism development) levels" (Hu et al., 2024, p. 143).

SET remains the fundamental theory to understand resident attitudes toward tourism development although it has integrated other notions and theories (Andereck

& Nyaupane, 2011; Chhabra & Gursoy, 2009; Su & Swanson, 2020). As mentioned by Hu et al., "the framework of SET has been combined with bottom-up spillover theory, TALC, and TCC, alongside with newly introduced concepts and theories such as emotional solidarity, place attachment, social responsibility, and value co-creation" (Hu et al., 2024, p. 145). Myriad integration of other concepts such as sustainability, psychological stress, and the livability theory has also been suggested to expand the path for future deliberations on QOL. Hu et al. suggested future paths of research:

- Devising a conceptual framework linking tourists' and residents' QOL.
- Combining subjective and objective scales to improve results' generalizability.
- Longitudinal research involving innovative methods to clarify the dynamics of the TD–QOL nexus over time.
- Broader investigations of QOL/well-being from the eudaimonic tradition to accommodate both the self and others as well as short-and long-term elements of well-being with a focus on self-realization and community development (Hu et al., 2024, p. 145).

Some studies have pondered on the impact of the pandemic on tourism development and the resident QOL (Lindberg et al., 2021; Ramkissoon, 2021), and it has been suggested that future research should optimize resources to nurture resilient and equitable sustainable tourism development. Andersson et al. (2018) highlight the significance of transformation to understand the dynamic association between tourism destination and the general public. The authors opine that transformation can be meaningful if an open system approach is embraced.

Several notable points of inquiry emerge from this book such as developing appropriate mechanisms to promote SDGs, non-pharmaceutical interventions as remedies and wellness facilitators, and surge of wellness tourism. Areas of further deliberation relate to important stakeholders of tourism beyond the tourists and hosts such as the tourism industry, inculcating a sense of responsibility toward all living beings, importance of local voices in policy decision-making processes, and designing transformative cyclical models, to ensure sustained transformative behavior and suggest pathways toward building sustainable and resilient tourism systems. Recent studies have focused on the hedonic and eudaimonic frameworks of well-being particularly in the context of transformative tourism experiences with important implications for the tourists and the host communities (Chhabra, 2022; Smith & Diekmann, 2017; Uysal & Sirgy, 2023; Volo, 2017).

Transformative experiences have strong personal and social implications and are of great importance to the individual, the broader society, and tourism (Sheldon, 2020). Participation in transformative experiences can bring various tangible (such as acquiring new knowledge and competencies) and intangible (changes in way of thinking and beliefs) changes in tourists. Becker (2020) notes that significant social changes can also happen in the way we view others, developing toleration, trust, altruistic disposition, and understanding of other cultures. Transformative moments can be described as "emotionally intensive, sensorially impressive and cognitively stimulating" with potential to inspire tourists to move beyond their comfort zones and revisit their core values and begin a "deeper awakening to a new way of being in the world" (Sheldon, p. 4). Describing the prerequisites for personal transformation, it is important for tourists to take solo trips with a keen desire to reach a transformation state of mind and enhanced cognitive state through learning opportunities. Enlightened and inspired individuals are "more likely to respect the ecosystem and behave responsibly toward the environment. Consequently, the change will have significant implications for the sustainability of the tourist destination itself" (Seočanac, 2022, p. 6). Sheldon writes that "transformed tourists live by a higher set of values, are more aware and sensitive to their impact on the destination, and moderate their behavior accordingly, they can transform destinations while transforming themselves" (Sheldon, 2020, p. 10).

Well-being aligns with SDG 3 (UN 2019) and is regarded as an essential pillar in gauging wealth of nations. There is surging interest in well-being (a subset of health), particularly from the standpoint of tourist and host experiences. As noted earlier in this book, the notion of well-being has also been associated with subjective well-being which is specifically focused on happiness and other positive psychology attributes (McCabe, 2009; Pearce & Packer, 2013). Some scholars suggest that well-being has also been scrutinized in association with later-life (Morgan et al., 2015) and QOL (Dolnicar et al., 2012) standpoints. Evidently well-being has been used a proxy to happiness, QOL and satisfaction by several studies. Specific landscapes, such as nature-based, hold potential to promote transformative behavior resulting in empathy toward other people and living beings and the environment. Hanna et al. write that "we are all part of nature. It is through this distinction that contemporary living for many has served as the rooted link where 'at the deepest level the psyche remains sympathetically bonded to the earth that mothered us into existence" (2019, p. 5).

Hanna et al. further point out that "eco-psychology and environmental education research continue to build evidence of the links between a child's active engagement in outdoor activities and environmental responsibility…." (2019, p. 6). Psychological well-being and understanding of sustainability are enhanced. Hanna draws from the work of Fisher and contends that nature-based activities help in "recalling how our human psyches are embedded in and nurtured by the larger psyche of nature and at relearning the essentially human art of revering, giving back to, and maintaining reciprocal relations with an animate natural world" (Fisher, 2012, p. 13).

Psychological health crisis has been noted by several studies in the contemporary times due an unprecedented increase in the diagnoses of mental health conditions all over the world. A plethora of literature also notes that mental illness results are often pathologized in an exaggerated medical paradigm of mental health and well-being which distances the sick people from activities and engagement which in fact can benefit the individual with mental health conditions (Walker et al., 2017). Examples of conducive activities can be group singing (Walker et al., 2017), group exercise (Chhabra, 2022), walking, and outdoor activities (Brymer et al., 2009; Rosa et al., 2018). It is interesting to note that such activities are overlooked as health, healing, and well-being resources (Walker et al., 2017). Human-nature connections offer therapeutic benefits albeit these are often referred as unsustainable because they can produce a negative impact on the remote and pristine environments (Humberstone, 2009; Page et al., 2005). However, emerging literature suggests that nature-based engagements and activities instill a deep sense of appreciation and reverence for nature which can instill responsible behavior and more nurturing of the natural environment (Brymer et al., 2009; Rosa et al., 2018; Thorpe & Rinehart, 2010). Thorpe and Rinehart (2010) note the role of embodied connections and performances which induce a sense of connectedness and spiritual associations with the natural environment in the context of earth's forces such as gravity. As an example, Brymer and Gray write that the outdoor adventurers "describe themselves as being at tune with the natural world or connected through life enhancing energy" (Brymer & Gray, 2009, p. 193). Along similar lines, Csikszentmihalyi and Csikszentmihalyi (1992) contend that outdoor participants become mystical and develop transcendental sentiments as they get immersed in the activity and the natural surroundings.

Undeniably, establishing a communion with nature and a reciprocal association is healing for "both the natural space and the individual psyche (Davis, 1998). Understanding one's position in the ecological system results in what Foucault's (1988) calls as 'affective reciprocity and reciprocal dependence' between the self and the other and this serves as an awakening call triggering 'appreciation for the unconscious and deeply rooted bonds between the psyche and 'Gaia' (Roszak, 1995)' (Hanna et al., 2019, p. 14). This relationship inspires contemplation on the 'general problems of the earth' and also a care for 'Mother Nature' offering benefits to both the participants well-being and also potentially their environmental reflections/practices" (2019, p. 9). In summary, Hanna et al. (2019) suggest that outdoor adventure activities strengthen the awareness and facilitation of mindful/sustainable behavior and generate healing thereby contributing to the overall well-being. Consequently, recent trends note the emergence of conscious consumers, who tend to "balance between experiencing the authentic offerings from a destination and reciprocating back equitably" (Mondal & Samaddar, 2021, p. 6). It is purported that conscientious visitors behave in a responsible manner and have sustained ethical habits such as pursuing healthy and environment-friendly activities at home as well as during travel (Mondal & Samaddar, 2021).

Undeniably, mobilizing resources and competencies are key success factors to boost community well-being which can be described based on dimensions such as QOL, individual welfare, community vitality, sociocultural, economic, and political welfare (Uysal & Sirgy, 2023). QOL dimensions include decent income and adequate infrastructure that supports community activity. Individual welfare is supported by people enjoying daily work and high motivations to achieve personal goals. Community vitality indicators refer to how proud the community is and its active participation. Political welfare is determined by community's ability to influence decision-making. Economic welfare refers to the ability to manage local economic linkages to ensure daily needs of life are met. Extant literature confirms that creative/circular economy initiatives significantly benefit QOL.

Positive impacts of tourism on public health cannot be denied. Travel holds potential to trigger positive emotions through immersive experiences and social bonding opportunities with the host community (Chhabra, 2022; Dillette et al., 2019). Fredrickson (2013) describes positive emotions transient mental states that spark pleasant feelings and enhance health and well-being. Tourism experiences trigger or are shaped by emotions of tourists and hold potential to shape physical health and overall well-being. Positive emotions reduce the magnitude or invasion of diseases such as related to heart, inflammation, and headaches (Seligman, 2011). Positive feelings also inspire healthy activities such as exercise, healthful eating, and good quality of sleep and help to conquer bad unhealthy habits (Godovykh & Ridderstaat, 2020). Upbeat emotions can offer relief from stress and promote positive, perseverance, and resilient disposition (Chhabra, 2022; Dillette et al., 2019). Social bonding offers an opportunity to facilitate a sense of belonging and generate happy feelings in residents as well as the tourists with positive health outcomes (Kansky & Diener, 2021).

Positive emotions and their effect in well-being have been scrutinized by several studies (Chhabra, 2022; Chhabra & Kim, 2023; Filep et al., 2016; Godovykh & Ridderstaat, 2020). Scholars have called for a detailed scrutiny of human flourishing in tourism environments (Godovykh & Ridderstaat, 2020). The authors report that the presence of tourists affects the health of the local people. As an instance, tourism development generates economic benefits such as increasing income of local residents, generating jobs, and tax revenue (Gursoy & Nunkoo, 2019; Liu & Var, 1986; Milman & Pizam, 1988; Zhou et al., 1997). Extant literature claims that income of local residents significantly affects their physical and psychological health (Ettner, 1996; Lang et al., 2019). Godovykh and Ridderstaat's (2020) study shows that the impact of tourism development tourism growth produces a negative effect on health of hosts in the short time but eventually improves their health and long life (Kenny et al., 2019). Godovykh and Ridderstaat report that "the lack of significant long-term effects of residents' income on health can be related to the combined influence of increased levels of stress and obesity .... traditionally associated with higher income and better healthcare opportunities, which also increase in tandem with rising income" (Godovykh & Ridderstaat, 2020, p. 6).

The report by the World Commission on Environment and Development in the late 1990s, conceptualizing the notion of sustainable development, stressed on key need to fulfil the needs of the contemporary societies without compromising the adeptness of future generations to meet sustainable goals (Butler, 1991; Murphy & Price, 2005; Weaver, 2014). Before the world was hit by the COVID-19 catastrophe, extant documented literature emphasized on three key pillars of sustainable development of tourism: socio/cultural, economic, and environmental (Chhabra, 2010; Sharpley, 2020; Weaver, 2014). The harmonious balance between these pillars, particularly from ecological and economic standpoint, has been emphasized by numerous scholars (Redclift, 1987; Sharpley, 2020). Furthermore, the crucial influence of the political environment in shaping and extending the level of support for sustainable development has also been highlighted. In a nutshell, the key purpose of sustainable development can be described as a form of development that supports responsible use of resources to enhance quality of life of people and holds capability to resolve, particularly, contemporary pandemic-related social issues such as stress from isolation and loss of loved ones (from lockdowns) (Pietromonaco & Overall, 2022; Ramkissoon, 2021; Saltzman et al., 2020). Sharpley (2020) points out that sustainable development is a process of collaborative development between communities, economy, resources (both finite and infinite), and ecological environment. According to Qiu, Zhou, and Kim, "the concept of 'people-oriented' development is to seek and realize people's freedom and all-round development, so that people can obtain rich material wealth and spiritual enjoyment, which is the goal of sustainable development. It is the essential requirement of sustainable development to achieve comprehensive and coordinated development" (Qiu et al., 2021, p. 4). Harmonious collaboration between numerous stakeholders has become crucial to plan pathways for integrated development. In fact, the pandemic has resulted in adding one more key set of stakeholders (health-related) to the sustainable development equation.

Sustained transformations and enhanced quality of life of both visited and visiting communities are key to strengthening resilience capacity (Dolnicar & Leisch, 2008; Mondal & Samaddar, 2021; Moscardo et al., 2017; Taylor & Norman, 2019). There is a surge of mindful visitors who are becoming more aware and concerned of others in their home and travel environments and practice ethical consumption to enhance their quality of life (Barber & Deale, 2014; Sheth et al., 2011). In academic literature, there is compelling need to synthesize previous and current literature to develop compatible sustainable development and transformative quality-of-life models (Chhabra, 2022; Sheldon, 2020). Also, it is necessary to consider emerging trends to develop flexible strategies.

Trends

Based on the perusal of literature and numerous other sources such as websites of leading travel agencies, Euromonitor, and tourism research institutes (such as Good Tourism Institute, camping agencies, Forbes, and TravelPulse), numerous trends can be noted. Some trends indicate mounting demand for eco-conscious and deep transformative experiences, while other trends suggest demand for travel that is likely to generate externalities for the visited communities and generate minimum transformative benefits for the tourists. The latter calls for proactive strategies such as developing ethical codes of conduct and mindful knowledge to minimize negative impacts, promote sustainable use of resources, and inspire all relevant stakeholders (particularly the tourists) to inculcate a higher sense of consciousness and seek a transformative state of mind. Table 12.1 presents trends that are likely to stimulate self-growth, B-cognition moments (triggered by disorientated/out-of-the ordinary experiences and/or positive emotions), and sustained demand for responsible and transformative experiences.

Other trends include home swapping, private group travel, skip-gen travel, and increased demand for shoulder seasons (Altschuler, 2024; Travel Pulse, 2024). Home swapping refers to travelers who are looking for longer stays as the digital

Table 12.1 Emerging trends

Trends	
Astro-tourism	This type of tourism is patronized to view astronomical phenomena. Astro-tourists visit unpolluted and uncrowded places to gaze at the sky. They prefer to spend a lot of time gazing at the stars, planets, and constellations overhead. To meet this growing demand, hotels focused on wellness offerings, and spas are offering spaces to guests to gaze and watch for comets, other planets, eclipses, and patterns in the galaxy. Telescopes are offered. This type of travel can be transformative
Wild feasting	Demand is noted for artistically and mindfully organized culinary experiences in natural settings by integrating locally searched ingredients. As an instance, in Sweden, visitors can patronize "do-it-yourself outdoor restaurants" where they can book a table in a scenic spot. Guests hunt for local plants and vegetables such as nettles, birch leaves, and trumpet chanterelles. They are given a chef's recipe card and invited to cook them on an open fire. These types of experiences promote responsible and local consumption
Silent travel	In an age of overstimulation and digitalization, particularly in the post-pandemic times, silence is being valued. Offering a chance to restore and reset, silent travel represents a more mindful kind of trip. Examples include silent meditation retreats which have become a popular wellness trend. It is also about traveling to secluded nature resorts and seeking solitary hotels, silent walking tours, etc.
Calmcation	More than 50% of the travelers desire calm and relaxing moments. They are still fatigued from their pandemic experiences and are seeking ways to unwind. "Calmcation" is trending and is a perfect answer to distancing oneself away from stresses and unpredictability of everyday life. Travelers are seeking comfortable camping experiences so that they can relax in natural settings
Low-carbon adventure travel	Preference for cooler summer destinations facilitating demand for low-carbon adventure travel. More travelers are opting to spend time outdoors and enjoy connecting with nature. Focus by this travel segment is also to reduce carbon emissions and seek emission-free immersive experiences in natural settings. Example of low-carbon adventure activities being sought include hiking, mountain-climbing, trekking, rock climbing, scuba diving, snorkeling, cycling
Long-distance train travel	According to a travel survey conducted by Euromonitor, rail travel is predicted to soar. Demand is in fact growing for long-distance and luxury train travel. More and more travelers are becoming eco-conscious and are looking for sustainable modes of travel. Japan, India, and Canada have developed extensive rail networks
Responsible revenge travel	Revenge travel continues to trend from last year. People are still trying to make up for their missed adventure time during the pandemic. Last year, they indulged in fast-paced and unsustainable trips which in fact added to their fatigue. Difference noted this year is that travelers are still keen to take their time back, but they are planning their trips in a more responsible and slow manner

nomadism continues to trend. Appeal has grown for working and living abroad. Home swapping, in fact, is being considered a perfect way to secure a (free) home overseas in exchange for offering one's own home (time duration can be weeks or months). As travel costs continue to escalate, home swapping has become a more economically viable option compared to expensive hotels or Airbnbs. Next, private group travel and skip gen travel have become trendy. The post-pandemic appeal to travel with friends or family on a joint vacation experience has picked up. In fact, this trend has led to the surge in demand for luxury travel among family groups across three generations and middle-aged empty nesters who are keen to revive long-lasting friendships and family bonds. Skip-gen (as mom or dad do not join and one generation skips, hence the name) travel relates to grandparents vacationing with their grandchildren, and they seek nostalgic places. Examples of packages designed by travel agencies include skip-gen safaris. Demand has also increased for travel in off-peak (shoulder seasons). Destinations in high demand during peak seasons are being sought during shoulder seasons. Advantages for traveling during off-peak times include less crowds and better prices. Furthermore, as mentioned in Table 12.1, train stations and train rides are becoming popular. Today, as visitors seek more local experiences, train stations are becoming more friendly and hospitable. They are offering a variety of shops, restaurants, and bars for enjoyment and exploration. Historic train stations are being renovated to offer tailored food and drink venues.

Another emerging trend is set-jetting (Travel Pulse, 2024). People are planning travel based on the filming locations of their favorite television shows and movies. Examples of jet-setting locations include Paris (as viewed in "Emily in Paris") and Florida Keys (as viewed in the new show "Bad Monkey"). Slow travel experiences remain in demand. Digital nomadism continues to trend. Destination dupes refers to opting to travel to a locale which is a look-alike of the favorite destination but less expensive. Mindfulness and tangible efforts to minimize carbon footprint can be evidenced today. Travelers are reducing overtourism in popular destinations by traveling to less-popular destinations. Furthermore, as mentioned by numerous contributors of this book and noted by different travel agencies and media, wellness travel continues to surge thereby prioritizing personal health and the desire to find time to relax, destress, and detoxify. Evidently, shift toward more responsible and transformative consumption is noted. To ensure it translates into a win-win situation for the guests, hosts, and the hospitality businesses, a sustained transformative roadmap needs to be pursued so that the QOL of life of all involved is robust and the entire tourism system becomes and stays resilient.

In summary, while being cognizant of contemporary trends, this book highlights topics to consider and offers a roadmap to boost QOL and sustainability. As a case in point, herbal health remedies guided by traditional knowledge and scientifically tested mechanisms and nutritional and nourishing experiences such as by immersing in therapeutic and natural landscapes can be powerful transformation catalysts and hold tremendous potential to boost QOL of both visitors and hosts. From the human resource standpoint, Ayazlar and Ayazlar (2024) argue that mindfulness is valuable in fostering mental health and overall well-being of employees and also boosts productivity, through meditation exercises. The authors illustrate that mindfulness enhances emotional intelligence which in turn promotes well-being and sustainable behavior. Community-empowered models in niche forms of tourism are crucial, as illustrated by Atal and Atal (2024) and Gogoi and Maheshwari (2024), which support local and indigenous livelihoods. The authors also reiterate significance of experiential and knowledge exchange platforms to promote inclusive and mindful agenda in fragile and indigenous destinations. Similar message is echoed by Cohen (2024) in his study on marine diving tourism. The author stresses on the efficacy of interpretation programs and underwater citizen science projects for successful immersive educational experiences that can facilitate sustained transformation and personal growth. All contributors suggest that transformation of the host community is equally important to minimize overuse their resources and garner resources to promote local economic interests and conservation strategies (Pham et al., Pham, Vogt, & Andereck, 2024a, b). Environmental interpretation and peak experiences play important role in promoting transformation. Kim (2024) calls for synergistic relationships between natural areas, locals, and tourism and stresses on the need to tap into social media applications to generate mindfulness. Economic benefits and government support are crucial for ongoing success of sustainable initiatives. In summary, this book underlines the significance of mindfulness, inclusiveness, knowledge-building initiatives and platforms, interaction with natural resources, local community empowerment, and co-created experiences between the guests and the hosts. Going forward, Fig. 12.1 presents a transformative quality-of-life model to underline the significance of transformation and heightened state of consciousness. Awakened and enlightened minds inspire connections, mindfulness, and sustainable use of resources to build and strengthen resilient tourism systems. The ultimate goal of ongoing journeys of self-inquiry and self-growth of guests and hosts should be to build and strengthen resilient tourism systems.

The model in Fig. 12.1 captures the key messages conveyed by all contributors of this book. It illustrates that niche types of tourism (such as marine dive tourism, motorcycle tourism, and off-highway vehicle recreation) and wellness experiences in natural therapeutic natural (Little, 2012) and spiritual spaces (such as natural landscapes, retreat centers, and herbal gardens) offer knowledge enhancement (learning) and social opportunities (Chhabra, 2022; Gill et al., 2019; Hanna et al., 2019; Sheldon, 2020). In the words of Sheldon, "locations in nature and the

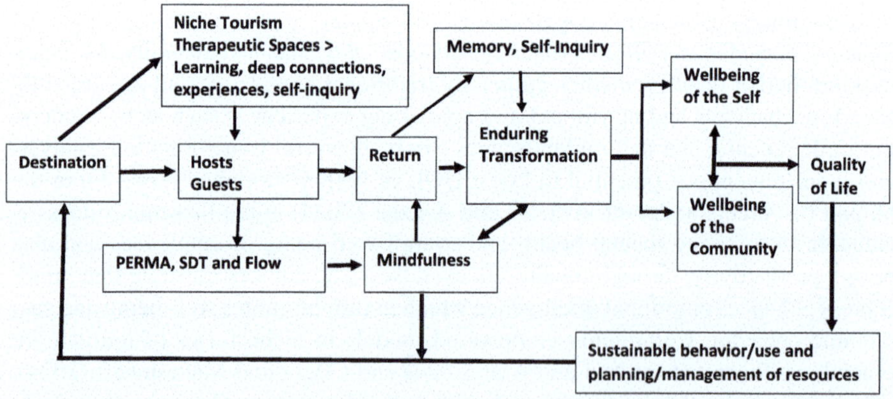

Fig. 12.1 Sustainable transformative quality of life paradigm. (Source: After Chhabra (2022))

wilderness naturally create mindfulness and are fertile places for transformation to occur" (2020, p. 5).

The niche tourism stakeholders in the destination and the key facilitators such as tour guides, instructors of retreat or healing centers (Chhabra, 2022), and horticulturists at herbals gardens (Atal & Atal, 2024) or cultural custodians of indigenous areas should aim to design co-created memorable and deep experiences for guests, local residents, and other important stakeholders. These co-created experiences hold potential to generate both hedonic and eudaimonic as evidenced by several studies, based on PERMA (based on five dimensions: positive emotions, engagement, relatedness, meaningfulness, and accomplishment) model of well-being (Chhabra, 2022; Khaw & Kern, 2014; Seligman, 2011, 2018), social determination theory (Buzinde, 2020), and flow experience theory (Csikszentmihalyi, 1990; Gill et al., 2019). The main goal should be to aim for transformative experiences which can inspire mindfulness at the tourism destinations and home destinations of tourists (Ponder & Holladay, 2013). In regard to the hosts (stakeholders of niche tourism and services), SET can inspire mindfulness, as they anticipate/receive economic, social, hedonic, and eudaimonic benefits. In the postexperience phase, for tourists, fond memories and quest for self-actualization can inspire the desire for sustained transformation and advance the well-being of the self and the community.

Remembrance and continued engagements can happen through word of mouth (such as sharing of learning experiences, spearheading wellness programs or traveling again to continue the quest to reach a higher sense of consciousness). This roadmap can continue to open new paths for enduring transformation. As indexed by recent studies, having experienced/nurtured the optimum point of physiological, psychological, emotional, and/or spiritual state, an ongoing desire to preserve a contemplative/self-actualizing state of "being" can serve as a shock buffer against predictable and unpredictable catastrophes or micro-level disturbances (Buzinde, 2020; Chhabra, 2022; Öznalbant & Alvarez, 2020). Ultimate outcome should be cyclical pursuit of transformational journeys facilitating enhanced quality of life and mindful cognizance of resources by promoting various SDGs (Sustainable Development Goals) to minimize impacts (of, for instance, overtourism), holistic well-being, and healthfulness. This roadmap will remain an ongoing process nourished by endurance and aspiration to nurture the self, broader community, and the natural environment (Chhabra, 2022). Future studies should empirically examine intersections between sustained transformation and quality of life from health and wellness viewpoints. Sustained transformation at the individual and the community level can serve as a catalyst and strengthen resilience capacity of both visited and residential destinations. This line of inquiry also calls for further deliberations and may spur future research.

References

Altschuler, W. 2024's most surprising travel trends. Retrieved from: https://www.forbes.com/sites/wendyaltschuler/2024/02/21/the-hottest-travel-trends-that-will-surprise-you/?sh=6f5c9cd18194

Andersson, V., Clausen, H. B., & García, M. V. (2018). Transformative change: Resilience and sustainable tourism development for communities. *Diálogos Latinoamericanos, 19*(27), 117–126.

Andereck, K. L., & Nyaupane, G. P. (2011). Exploring the nature of tourism and quality of life perceptions among residents. *Journal of Travel Research, 50*(3), 248–260.

Atal, N., & Atal, C. (2024). Healing potential of herbs and herbal tourism. In D. Chhabra, N. Atal, & A. Maheshwari (Eds.), *Sustainable development and resilience of tourism: Wellbeing and quality of life perspectives*. Springer. In Press.

Ayazlar, R., & Ayazlar, G. (2024). The importance of mindfulness for tourism employees. In D. Chhabra, N. Atal, & A. Maheshwari (Eds.), *Sustainable development and resilience of tourism: Wellbeing and quality of life perspectives*. Springer. In Press.

Barber, N. A., & Deale, C. (2014). Tapping mindfulness to shape hotel guests' sustainable behavior. *Cornell Hospitality Quarterly, 55*(1), 100–114.

Becker, J. C. (2020). Ideology and the promotion of social change. *Current Opinion in Behavioral Sciences, 34*, 6–11.

Brymer, E., & Gray, T. (2009). Dancing with nature: Rhythm and harmony in extreme sport participation. *Journal of Adventure Education & Outdoor Learning, 9*(2), 135–149.

Brymer, E., Downey, G., & Gray, T. (2009). Extreme sports as a precursor to environmental sustainability. *Journal of Sport & Tourism, 14*(2–3), 193–204.

Butler, R. W. (1980). The concept of a tourist area cycle of evolution: Implications for management of resources. *Canadian Geographer / Le Géographe Canadien, 24*(1), 5–12. https://doi.org/10.1111/j.1541-0064.1980.tb00970.x

Butler, R. W. (1991). Tourism, environment, and sustainable development. *Environmental Conservation, 18*(3), 201–209.

Buzinde, C. N. (2020). Theoretical linkages between well-being and tourism: The case of self-determination theory and spiritual tourism. *Annals of Tourism Research, 83*, 102920.

Chhabra, D. (2010). *Sustainable marketing of cultural and heritage tourism*. Routledge.

Chhabra, D. (2022). Transformational model of well-being for serious travellers. *International Journal of Spa and Wellness, 5*(2), 129–146.

Chhabra, D., & Gursoy, D. (2009). Life dynamism explorations on perceived quality of life and social exchange paradigms in casino settings. *Leisure Sciences, 31*(2), 136–157.

Chhabra, D., & Kim, E. G. (2023). Sustaining inner transformation through spiritual tourism. *International Journal of Spa and Wellness*, 1–22.

Cohen, D. (2024). Transformative potential of sustainable scuba dive tourism. In D. Chhabra, N. Atal, & A. Maheshwari (Eds.), *Sustainable development and resilience of tourism: Wellbeing and quality of life perspectives*. Springer. In Press.

Csikszentmihalyi, M. (1990). Flow: the psychology of optimal experience. New York: Harper & Row.

Csikszentmihalyi, M., & Csikszentmihalyi, I. S. (1992). *Optimal experience: Psychological studies of flow in consciousness*. Cambridge University Press.

Dann, G. M. S. (2012). Tourist motivation and quality-of-life: In search of the missing link. In M. Uysal, R. R. Perdue, & M. J. Sirgy (Eds.), Handbook of tourism and quality-of-life research: Enhancing the lives of tourists and residents of host communities (pp. 233–250). Springer.

Davis, J. (1998). The transpersonal dimensions of ecopsychology: Nature, nonduality, and spiritual practice. *The Humanistic Psychologist, 26*(1–3), 69–100.

Dillette, A. K., Douglas, A. C., & Andrzejewski, C. (2019). Yoga tourism–a catalyst for transformation? *Annals of Leisure Research, 22*(1), 22–41.

Dolnicar, S., & Leisch, F. (2008). Selective marketing for environmentally sustainable tourism. *Tourism Management, 29*(4), 672–680.

Dolnicar, S., Yanamandram, V., & Cliff, K. (2012). The contribution of vacations to quality of life. *Annals of Tourism Research, 39*(1), 59–83.

Dwyer, L. (2022). Destination competitiveness and resident well-being. *Tourism Management Perspectives, 43*, 100996.

Dwyer, L. (2023). Tourism development and sustainable well-being: A beyond GDP perspective. *Journal of Sustainable Tourism, 31*(10), 2399–2416.

Ettner, S. L. (1996). New evidence on the relationship between income and health. *Journal of Health Economics, 15*(1), 67–85.

Ferriss, A. L. (2006). A theory of social structure and the quality of life. *Applied Research in Quality of Life, 1*(1), 117–123. https://doi.org/10.1007/s11482-006-9003-1

Filep, S., Laing, J., & Csikszentmihalyi, M. (2016). What is positive tourism? Why do we need it? In *Positive tourism* (pp. 17–30). Routledge.

Fisher, A. (2012). What is ecopsychology? A radical view. In *Ecopsychology: Science, totems, and the technological species* (pp. 79–114).

Foucault, M. (1988). Technologies of the Self. In L. H. Martin, H. Gutman, & P. Hutton (Eds.), *Technologies of the Self: A seminar with Michel Foucault*. University of Massachusetts Press.

Fredrickson, B. L. (2013). Positive emotions broaden and build. In *Advances in experimental social psychology* (Vol. 47, pp. 1–53). Academic Press.

Genç, R. (2012). Subjective aspects of tourists' quality-of-life (QOL). In *Handbook of tourism and quality-of-life research: Enhancing the lives of tourists and residents of host communities* (pp. 149–167). Springer Netherlands.

Genç, R. (2023). Subjective aspects of quality of life (QOL). In *Handbook of tourism and quality-of-life research II: Enhancing the lives of tourists, residents of host communities and service providers* (pp. 121–138). Springer International Publishing.

Gill, C., Packer, J., & Ballantyne, R. (2019). Spiritual retreats as a restorative destination: Design factors facilitating restorative outcomes. *Annals of Tourism Research, 79*, 102761.

Godovykh, M., & Ridderstaat, J. (2020). Health outcomes of tourism development: A longitudinal study of the impact of tourism arrivals on residents' health. *Journal of Destination Marketing & Management, 17*, 100462.

Gogoi, P., & Maheshwari, A. (2024). Mapping the impact of motorcycle tourism in trans-Himalayas: A study on alternative livelihood creation in rural Ladakh, India. In D. Chhabra, N. Atal, & A. Maheshwari (Eds.), *Sustainable development and resilience of tourism: Wellbeing and quality of life perspectives*. Springer. In Press.

Gursoy, D., & Nunkoo, R. (Eds.). (2019). *The Routledge handbook of tourism impacts: Theoretical and applied perspectives*. Routledge.

Hanna, P., Wijesinghe, S., Paliatsos, I., Walker, C., Adams, M., & Kimbu, A. (2019). Active engagement with nature: Outdoor adventure tourism, sustainability and wellbeing. *Journal of Sustainable Tourism., 27*, 1355.

Hu, R., Li, G., Liu, A., & Chen, J. L. (2024). Emerging research trends on residents' quality of life in the context of tourism development. *Journal of Hospitality & Tourism Research, 48*(1), 131–152.

Humberstone, B. (2009). Inside/outside the Western 'Bubble': The nexus of adventure, adventure sports and perceptions of risk in UK and Mauritius. In J. Ormrod & B. Wheaton (Eds.), *On the edge: Leisure, consumption and the representation of adventure sports*. University of Michigan: Leisure Studies Association.

Kansky, J., & Diener, E. (2021). Science of well-being: Notable. *Positive Psychology: An International Perspective*, 43.

Karagöz, D., & Kama, S. (2023). Self-identification with tourism experiences, hedonic–eudaimonic consumption. In handbook of tourism and quality-of-life research II: enhancing the lives of tourists, residents of host communities and service providers (pp. 211–223). Cham: Springer International Publishing.

Kenny, K., Sandercock, M., & Webster, J. (2019). Case report: Identification of elemental mercury. *Canadian Society of Forensic Science Journal, 52*(3), 122–128.

Kesgin, M. (2023). Experiencing wellbeing and life satis-faction through gastronomy tourism. In M. Uysal & M. J. Sirgy (Eds.), *Handbook of tourism and quality-of-life research II: Enhancing the lives of tourist, residents of host communities and service providers*. Springer Nature.

Khaw, D., & Kern, M. (2014). A cross-cultural comparison of the PERMA model of well-being. *Undergraduate Journal of Psychology at Berkeley, University of California, 8*(1), 10–23.

Kim, K. (2002). The effect of tourism impacts upon quality-of-life of residents in the community. Unpublished Dissertation. Unpublished doctoral dissertation, Virginia Polytechnic Institute and State University, Blacksburg, Virginia.

Kim, K., Uysal. M., & Sirgy. J. (2003, June). The effects of tourism impacts upon quality of life of residents in the community. TTRA' s 34th annual conference proceedings, pp. 16–20. St. Louis, MO, USA.

Kim, E. (2024). Impact of hybrid word of mouth on consumer behavior: Case of off-highway vehicle recreation markets. In D. Chhabra, N. Atal, & A. Maheshwari (Eds.), *Sustainable development and resilience of tourism: Wellbeing and quality of life perspectives*. Springer.

Koc, E., & Ayyildiz, A. Y. (2023). Well-being and aesthetic journeys: Motivations for cosmetic surgery tourism. In *Handbook of tourism and quality-of-life research II: Enhancing the lives of tourists, residents of host communities and service providers* (pp. 197–210). Springer International Publishing.

Lang, M., McManus, T. C., & Schaur, G. (2019). The effects of import competition on health in the local economy. *Health Economics, 28*(1), 44–56.

Lew, A. (2020). The global consciousness path to sustainable tourism: A perspective paper. *Tourism Review., 75*, 69.

Lindberg, K., Munanura, I. E., Kooistra, C., Needham, M. D., & Ghahramani, L. (2021). Understanding effects of tourism on residents: A contingent subjective well-being approach. *Journal of Travel Research*. https://doi.org/10.1177/0047287520988912

Little, J. (2012). Transformational tourism, nature and wellbeing: New perspectives on fitness and the body. *Sociologia Ruralis, 52*(3), 257–271.

Liu, J. C., & Var, T. (1986). Resident attitudes toward tourism impacts in Hawaii. *Annals of Tourism Research, 13*(2), 193–214.

Marques, I., Serrasqueiro, Z., & Nogueira, F. (2021). Managers' competences in private hospitals for investment decisions during the COVID-19 pandemic. *Sustainability, 13*(4), 1757.

McCabe, S. (2009). Who needs a holiday? Evaluating social tourism. *Annals of Tourism Research, 36*(4), 667–688.

Milman, A., & Pizam, A. (1988). Social impacts of tourism on central Florida. *Annals of Tourism Research, 15*(2), 191–204.

Mitta, O., Shahvali, M., Ward, P., Duerden, M., & Bastiaansen, M. (2023). Human connection: A crucial ingredient in vacation experience. In M. Uysal & M. J. Sirgy (Eds.), Handbook of tourism and quality-of-life research II: Enhancing the lives of tourist, residents of host communities and service providers. Springer Nature.

Mondal, S., & Samaddar, K. (2021). Responsible tourism towards sustainable development: Literature review and research agenda. *Asia Pacific Business Review, 27*(2), 229–266.

Morgan, N., Pritchard, A., & Sedgley, D. (2015). Social tourism and well-being in later life. *Annals of Tourism Research, 52*, 1–15.

Moscardo, G. (2009). Tourism and quality of life: Towards a more critical approach. *Tourism and Hospitality Research, 9*(2), 159–170.

Moscardo, G., Konovalov, E., Murphy, L., McGehee, N. G., & Schurmann, A. (2017). Linking tourism to social capital in destination communities. *Journal of Destination Marketing and Management, 6*(4), 286–295. https://doi.org/10.1016/j.jdmm.2017.10.001

Murphy, P. E., & Price, G. G. (2005). Tourism and sustainable development. *Global Tourism, 3*, 167–193.

Öznalbant, E., & Alvarez, M. D. (2020). A socio-cultural perspective on yoga tourism. *Tourism Planning & Development, 17*(3), 260–274.

Page, S. J., Bentley, T. A., & Walker, L. (2005). Scoping the nature and extent of adventure tourism operations in Scotland: how safe are they?. *Tourism Management, 26*(3), 381–397.

Pearce, P. L., & Packer, J. (2013). Minds on the move: New links from psychology to tourism. *Annals of Tourism Research, 40*, 386–411.

Pham, K., Vogt, C., & Andereck, K. (2024a). Overtourism; A potential outcome in contemporary tourism, part I: Concepts, indicators and impacts. In D. Chhabra, N. Atal, & A. Maheshwari (Eds.), *Sustainable development and resilience of tourism: Wellbeing and quality of life perspectives*. Springer. In Press.

Pham, K., Andereck, K., & Vogt, C. (2024b). Overtourism: A potential outcome in contemporary tourism. Part 2: Causes, solutions and management challenges. In D. Chhabra, N. Atal, & A. Maheshwari (Eds.), *Sustainable development and resilience of tourism: Wellbeing and quality of life perspectives*. Springer. In Press.

Pietromonaco, P., & Overall, N. (2022). Implications of social isolation, separation and loss during the COVID-19 pandemic for couples' relationships. *Current Opinion in Psychology, 43*, 189.

Pocinho, M., Garcês, S., & de Jesus, S. N. (2022). Wellbeing and resilience in tourism: A systematic literature review during COVID-19. *Frontiers in Psychology, 12*, 748947. https://doi.org/10.3389/fpsyg.2021.748947

Ponder, L. M., & Holladay, P. J. (2013). The transformative power of yoga tourism. *Transformational Tourism: Tourist Perspectives, 1*(7).

Qiu, P., Zhou, Z., & Kim, D. J. (2021). A new path of sustainable development in traditional agricultural areas from the perspective of open innovation—A coupling and coordination study on the agricultural industry and the tourism industry. *Journal of Open Innovation: Technology, Market, and Complexity, 7*(1), 16.

Ramkissoon, H. (2021). Pro-sociality in times of separation and loss. *Current Opinion in Psychology*.

Redclift, M. (1987). Sustainable development: Exploring the contradictions. Routledge.

Rosa, C. D., Profice, C. C., & Collado, S. (2018). Nature experiences and adults' self-reported pro-environmental behaviors: The role of connectedness to nature and childhood nature experiences. *Frontiers in Psychology, 9*, 1055.

Roszak, T. (1995). Where psyche meets Gaia. In *Ecopsychology: Restoring the earth, healing the mind* (pp. 1–17).

Saarinen, J. (2021). Is being responsible sustainable in tourism? Connections and critical differences. *Sustainability, 13*(12), 6599. https://doi.org/10.3390/su13126599

Saltzman L., Hansel, T., & Bordnick, P. (2020). Loneliness, isolation, and social support factors in post-COVID-19 mental health. *Psychol Trauma: Theo, Res., Prac, and Policy*.

Santos, A., Gonzalez, C., Haegeman, K., & Rainoldi, A. (2020). *Behavioural changes in tourism in times of COVID-19*. Publications Office of the European Union.

Seligman, M. (2011). *Flourish: A visionary new understanding of happiness and well-being*. Free Press.

Seligman, M. (2018). PERMA and the building blocks of well-being. *The Journal of Positive Psychology, 13*(4), 333–335.

Seočanac, M. (2022). Transformative experiences in nature-based tourism as a chance for improving sustainability of tourism destination. *Economics of Sustainable Development, 6*(1), 1–10.

Sevuk, D. (2023). Gender-based differences in medical tourists' destination preferences and their perception of quality of life. In M. Uysal & M. J. Sirgy (Eds.), *Handbook of tourism and quality-of-life research II: Enhancing the lives of tourist, residents of host communities and service providers*. Springer Nature.

Sharpley, R. (2020). Tourism, sustainable development and the theoretical divide: 20 years on. *Journal of Sustainable Tourism, 28*(11), 1932–1946.

Sheldon, P. J. (2020). Designing tourism experiences for inner transformation. *Annals of Tourism Research, 83*, 102935.

Sheth, J. N., Sethia, N. K., & Srinivas, S. (2011). Mindful consumption: A customer-centric approach to sustainability. *Journal of the Academy of Marketing Science, 39*(1), 21–39.

Sirgy, M. J., & Uysal, M. (2023). Putting Things Together to Have a Better Understanding of the Big Picture. In Handbook of Tourism and Quality-of-Life Research II: Enhancing the Lives of Tourists, Residents of Host Communities and Service Providers (pp. 575–602). Cham: Springer International Publishing.

Smith, M. K., & Diekmann, A. (2017). Tourism and wellbeing. *Annals of Tourism Research, 66*, 1–13.

Streimikiene, D., Svagzdiene, B., Jasinskas, E., & Simanavicius, A. (2021). Sustainable tourism development and competitiveness: The systematic literature review. *Sustainable Development, 29*(1), 259–271.

Su, L., & Swanson, S. R. (2020). The effect of personal benefits from, and support of, tourism development: The role of relational quality and quality-of-life. *Journal of Sustainable Tourism, 28*(3), 433–454.

Suess, C., Woosnam, K., Mody, M., Dogru, T., & Sirakaya Turk, E. (2021). Understanding how residents' emotional solidarity with Airbnb visitors influences perceptions of their impact on a community: The moderating role of prior experience staying at an Airbnb. *Journal of Travel Research., 60*, 1039. https://doi.org/10.1177/0047287520921234

Taylor, L. L., & Norman, W. C. (2019). The influence of mindfulness during the travel anticipation phase. *Tourism Recreation Research, 44*(1), 76–90.

Thorpe, H., & Rinehart, R. (2010). Alternative sport and affect: Non-representational theory examined. *Sport in Society, 13*(7–8), 1268–1291.

Travel Pulse. (2024). Travel Trends that will define travel in 2024. Retrieved from: https://www.travelpulse.com/news/features/the-travel-trends-that-will-define-2024

Um, J., & Yoon, S. (2021). Evaluating the relationship between perceived value regarding tourism gentrification experience, attitude, and responsible tourism intention. *Journal of Tourism and Cultural Change, 19*(3), 345–361.

Uysal, M., & Sirgy, M. J. (2023). Prologue: Tourism and quality of life (QoL) research II. In *handbook of tourism and quality-of-life research II: enhancing the lives of tourists, residents of host communities and service providers* (pp. 1–8). Springer International Publishing.

Uysal, M., Sirgy, M. J., Woo, E., & Kim, H. (2016). Quality of life (QOL) and well-being research in tourism. *Tourism Management, 53*, 244–261. https://doi.org/10.1016/j.tourman.2015.07.013

Uysal, M., Berbekova, A., & Kim, H. (2020). Designing for quality of life. *Annals of Tourism Research, 83*, 102944.

Volo, S. (2017). Eudaimonic well-being of islanders: Does tourism contribute? The case of the Aeolian Archipelago. *Journal of Destination Marketing and Management, 6*(4), 465–476. https://doi.org/10.1016/j.jdmm.2016.08.002

Walker, C., Hart, A., & Hanna, P. (2017). *Building a new community psychology of mental health.* Palgrave Macmillan UK.

Weaver, D. (2014). The sustainable development of tourism. *The Wiley Blackwell companion to tourism*, 524–534.

Williams, D. R., & McIntyre, N. (2011). Place affinities, lifestyle mobilities, and quality-of-life. In handbook of tourism and quality-of-life research: Enhancing the lives of tourists and residents of host communities (pp. 209–231). Dordrecht: Springer Netherlands.

Zhou, D., Yanagida, J. F., Chakravorty, U., & Leung, P. (1997). Estimating economic impacts from tourism. *Annals of Tourism Research, 24*(1), 76–89.

Zielińska-Szczepkowska, J. (2023). Travel behavior of the elderly and quality-of-life. In *Handbook of tourism and quality-of-life research II: Enhancing the lives of tourists, residents of host communities and service providers* (pp. 153–165). Springer International Publishing.

Index

D. Chhabra et al. (eds.), *Sustainable Development and Resilience of Tourism*,
https://doi.org/10.1007/978-3-031-63145-0

The manufacturer's authorised representative in the EU is Springer Nature Customer Service Centre GmbH, Europaplatz 3, 69115 Heidelberg, Germany. If you have any concerns regarding our products, please contact ProductSafety@springernature.com

Printed and bound by CPI Group (UK) Ltd, Croydon, CR0 4YY
01/10/2026
02225965-0004